KETAMINE

Use and Abuse

KETAMINE

Use and Abuse

Edited by David T. Yew

CRC Press
Taylor & Francis Group
Boca Raton   London   New York

CRC Press is an imprint of the
Taylor & Francis Group, an **informa** business

CRC Press
Taylor & Francis Group
6000 Broken Sound Parkway NW, Suite 300
Boca Raton, FL 33487-2742

© 2015 by Taylor & Francis Group, LLC
CRC Press is an imprint of Taylor & Francis Group, an Informa business

No claim to original U.S. Government works

Printed on acid-free paper
Version Date: 20150130

International Standard Book Number-13: 978-1-4665-8339-9 (Hardback)

Visit the Taylor & Francis Web site at
http://www.taylorandfrancis.com

and the CRC Press Web site at
http://www.crcpress.com

Contents

Preface

With the emergency of diversified categories of abusive drugs in the world, ketamine has been considered as one of the most prevalent and widely employed abusive agents in Hong Kong and southern Asia. In Hong Kong, ketamine abuse became problematic for the local society in the late 1990s and remained as such to this day. It is one of the three frequently abused agents locally, and for the addicts, ketamine is no longer considered a recreational drug. Ketamine is now used by the addicts at an average of three times a week and with dosages of often greater than 1 g per night. Ketamine and its analogue as abusive agents have since spread to Europe and America. In England, Paul I. Dargan and his group reported methoxetamine, a ketamine analogue, induced renal damages and neurological lesions. Ketamine addiction is now becoming obvious in other parts of the world. In this book, we tried to present both the misuse (toxicity) and the normal use of ketamine clinically at acceptable dosage, along with its possible future contribution to the treatment of depression. In this volume, there are eight chapters on the toxicity of ketamine as an acute or chronic abusive agent, including an interesting chapter on postmortem toxicity and another one on developmental toxicity. In addition, there is a chapter on the pharmacology of ketamine and a chapter on neuroimaging. On the other hand, five chapters are targeted on the usage of ketamine and its clinical testing. Finally, representing the angle of the social scientists is a chapter on the psychosocial factors of ketamine addicts. During the course of our ketamine research, the editor and associates were fortunate to obtain the support of the Beat Drug Fund of the Hong Kong Government and the endowment fund of Wai Yin Foundation, Hong Kong, without which some of the results described here could not have come to light. It is hoped that this book represents an initiative to investigate the different facets of ketamine and would promote more attention on the subject in this direction.

Editor

Professor David T. Yew is the Professor Emeritus of Anatomy in the School of Biomedical Sciences, Chinese University of Hong Kong. He graduated in 1969 from the Chinese University with a Bachelor of Science degree in Zoology and he received in 1974 a PhD degree in Anatomy from Wayne State University in Detroit, the United States. He also received in 1988 a Doctor of Medical Sciences degree and in 1995 a Doctor of Medicine (habil) from the University of Rostock in Germany.

He is an elected fellow of the Society of Biology, United Kingdom; Royal Society of Public Health, United Kingdom; Anatomical Society, United Kingdom; and American Association of Anatomy, USA. During the past 30 plus years, Professor Yew has published more than 300 papers in scientific journals and has authored/edited of 12 books and authored more than 25 review chapters. He is presently on the editorial boards of more than 10 journals in the fields of anatomy, neuroscience, aging, and radiology. In the past 10 years, Professor Yew has been involved in research on neurodegeneration, particularly damage in the nervous system brought about by ketamine abuse, and is now one of the major investigators in this area globally.

Contributors

Wai-Chi Chan
Department of Psychiatry
The University of Hong Kong
Hong Kong SAR, China

Sherry K.W. Chan
Department of Psychiatry
The University of Hong Kong
Hong Kong SAR, China

Eric Y.H. Chen
Department of Psychiatry
and
The State Key Laboratory of Brain
and Cognitive Sciences
The University of Hong Kong
Hong Kong SAR, China

Yuet-Wah Cheung
Department of Sociology
The Chinese University of
Hong Kong
Shatin, New Territories,
Hong Kong, China

Cristiano Chiamulera
Department of Public Health and
Community Medicine
University of Verona
Verona, Italy

Doris C.K. Ching
Toxicology Reference Laboratory
Hospital Authority
Hong Kong, China

Eric Yu-Pang Cho
Faculty of Medicine
School of Biomedical Sciences
The Chinese University of
Hong Kong
Shatin, New Territories,
Hong Kong, China

Calvin Y.K. Chong
Toxicology Reference Laboratory
Hospital Authority
Hong Kong, China

Peggy Sau Kwan Chu
Division of Urology
Department of Surgery
Tuen Mun Hospital
Hong Kong, China

Ginetta Collo
Department of Public Health and
Community Medicine
University of Verona
Verona, Italy

Paul I. Dargan
Clinical Toxicology
Guy's and St. Thomas' NHS
Foundation Trust
and
King's College London
London, United Kingdom

Chaoxuan Dong
Department of Pediatrics
University of Tennessee Health
Science Center
Memphis, Tennessee

Yue Hao
Department of Pharmacy
Shenzhen University Health
Science Centre
Shenzhen, China

Lawrence K. Hui
School of Biomedical Sciences
The Chinese University of
Hong Kong
Shatin, New Territories,
Hong Kong, China

Ismail Laher
Department of Anesthesiology,
Pharmacology and Therapeutics
The University of British Columbia
Vancouver, British Columbia,
Canada

Jacqueline C. Lam
Faculty of Pharmacy
University of Sydney
Sydney, Australia

Lok Hang Lam
Brain Research Center
Faculty of Medicine
School of Biomedical Sciences
The Chinese University of
Hong Kong
Shatin, New Territories,
Hong Kong, China

Phoebe Y.H. Lam
Oxford Institute for Radiation
Oncology and Biology
University of Oxford
Oxford, United Kingdom

Wai Ping Lam
Brain Research Center
Faculty of Medicine
School of Biomedical Sciences
The Chinese University of
Hong Kong
Shatin, New Territories,
Hong Kong, China

Ping Chung Leung
Institute of Chinese Medicine
The Chinese University of
Hong Kong
Shatin, New Territories,
Hong Kong, China
and
State Key Laboratory of
Phytochemistry and Plant
Resources in West China (Partner
Laboratory in the Chinese
University of Hong Kong)
Hong Kong, China

Qi Li
Department of Psychiatry
The University of Hong Kong
Hong Kong SAR, China

Willmann Liang
School of Biomedical Sciences
The Chinese University of
Hong Kong
Shatin, New Territories,
Hong Kong, China

Wai Kit Ma
Division of Urology
Department of Surgery
Queen Mary Hospital
The University of Hong Kong
Hong Kong, China

Tony W.L. Mak
Toxicology Reference Laboratory
Hospital Authority
Hong Kong, China

Chi Fai Ng
Division of Urology
Department of Surgery
The Chinese University of
Hong Kong
Shatin, New Territories,
Hong Kong, China

Andrew M. Perez
Department of Anesthesiology
The ICAHN School of Medicine at
Mount Sinai
New York, New York

Shwetha S. Rao
Clinical Toxicology
Guy's and St. Thomas' NHS
Foundation Trust
London, United Kingdom

Ou Sha
Department of Medicine
Shenzhen University Health
Science Centre
Shenzhen, China

Pak C. Sham
Department of Psychiatry
and
The State Key Laboratory of Brain
and Cognitive Sciences
and
Centre for Genomic Sciences
LKS Faculty of Medicine
The University of Hong Kong
Hong Kong SAR, China

Tan Sijie
Brain Research Center
Faculty of Medicine
School of Biomedical Sciences
The Chinese University of
Hong Kong
Shatin, New Territories,
Hong Kong, China

Lin Sun
Weifang Medical University
Baotong West Street
Shandong, China

Hong Chai Tang
Institute of Chinese Medicine
The Chinese University of
Hong Kong
Shatin, New Territories,
Hong Kong, China

Magdalene H.Y. Tang
Toxicology Reference Laboratory
Hospital Authority
Hong Kong, China

Vincenzo Tedesco
Department of Public Health and
Community Medicine
University of Verona
Verona, Italy

Maria S.M. Wai
School of Biomedical Sciences
The Chinese University of
Hong Kong
Shatin, New Territories,
Hong Kong, China

Chun-Mei Wang
Brain Research Center
Faculty of Medicine
School of Biomedical Sciences
The Chinese University of
Hong Kong
Shatin, New Territories,
Hong Kong, China

Nan Wang
Department of Anesthesiology
Jinling Hospital
School of Medicine
Nanjing University
Nanjing, China

James Watterson
Department of Forensic Science
Laurentian University
Ontario, Canada

Yeak Wan Wong
Brain Research Center
Faculty of Medicine
School of Biomedical Sciences
The Chinese University of
Hong Kong
Shatin, New Territories,
Hong Kong, China

David M. Wood
Clinical Toxicology
Guy's and St. Thomas' NHS
Foundation Trust
and
King's College London
London, United Kingdom

Jian-Jun Yang
Department of Anesthesiology
Jinling Hospital
School of Medicine
Nanjing University
Nanjing, China

David T. Yew
School of Biomedical Sciences
and
Institute of Chinese Medicine
The Chinese University of
Hong Kong
Shatin, New Territories,
Hong Kong, China

Xin Zhang
Institute of Chinese Medicine
The Chinese University of
Hong Kong
Shatin, New Territories,
Hong Kong, China

Li Zhou
Department of Histology and
Embryology
Basic Medical School of
Jilin University
Changchun, China

chapter one

Ketamine use and misuse— Impacts on the nervous system
An overview

David T. Yew

Like all abusive agents, ketamine acts on the central nervous system (CNS). Categorized as a dissociative anesthetic, this drug is known to interfere with the reception of sensory input such that interpretation by the association area of the brain is disrupted (Mion and Villevielle 2013). The drug primarily works on glutamatergic neurons, specifically as an N-methyl-D-aspartate (NMDA) receptor noncompetitive antagonist (Mion and Villevielle 2013). The active molecule of this drug is (S)-ketamine, which has a greater binding affinity for NMDA receptors than its (R) enantiomers, and it can be metabolized to norketamine (Mion and Villevielle 2013).

Today, ketamine is typically used as a general anesthetic. The application of this drug in routine surgical practice has been determined to be safe in normal clinical dosages. It has been employed in both veterinary and human surgery because of its quick induction and the rapid recovery associated with its use. In the past 12 years, ketamine has mostly been used in animal surgery as a rapid anesthetic (Carter and Story 2013), either singly or in combination with other tranquilizers and hypnotics. In animals, propofol is commonly employed along with ketamine during anesthesia (Lerche et al. 2000), often as a continuous intravenous anesthetic (Seliskar et al. 2007). Similarly, in the last three decades, ketamine has been popular for acute use in humans, especially for children, where continuous intravenous infusion with a clinical dosage of usually 20–60 mg/min is implemented. Despite risks concerning its misuse, ketamine is still regarded as a safe anesthetic, and while hallucinations have been reported in patients, these episodes can be controlled by diazepam (Youssef-Ahmed et al. 1996). To minimize the deleterious effects of ketamine, short-term administration is preferred in older human patients (Hosseinzadeh et al. 2013) and during obstetric procedures such as Cesarean sections (Behdad et al. 2013). In addition, during short surgical procedures such as dental surgery, ketamine is often the general anesthetic of choice (Bahetwar et al. 2011; Braidy et al. 2011; Cillo 2012). For longer procedures in humans such as abdominal surgery, ketamine is normally used with other analgesics to maintain a longer period of sedation (Khajavi et al. 2013; Singh et al. 2013). In small clinics, ketamine is typically chosen as an agent for the relief of sporadic pain, often after operative intervention (Eghbal et al. 2013; Javid et al. 2012; Mendola et al. 2012; Patil and Anitescu 2012; Safavi et al. 2011). Some clinics have even used ketamine as a cough suppressant in general practice (Honarmand et al. 2013). Despite its widespread use in the clinical setting, prolonged and chronic use of ketamine either alone or in combination with other CNS agents can be dangerous for the nervous system (Mellon et al. 2007). The situation is true for all ages of humans including adults and children, as well as for embryos and neonates (Félix et al. 2014; Turner et al. 2012).

Despite the beneficial and utilizable effects of ketamine as a one-off or sporadic anesthetic, long-term use has been shown to produce serious health consequences. Chronic addiction of ketamine undoubtedly results in degenerative changes of the nervous system (Chan et al. 2012; Featherstone et al. 2012; Félix et al. 2014; Kanungo et al. 2013; Mak et al. 2010; Olney et al. 2002; Roberts et al. 2014; Sun et al. 2012; Tan et al. 2012; Yeung et al. 2010). The deleterious changes can, in general, be divided into two aspects: intermediate effects and long-term changes. The intermediate effects include cell death and degeneration (Figures 1.1 and 1.2) after several months of use. The long-term changes refer to the production of toxic materials or mutations that can induce further neurodegenerative changes in the years to come. In this latter category, two possible toxic end products that are generated in the CNS are amyloids (Figure 1.3) and mutated tau proteins (Yeung et al. 2010).

Ketamine is known to initiate apoptosis even within simple cell cultures. For example, this drug has been shown to have a proapoptotic effect on SH5Y5Y neuroblastoma cells, where it appears to act through the Bax/Bcl2 ratio as well as the eventual elevation of the activated form of the caspase-3 enzyme. Interestingly, the susceptibilities of cells differed after ketamine treatment based on cell maturity, with differentiated cells being more resistant to ketamine-induced apoptosis than immature and undifferentiated cells (Mak et al. 2010). This observation confirmed the hypertoxicity of ketamine on developing tissues and led researchers to question whether ketamine would have an effect on undifferentiated neoplastic cells.

The effect of ketamine treatment on apoptosis in the nervous system has been well documented in various mammals. In the primate, for example, short-term ketamine exposure caused widespread neuronal apoptosis, similar to that observed after exposure to isoflurane and propofol (Creeley et al. 2013). Ketamine resembles nitrous oxide, barbiturates, phencyclidine, and ethanol, in that it may either directly or indirectly act as an

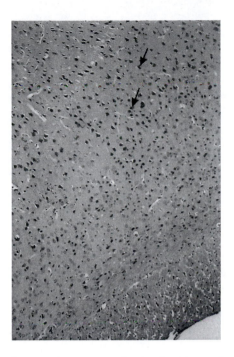

Figure 1.1 Region of cortex in a mouse brain showing pyknotic and irregular nuclei (arrows) after ketamine treatment of 3 months. Magnification, ×200.

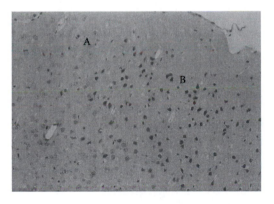

Figure 1.2 Prefrontal cortex of a mouse after ketamine treatment of 6 months. Note, area A has fewer cells than area B. Magnification, ×400.

NMDA antagonist or GABA mimetic, thus exerting toxicity on humans (Olney et al. 2002). In rats, it has been shown that ketamine-induced neuro-apoptosis is accompanied by the activation of glycogen synthase kinase 3 beta (GSK3B), whereas treatment with lithium, a GSK3B inhibitor, attenuated the neuroapoptotic effect of the drug (Liu et al. 2013).

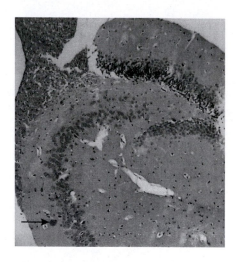

Figure 1.3 A possible amyloid body (arrow) in the hippocampus of a mouse treated with ketamine for 3 months. Magnification, ×200.

Recently, ketamine has also been shown to have a substantial effect on the neurons of lower vertebrates. In zebrafish, for example, ketamine has been shown to be neurotoxic to the motor neurons, and it down-regulates or suppresses a number of genes involved in neurogenesis (Kanungo et al. 2013). In addition, the profound effect of this drug on the nervous system changed the behavior and physiology of the adult fish (Riehl et al. 2011). Motor damage after ketamine use has also been reported in humans and the mouse (Wang et al. 2013; Figures 1.1 through 1.3), with lesions clearly present in the cortex, cerebellum, and the striatum (Wang et al. 2013). In addition, Fang et al. (2013) confirmed the down-regulation of L-DOPA uptake in the striatum after ketamine treatment. Animal studies performed by other groups further outlined additional mechanistic changes after ketamine use, aligning with our reports in the human, monkey, and mouse (Chan et al. 2012; Sun et al. 2012; Tan et al. 2011, 2012; Wang et al. 2013; Yeung et al. 2010; Yu et al. 2012). In addition, substantial EEG changes as well as astrocytic EAAT2 transporter changes were recorded by Featherstone et al. (2012) in the mouse.

Memory loss is not unique for humans or mammals after chronic addiction of ketamine (Tan et al. 2011). Figure 1.4a displays an area of the mouse hippocampus after chronic treatment of ketamine showing more spindle-shaped cells with dense nuclei than those in the equivalent region of the control group (Figure 1.4b). Tan et al. (2011) reported a decline in learning and memory performance via the Morris Maze test in mice addicted to ketamine for 3 months. He also reported that 110 genes were up-regulated while 136 genes were down-regulated in the prefrontal cortex of the ketamine-treated mice. The up-regulated genes

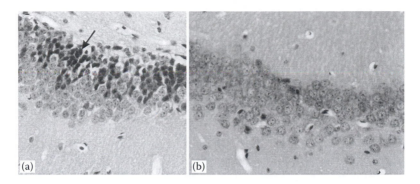

Figure 1.4 Comparison of the hippocampus in a mouse after ketamine treatment for 3 months and an untreated control. (a) In the ketamine-treated mouse, the hippocampus has more spindle-shaped cells with irregular and dense nuclei (arrow) than (b) in the ketamine-free control. Magnification, ×400.

were related to mRNAs and proteins of the alpha-5 subunits of GABA (A) receptors, an interactive site of ketamine. After 6 months of addiction, the mice exhibited significant deterioration in muscle strength and nociceptive response (Sun et al. 2011). In addition, cell death was clearly depicted by pyknotic and irregular nuclei in many parts of the cortex as well as by TUNEL staining (Yeung et al. 2010). In histological studies, the loss in density observed in damaged loci was frequently observed adjacent to normal areas (Figure 1.2), indicating that damage was initially localized, only appearing in many regions of the gray matter after lengthy episodes of addiction. Interestingly, not only were the lesions cortical, they were also seen in the diencephalon (Figure 1.5). Although thalamic lesions were

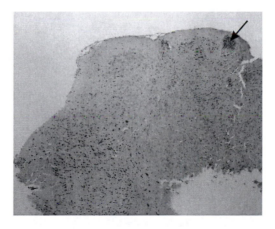

Figure 1.5 A lesion (arrow) in the diencephalon of a mouse treated with ketamine for 6 months. Magnification, ×50.

also sometimes seen in human addicts, diencephalic lesions were rarely observed in monkeys. This supports the idea that ketamine may (as demonstrated in the mouse) affect sensory areas of the brain and other areas of the diencephalon.

In the zebrafish, spatial memory loss was induced by putting the fish in a tank dosed with ketamine (0.01 mg/ml of freshwater) for 2 h. Some 6 h after the end of the treatment, the fish had still not moved from the quadrant of the aquarium that they had initially occupied. In mammalian models, prediction error, emotional learning, and inference are usually impaired, along with a loss of simple spatial memory after a long period of ketamine treatment (Bolton et al. 2012; Corlett et al. 2011).

The monkey, being a close "relative" of the human, is another model for exploring chronic ketamine toxicity. In the early phase of chronic ketamine treatment, the monkey's brain, much like that of a human in the early stages of addiction, showed episodes of diffuse hyperactivity over the cortex (Figure 1.6). This sudden burst of temporary cortical hyperactivity was not confined to particular areas and usually lasted between 2 weeks and 1 month. In addition, even though the animals being used in experiments were young adults, the episodes of hyperactivity recorded were suggested to more resemble a naïve and developing brain suddenly exposed to stimulation (Fang et al. 2005). After 1 month of treatment, cell death and diminished motor activities began to be established (Sun et al. 2012). Furthermore, in contrast to the increase of catecholaminergic excitatory neurons along the tectum and midline of the mesencephalon of the mice, in the monkey model, there was a decrease in fMRI activity in the ventral tegmentum, substantia nigra, posterior cingulate, and visual cortex, and hyperfunction was evident in the entorhinal and the

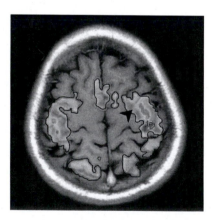

Figure 1.6 fMRI of the brain of a human addict on ketamine for 6 months showing diffuse hypersensitivities (arrowhead) after a simple sensory stimulation.

striatal areas instead (Yu et al. 2012). Two explanations for the neuronal reactivity differences observed in the mouse and monkey have been suggested. The first is simply the variation in the different species. However, the more likely answer is that the differences in hyperfunction are independent of the catecholaminergic neurons or their projections. Despite the differences observed, it is essential to note that the midbrain, the limbic system, and the striatal axis are all instrumental to hallucination, psychosis, and tuning of motor activities (especially the striatum). Another crucial area for motor coordination, the cerebellum, has been shown to be down-regulated in humans after ketamine addiction (Chan et al. 2012). Thus, there appears to be a correlation between the perturbing behavior of ataxia and loss of coordination in these subjects.

Psychological studies on human ketamine addicts are difficult to tackle in that subjects are typically unwilling to expose his or her identity because of the stigma surrounding illicit drug use. Even those willing to participate do not talk much about personal neurological signs and symptoms for fear of being labeled as "psychotic." As a consequence, many studies on how ketamine affects the human CNS are limited to a small number of participants. Table 1.1 illustrates the relatively small number of results

Table 1.1 Profiles of CNS Derangements According to Psychiatric Examinations of 42 Human Ketamine Addicts

	≤2	4–8	>10
Duration of addiction (years)	≤2	4–8	>10
Number of patients	3	34	5
Categories of derangement			
Loss of memory			
Sporadic loss of recent memory	✓	✓	✓
Loss of some retrograde memory	—	—	✓
Difficulties in logical deduction of reasonably complex problems (multiplication followed by addition)		✓[a]	✓
Autonomic disturbances (including sweating, irregular bowel movement, stomach dyspepsia, palpitation)	✓	✓	✓
Neurosis, depression, and uncontrollable temper		✓[b]	✓[b]

Note: The average patient dose of ketamine was at least 2 g/day and at least 3 days/week usually consecutive and bridging over weekends.

[a]   In this category, 3 patients with addiction of >7 years exhibited difficulties.
[b]   In this category, 1 patient with addiction of 4 years and 35 patients with addiction of >5 years exhibited neurosis, depression, and uncontrollable temper.

obtained from one of these CNS studies. In a sequence of interviews with these subjects, researchers observed a general pattern of memory loss, autonomic disturbance, and emotional problems (Yew et al. unpublished report). As most of these volunteers met with researchers in discussion groups where they were sitting, it was not possible to assess any problem in mobility (Table 1.1).

Equally challenging are studies on the brains of ketamine addicts. Wang et al. (2013) were fortunate enough to gain consent of a small number of ketamine addicts with between 6 months and 12 years of addiction and who were willing to have their brains examined by magnetic resonance imaging (MRI). This experiment yielded some preliminary observations on the relationship between the sites of damage, the progression of damage over time, and the degree of damage of these addicts' brains. A detailed evaluation of an average dose consumption revealed that the majority of subjects were on high dosages of nothing less than 1 g of ketamine per day. This was true not only for this group but also for other groups that subsequently visited. The subjects were also taking ketamine at least 3 days/week. Damage of the nervous system in these addicts began with vesicle formation (either diffuse or localized) in the white matter of the brain. If the vesicles were localized, they were usually present in the anterior one-third of the brain (Wang et al. 2013). Damage to the white matter seems to be a common feature as it has been observed in the mouse as well as in primates and humans. In mice, the damage observed often involves edema of the white matter loci and the loss of axons (Figure 1.7). In humans, most of the vesicular lesions of the white matter were documented by MRI at early stages (i.e., typically within 1 year) of ketamine addiction. Subsequently, the gray matter is affected and a high level of

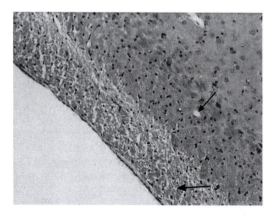

Figure 1.7 Degeneration of white fibers (arrows) in the white matter of a mouse treated with ketamine for 6 months. Magnification, ×400.

damage is sustained after addiction of between 4 and 7 years (Wang et al. 2013). A high number of different areas of the brain have been recorded as having lesions; these include the cortex, hippocampus, basal forebrain, striatum, and the cerebellum, with the brainstem showing damage during the later years of addiction (Wang et al. 2013). Lesions in the brainstem were usually found deep to the surface and were close to the midline, possibly involving reticular formation (Wang et al. 2013). Figure 1.8a through d show representative examples of some of the degenerative changes that occur at around 6 to 7 years of addiction in humans.

Although ketamine-induced lesions in the human nervous system have been well documented, detailed mechanisms of the defects still remain elusive. A decisive prognosis to understand and counteract the neurodegenerative effects of ketamine addiction depends on further research. At the moment, it is clear that ketamine use leads to major neuronal lesions and that most lesions appear by 4 years of addiction. However, for polydrug users, these lesions are likely to appear much earlier, even as early as 1 year of addiction (Wang et al. 2013).

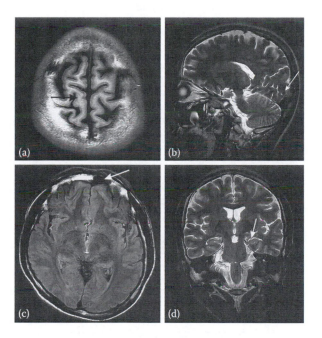

Figure 1.8 MRI images of the brains of human ketamine addicts. Atrophy of (a) the cortex and (b) the occipital cortex (arrow) of a human addict on ketamine for 7 years. Note the thin gyrus (arrow) in (a). (c) An area of the brain of a human addict on ketamine for 6 years, which is partially devoid of the meninges (arrow). (d) Start of the degenerative changes (e.g., ruffling of cortex; arrow) observed in the limbic system of a human addict on ketamine for 7 years.

References

Bahetwar, S.K., Pandey, R.K., Saksena, A.K., and Chandra, G. 2011. A comparative evaluation of intranasal midazolam, ketamine and their combination for sedation of young uncooperative pediatric dental patients: A triple blind randomized crossover trial. *J. Clin. Pediatr. Dent.* 35:415–420.

Behdad, S., Hajiesmaeili, M.R., Abbasi, H.R., Ayatollahi, V., Khadiv, Z., and Sedaghat, A. 2013. Analgesic effects of intravenous ketamine during spinal anesthesia in pregnant women undergone caesarean section: A randomized clinical trial. *Anesth. Pain Med.* 3:230–233.

Bolton, M.M., Heaney, C.F., Sabbagh, J.J., Murtishaw, A.S., Magcalas, C.M., and Kinney, J.W. 2012. Deficits in emotional learning and memory in an animal model of schizophrenia. *Behav. Brain Res.* 233:35–44.

Braidy, H.F., Singh, P., and Ziccardi, V.B. 2011. Safety of deep sedation in an urban oral and maxillofacial surgery training program. *J. Oral Maxillofac. Surg.* 69:2112–2119.

Carter, J., and Story, D.A. 2013. Veterinary and human anesthesia: An overview of some parallels and contrasts. *Anesth. Intensive Care* 41:710–718.

Chan, W.M., Xu, J., Fan, M., Jiang, Y., Tsui, T.Y., Wai, M.S., Lam, W.P., and Yew, D.T. 2012. Downregulation in the human and mice cerebella after ketamine versus ketamine plus alcohol treatment. *Microsc. Res. Tech.* 75:258–264.

Cillo, J.E. Jr. 2012. Analysis of propofol and low-dose ketamine admixtures for adult outpatient dentoalveolar surgery: A prospective, randomized, positive-controlled clinical trial. *J Oral Maxillofac Surg.* 70(3):537–546.

Corlett, P.R., Honey, G.D., Krystal, J.H., and Fletcher, P.C. 2011. Glutamatergic model psychosis: Prediction error, learning, and inference. *Neuropsychopharmacology* 36:294–315.

Creeley, C., Dikranian, K., Dissen, G., Martin, L., Olney, J., and Brambrink, A. 2013. Propofol-induced apoptosis of neurones and oligodendrocytes in fetal and neonatal rhesus macaque brain. *Br. J. Anesth.* 110 suppl 1:i29–i38.

Eghbal, M.H., Taregh, S., Amin, A., and Sahmeddini, M.A. 2013. Ketamine improves postoperative pain and emergence agitation following adenotonsillectomy in children. A randomized clinical trial. *Middle East J. Anesthesiol.* 22:155–160.

Fang, C.K., Chen, H.W., Wang, W.H., Liu, R.S., and Hwang, J.J. 2013. Acute effects of three club drugs on the striatum of rats: Evaluation by quantitative autoradiography with [18F] FDOPA. *Appl. Radiat. Isot.* 77:153–159.

Fang, M., Lorke, D.E., Li, J., Gong, X., Yew, J.C., and Yew, D.T. 2005. Postnatal changes in functional activities of the pig's brain: A combined functional magnetic resonance imaging and immunohistochemical study. *Neurosignals* 14:222–233.

Featherstone, R.E., Liang, Y., Saunders, J.A., Tatard-Leitman, V.M., Ehlichman, R.S., and Siegel, S.J. 2012. Subchronic ketamine treatment leads to permanent changes in EEG, cognition and the astrocytic glutamate transporter EAAT2 in mice. *Neurobiol. Dis.* 47:338–346.

Félix, L.M., Antunes, L.M., and Coimbra, A.M. 2014. Ketamine NMDA receptor-independent toxicity during zebrafish (*Danio rerio*) embryonic development. *Neurotoxicol. Teratol.* 41:27–34.

Honarmand, A., Safavi, M., and Khalighinejad, F. 2013. A comparison of the effect of pretreatment with intravenous dexamethasone, intravenous ketamine, and their combination for suppression of remifentanil induced cough: A randomized, double-blind placebo-controlled clinical trial. *Adv. Biomed. Res.* 2:60. doi: 10.4103/2277-9175.115808.

Hosseinzadeh, H., Eidy, M., Golzari, S., and Vasebi, M. 2013. Hemodynamic stability during induction of anesthesia in elderly patients: Propofol + ketamine versus propofol + etomidate. *J. Cardiovasc. Thorac. Res.* 5:51–54.

Javid, M.J., Hajijafari, M., Hajipour, A., Makarem, J., and Khazaeipour, Z. 2012. Evaluation of a low dose ketamine in post tonsillectomy pain relief: A randomized trial comparing intravenous and subcutaneous ketamine in pediatrics. *Anesth. Pain Med.* 2:85–89.

Kanungo, J., Cuevas, E., Ali, S.F., and Paule, M.G. 2013. Ketamine induces motor neuron toxicity and alters neurogenic and proneural gene expression in zebrafish. *J. Appl. Toxicol.* 33:410–417.

Khajavi, M., Emami, A., Etezadi, F., Safari, S., Sharifi, A., and Shariat Moharari, R. 2013. Conscious sedation and analgesia in colonoscopy: Ketamine/propofol combination has superior patient satisfaction versus fentanyl/propofol. *Anesth. Pain Med.* 3:208–213.

Lerche, P., Nolan, A.M., and Reid, J. 2000. Comparative study of propofol or propofol and ketamine for the induction of anesthesia in dogs. *Vet. Rec.* 146:571–574.

Liu, J.R., Baek, C., Han, X.H., Shoureshi, P., and Soriano, S.G. 2013. Role of glycogen synthase kinase 3β in ketamine-induced developmental neuroapoptosis in rats. *Br. J. Anesth.* 110 suppl 1:i3–i9.

Mak, Y.T., Lam, W.P., Lü, L., Wong, Y.W., and Yew, D.T. 2010. The toxic effect of ketamine on SH-SY5Y neuroblastoma cell line and human neuron. *Microsc. Res. Tech.* 73:195–201.

Mellon, R.D., Simone, A.F., and Rappaport, B.A. 2007. Use of anesthesia in neonates and young children. *Anesth. Analg.* 104:509–520.

Mendola, C., Cammarota, G., Netto, R., Cecci, G., Pisterna, A., Ferrante, D., Casadio, C., and Della Corte, F. 2012. S+ ketamine for control of perioperative pain and prevention of post thoracotomy pain syndrome: A randomized, double–blind study. *Minerva Anestesiol.* 78:757–766.

Mion, G., and Villevielle, T. 2013. Ketamine pharmacology: An update (pharmacodynamics and molecular aspects, recent findings. *CNS Neurosci. Ther.* 19:370–380.

Olney, J.W., Wozniak, D.F., Jevtovic-Todorovic, V., Farber, N.B., Bittigau, P., and Ikonomidou, C. 2002. Drug-induced apoptotic neurodegeneration in the developing brain. *Brain Pathol.* 12:488–498.

Patil, S., and Anitescu, M. 2012. Efficacy of outpatient ketamine infusions in refractory chronic pain syndromes: A 5 year retrospective analysis. *Pain Med.* 13:263–269.

Riehl, R., Kyzar, E., Allain, A., Green, J., Hook, M., Monnig, L., Rhymes, K., Roth, A., Pham, M., Razavi, R., Dileo, J., Gaikwad, S., Hart, P., and Kalueff, A.V. 2011. Behavioral and physiological effects of acute ketamine exposure in adult zebrafish. *Neurotoxicol. Teratol.* 33:658–667.

Roberts, E.R., Curran, H.V., Friston, K.J., and Morgan, C.J. 2014. Abnormalities in white matter microstructure associated with chronic ketamine use. *Neuropsychopharmacology* 39:329–338.

Safavi, M., Honarmand, A., and Nematollahy, Z. 2011. Pre-incisional analgesia with intravenous or subcutaneous infiltration of ketamine reduces postoperative pain in patients after open cholecystectomy: A randomized, double-blind, placebo-controlled study. *Pain Med.* 12:1418–1426.

Seliskar, A., Nemec, A., Roskar, T., and Butinar, J. 2007. Total intravenous anaesthesia with propofol or propofol/ketamine in spontaneously breathing dogs premedicated with medetomidine. *Vet Rec.* 160(3):85–91.

Singh, H., Kundra, S., Singh, R.M., Grewal, A., Kaul, T.K., and Sood, D. 2013. Preemptive analgesia with ketamine for laparoscopic cholecystectomy. *J. Anesthesiol. Clin. Pharmacol.* 29:478–484.

Sun, L., Lam, W.P., Wong, Y.W., Lam, L.H., Tang, H.C., Wai, M.S., Mak, Y.T., Pan, F., and Yew, D.T. 2011. Permanent deficits in brain functions caused by long-term ketamine treatment in mice. *Hum. Exp. Toxicol.* 30:1287–1296.

Sun, L., Li, Q., Li, Q., Zhang, Y., Liu, D., Jiang, H., Pan, F., and Yew, D.T. 2012. Chronic ketamine exposure induces permanent impairment of brain functions in adolescent cynomolgus monkeys. *Addict Biol.* 19:185–194.

Tan, S., Lam, W.P., Wai, M.S., Yu, W.H., and Yew, D.T. 2012. Chronic ketamine administration modulates midbrain dopamine system in mice. *PLoS One* 7:e43947. doi: 10.1371/journal.pone.0043947.

Tan, S., Rudd, J.A., and Yew, D.T. 2011. Gene expression changes in $GABA_A$ receptors and cognition following chronic ketamine administration in mice. *PLoS One* 6:e21328. doi: 10.1371/journal.pone.0021328.

Turner, C.P., Gutierrez, S., Liu, C., Miller, L., Chou, J., Finucane, B., Carnes, A., Kim, J., Shing, E., Haddad, T., and Phillips, A. 2012. Strategies to defeat ketamine-induced neonatal brain injury. *Neuroscience* 210:384–392.

Wang, C., Zheng, D., Xu, J., Lam, W., and Yew, D.T. 2013. Brain damages in ketamine addicts as revealed by magnetic resonance imaging. *Front. Neuroanat.* 7:23. doi: 10.3389/fnana.00023 (Epub 2013).

Yeung, L.Y., Wai, M.S., Fan, M., Mak, Y.T., Lam, W.P., Li, Z., Lu, G., and Yew, D.T. 2010. Hyperphosphorylated tau in the brains of mice and monkeys with long-term administration of ketamine. *Toxicol. Lett.* 193:189–193.

Youssef-Ahmed, M.Z., Silver, P., Nimkoff, L., and Sagy, M. 1996. Continuous infusion of ketamine in mechanically ventilated children with refractory bronchospasm. *Intensive Care Med.* 22:972–976.

Yu, H., Li, Q., Wang, D., Shi, L., Lu, G., Sun, L., Wang, L., Zhu, W., Mak, Y.T., Wong, N., Wang, Y., Pan, F., and Yew, D.T. 2012. Mapping the central effects of chronic ketamine administration in an adolescent primate model by functional magnetic imaging (fMRI). *Neurotoxicol.* 33:70–77.

chapter two

Clinical applications and side effects of ketamine

Ou Sha, Yue Hao, Eric Yu-Pang Cho, and Li Zhou

Contents

2.1 Introduction

Ketamine is one of the core medicines recorded in the Essential Drugs List of the World Health Organization (World Health Organization 2011). It has various effects in humans, including anesthesia, analgesia, hallucinations, elevated blood pressure, and bronchodilation (Peck et al. 2008). Ketamine is primarily used for the induction and maintenance of general anesthesia, usually in combination with a sedative. Other clinical applications include analgesia, sedation, treatment of bronchospasm, and treatment of depression and bipolar disorder (Diazgranados et al. 2010). The potential therapeutic uses of ketamine are currently being investigated, particularly in treating heroin and alcohol addiction (Jovaisa et al. 2006; Krupitsky et al. 2007) and seizures (Erdogan Kayhan et al. 2012; Gaspard et al. 2013). Like other drugs of its class, ketamine induces a state referred to as "dissociative anesthesia" (Bergman 1999) and it is thus also popular as a recreational drug.

Initially known as CI-581, ketamine was first synthesized in 1962 as a replacement for phencyclidine (PCP), which had a range of adverse effects (Domino 2010). Like PCP, ketamine was shown to be a potent "dissociative anesthetic" that produced profound anesthesia and analgesia, but with a shorter duration of action and fewer psychotomimetic side effects (Corssen and Domino 1966). After Food and Drug Administration approval in 1970, ketamine anesthesia was first used on American soldiers during the Vietnam War. In the nearly 50 years since then, ketamine has gained popularity in clinical practice, being used for both veterinary and human clinical anesthesia, and it is still a popular topic in medical research (Golpayegani et al. 2012; Kranaster et al. 2014; Lizarraga and Chambers 2012; Wagner et al. 2012). On the other hand, soon after it was introduced to medicine, many young US individuals started to use it as a drug to make them dispirited so as to protest against the US war in Vietnam (Domino 2010). Indeed, ketamine is still one of the main recreational club drugs that is used worldwide today, because of its low expense and easy production (Brown and Melton 2011; Chakraborty et al. 2011; Joe-Laidler and Hunt 2008). However, ketamine might also result in a series of toxic effects in the urinary system, locomotor system, and nervous system (Chen et al. 2011; Wiley et al. 2011; Yu et al. 2012). Ketamine has gained a great deal of attention from researchers for decades not only because of its clinical importance but also because of the social and health problems accompanying its use. This chapter mainly reviews the clinical uses and adverse effects of ketamine, as well as the progress that has been made in recent years in the research of its use.

2.2 Mechanisms of action

Ketamine, or 2-(o-chlorophenyl)-2-(methylamino)cyclohexanone, is a PCP and cyclohexamine derivative (Bergman 1999). Pharmacologically, its main action is on glutamate and it has been classified as a noncompetitive N-methyl-D-aspartate (NMDA) receptor antagonist (Harrison and Simmonds 1985). NMDA receptor blockade appears to be the primary mechanism of the anesthetic and analgesic effects of ketamine (Pai and Heining 2007). Ketamine has also been found to bind to other receptors; for example, it blocks muscarinic acetylcholine receptors, descending monoaminergic pain pathways, and voltage-gated calcium channels (Pharmaceutical Society of Australia 2011). Ketamine binds to and acts as a weak agonist to opioid receptors with a preference for the mu and kappa receptors. Ketamine may potentiate the effects of gamma-aminobutyric acid (GABA) synaptic inhibition, induce activation of dopamine release, and reduce the presynaptic release of glutamate. Ketamine also has local anesthetic properties, possibly through its ability to inhibit neuronal sodium channels (Pai and Heining 2007).

Peripherally, ketamine inhibits the reuptake of catecholamines, stimulating the sympathetic nervous system and resulting in cardiovascular symptoms. Ketamine also inhibits neuronal uptake and increases serotonergic activity, thought to be the underlying basis of nausea and vomiting. In addition, ketamine induces catecholamine release and stimulation of β2 adrenergic receptors to cause bronchodilation (Aroni et al. 2009).

2.3 Enantiomers

Ketamine is a chiral compound whereby the two enantiomers exhibit pharmacological and clinical differences. S-(+)-Ketamine shows a three- to fourfold greater affinity for the NMDA receptor than R-(+)-ketamine. It is also more potent as an anesthetic and as an analgesic agent than R-(+)-ketamine and the racemic mixture (White et al. 1985). In contrast, R-(+)-ketamine is associated with posthypnotic stimulatory properties and agitated behavior (Goldberg et al. 2010). Animal studies have shown that R-(+)-ketamine is a more potent relaxant of acetylcholine-induced airway smooth muscle contraction than S-(+)-ketamine. This difference appears to be caused by differential actions on receptor-linked calcium channels (Pabelick et al. 1997). The two enantiomers have similar pharmacokinetic but different pharmacodynamic profiles. Clinical studies have shown that, compared with the racemic mixture, S-(+)-ketamine has fewer psychological side effects and a shorter recovery time when used in anesthesia (Pfenninger et al. 2002). While the S-form has been commercially available for several years (Schüttler 1992), the *R*-form is not, although it has been suggested that the use of the latter has advantages, at least at lower doses (Mathisen et al. 1995).

2.4 Clinical applications

2.4.1 Anesthesia

Ketamine was once considered to be an ideal anesthetic for general anesthesia, but soon it was found to possess a relatively high risk of psychological side effects (Berti et al. 2009; Muetzelfeldt et al. 2008; Persson 2010). Patients often reported a variety of unusual symptoms when recovering from ketamine anesthesia. These emergence phenomena included delusions, hallucinations, confusion, and sometimes "out-of-body" and "near-death" experiences. These phenomena led to ketamine being withdrawn from mainstream anesthetic use in humans. However, ketamine and its complexes or combinations are still widely used and studied for use in anesthesia in specialized clinics, especially in pediatrics, dentistry, and psychiatry (Kaviani et al. 2011).

High-risk patients with cardiorespiratory disorders represent one of the prime candidates for ketamine anesthesia. Considering that ketamine tends to increase or maintain cardiac output (Adams 1997), it is sometimes used in anesthesia for emergency surgery when the patient's fluid volume status is unknown. Extensive use of ketamine in pediatric cardiac catheterizations has shown it to be highly effective with fewer catheter-associated arrhythmias than other general anesthetics. On the other hand, ketamine might be deleterious in patients with limited right ventricular functional reserve and increased pulmonary vascular resistance (Pai and Heining 2007). However, ketamine suppresses breathing much less than many other available anesthetics (Heshmati et al. 2003). It is therefore the anesthetic of choice when reliable ventilation equipment is not available. It is possible to perform ketamine anesthesia without protective measures to the airways. Furthermore, in patients with reactive airway disease, ketamine can be useful as it produces bronchodilation and profound analgesia, allowing administration of an increased inspiratory oxygen concentration (Pai and Heining 2007).

Recent studies have mainly focused on regional anesthesia using ketamine, which is considered safer and more effective than general anesthesia. For example, in children undergoing cleft palate surgery, infiltration with either ketamine or bupivacaine at the surgical site provides adequate analgesia without major side effects. Ketamine is superior to bupivacaine in terms of the requirement for a rescue analgesic, peaceful sleep pattern, and early resumption of feeding (Jha et al. 2013). In addition, ketamine is also commonly used as an adjuvant in local and regional pediatric anesthesia (Mossetti et al. 2012). Furthermore, an increasing number of applications of ketamine that were previously considered to be contraindicated have now been reported. For example, ketamine anesthesia has usually not been used in schizophrenia patients for fear of causing disease exacerbation. However, a recent preliminary clinical evaluation (with six patients) demonstrated that ketamine is safe for use as an anesthetic for electroconvulsive therapy in schizophrenia, in that it does not provoke dissociative symptoms (Kranaster et al. 2014).

2.4.2 Analgesia

Recently, many efforts have been made to study the analgesic effects of ketamine (Abrishamkar et al. 2012; Guedes et al. 2012; Kajiume et al. 2012; Polson et al. 2012) as it has been reported to effectively relieve both acute and chronic pain (Cohen et al. 2011; Domino 2010; Swartjes et al. 2011). A number of studies have reported the acute analgesic effects of ketamine; for instance, a dose of 0.5 mg/kg given approximately 15 min before surgery by the intravenous route was shown to provide analgesia for 24 h after surgery in patients undergoing appendectomy (Honarmand et al. 2012). In addition, a low dose of orally administered ketamine not only relieved postoperative pain but also was shown to be beneficial in enhancing the effect of local anesthetics (Kaviani et al. 2011). The preventive administration of S-(+)-ketamine via 12-h infusion was safe and also had anti-hyperalgesic action after cesarean section (Suppa et al. 2012). In addition, ketamine is currently one of the most important adjuncts in helping to reach the desired effect when administered at drug-specific modes and at proven effective dosages throughout the perioperative period (Rakhman et al. 2011; Safavi et al. 2012; Weinbroum 2012). For example, a low dose of ketamine applied intravenously with fentanyl in pediatric patients could reduce pain scores after the Nuss procedure without increasing the side effects (Cha et al. 2012). The psychotropic side effects must be taken into account when ketamine is used to relieve acute pain. They can, however, usually be avoided by concomitant application of a sedative such as a benzodiazepine (Elia and Tramèr 2005). Indeed, acute postoperative pain that is not treated properly can transform into chronic pain, which may have a major negative impact on the quality of a patient's life (Weinbroum 2012). However, studies have shown that ketamine is also effective in relieving chronic pain. For example, in a recent case study, a low-dose infusion of ketamine in a patient who was receiving long-term, high-dose intrathecal hydromorphone therapy immediately relieved the painful myoclonus in the lower extremities associated with opioid-induced hyperalgesia (Forero et al. 2012). In another case, intractable neuropathic pain caused by ulnar nerve entrapment was treated with ketamine (Hesselink and Kopsky 2012). Another example of chronic pain is complex regional pain syndrome

(CRPS), which has been clinically recognized for more than 140 years, and still lacks a satisfactory treatment. CRPS is a severe chronic pain condition characterized by sensory, autonomic, motor, and dystrophic signs and symptoms. A dynamic change in the physiology and structure of central pain projecting neurons mediated through the NMDA receptor has been found in CRPS patients. Ketamine blocks central sensitization through its effects on the NMDA receptor and has been shown to be effective in treating severely ill patients with generalized CRPS. It is now being used as an experimental and controversial treatment for CRPS (Azari et al. 2012). Indeed, one study showed that a 4-h ketamine infusion that escalated from 40 to 80 mg over a 10-day period resulted in a significant reduction in pain with increased mobility and a tendency to decreased autonomic dysregulation in CRPS patients. In the study, a total of 33 patients diagnosed with CRPS were treated with a continuous subanesthetic dose of intravenous ketamine at least once. Due to relapse, 12 out of the 33 patients received a second course of therapy, and 2 of the 33 patients received a third course. The results showed that there was complete pain relief in 25 (76%), partial relief in 6 (18%), and no relief in 2 (6%) patients (Goldberg et al. 2005).

Ketamine is also a very effective analgesic for other chronic pain syndromes, including chronic pancreatitis pain, postherpetic neuralgia, migraine, burns, neuropathies, and fibromyalgia (Bouwense et al. 2011; Domino 2010). Recently, in a 5-year study to investigate whether outpatient intravenous ketamine infusions were satisfactory for pain relief in patients suffering from various chronic intractable pain syndromes, it was shown that subanesthetic ketamine infusions could improve pain scores on a visual analog scale, and in approximately half of the 49 patients investigated, relief lasted for up to 3 weeks with minimal side effects (Patil and Anitescu 2012). In addition, ketamine has also be used as an effective adjunct to epidural corticosteroid therapy for chronic pains, such as chronic lumbar radicular pain (Amr 2011).

Ketamine has also been used in small doses (0.1–0.5 mg/kg) for the treatment of pain associated with movement and neuropathic pain (Lynch et al. 2005). In addition, it has been used as an intravenous coanalgesic with opiates to manage otherwise intractable pain, particularly if this pain is neuropathic (pain attributed to vascular insufficiency or shingles). It has the added benefit of counteracting spinal sensitization or the wind-up phenomena experienced with chronic pain. At these doses, the psychotropic side effects are less apparent and well managed with benzodiazepines (Elia and Tramèr 2005). Ketamine is most effective when used in combination with a low dose of an opioid that itself has analgesic effects but that can cause disorienting side effects at the higher doses required if used alone (Elia and Tramèr 2005). The combined use of ketamine with an opioid is also particularly useful for the pain experienced during cancer (Saito et al. 2006).

When ketamine is repeatedly used in chronic pain management, it can show some side effects, such as urological toxicity and hepato-toxicity (Bell 2012; Noppers et al. 2011; Sear 2011). For better analgesic effects with fewer side effects, new analogs of ketamine are continually being synthesized and studied. Recently, 2-[*p*-methoxybenzylamino]-2-[*p*-methoxyphenyl] cyclohexanone, (i.e., ket-OCH3) and 2-[*p*-methylbenzyl-amino]-2-[*p*-methoxyphenyl] cyclohexanone (i.e., ket-CH3) as well as their intermediates were reported to be effective for decreasing pain in rats (Ahmadi et al. 2012). Thus, research to determine the optimal use of different ketamine derivatives as analgesics is still ongoing.

2.4.3 Sedation

One of the main challenges in pediatric dentistry is being able to treat uncooperative or combative young patients safely and effectively. Various sedative agents have been studied for this purpose including oral keta-mine, which has been reported to have side effects (Damle et al. 2008). Therefore, combinations of drugs have been designed to achieve better sedative effects and to minimize the side effects (Motamed et al. 2012). For example, a combination of ketamine and midazolam was reported to be safe and effective, and when administered by the intranasal route, this mixture could produce moderate sedation during dental care procedures to pediatric patients who would otherwise be given treatment under general anesthesia (Bahetwar et al. 2011). This combination also effectively decreased the anxiety scores of all the young patients observed, who were referred for treatment under general anesthesia (Golpayegani et al. 2012). Regarding its use in veterinary medicine, low-dose ketamine–diazepam is another combination used for short-duration chemical restraint; this was suggested to be suitable to assist in the physical restraint required during blood sampling for assessment of hematologic, serum biochemi-cal, and coagulation parameters in cats (Reynolds et al. 2012). In addition, ketamine was reported to be an effective and safe sedative for use in agi-tated human patients with psychiatric illness during aeromedical trans-port by the Royal Flying Doctor Service in Australia; after treatment, the drug prevented agitation in patients for the subsequent 72 h (Le Cong et al. 2012). However, it has been suggested that ideally ketamine should be tailored to individual patients with different conditions (Taylor et al. 2011).

2.4.4 Antidepression

The antidepressant effects of ketamine are well reported (Engin et al. 2009; Mathew et al. 2012; Zhou et al. 2012). For example, it has been shown to decrease the "behavioral despair" of rats in the forced swim test, a widely used animal model of antidepressant drug action. This effect was not confounded by side effects on general activity and was comparable to that of the standard antidepressant drug fluoxetine (Engin et al. 2009). In a randomized placebo-controlled clinical study, ketamine was found to significantly improve treatment-resistant major depression within hours of injection, and this improvement lasted for up to 1 week after just a single dose (Zarate et al. 2006). These findings were corroborated by Liebrenz et al. (2009) and Goforth and Holsinger (2007). Liebrenz et al. (2009) successfully treated a patient with treatment-resistant major depression and a co-occurring alcohol and benzodiazepine dependence by giving an intravenous infusion of ketamine over a period of 50 min, while Goforth and Holsinger (2007) demonstrated the marked improvement in a patient with severe, recurrent major depressive disorder within 8 h of receiving a preoperative dose of ketamine and one treatment of electroconvulsive therapy with bitemporal electrode placement. Ketamine has also been applied in depressed patients who have undergone orthopedic surgery where it was found to significantly decrease the depressed mood, suicidal tendencies, somatic anxiety, and hypochondriasis, when compared with the control group (Kudoh et al. 2002). More recently, a two-site randomized controlled clinical trial of ketamine treatment in patients with treatment-resistant depression found that 64% of the patients responded after 24 h according to the Montgomery–Åsberg Depression Rating Scale, compared with just 28% who responded to midazolam (Murrough et al. 2013). These findings led to the current focus in both academia and industry of ketamine being a treatment for depression. Extensive preclinical characterization of the effects of ketamine has partly illuminated its mechanism of action.

It has been suggested that the glutamatergic system plays an important role in the treatment of major depressive disorder in that different types of glutamate receptors, including NMDA, α-amino-3-hydroxy-5-methyl-4-isoxazolepropionic acid (AMPA), and metabotropic glutamate receptors or transporters, appear to be involved in the etiology of depression and in the mechanisms of action of antidepressants (Angelucci et al. 2012; Murrough 2012; Tokita et al. 2012). This is a departure from the previous theories, which focused on serotonin and norepinephrine. With ketamine being an NMDA antagonist, it is being considered as a new generation of antidepressants with fast-onset effects (Autry et al. 2011). Indeed, the antidepressive effects of ketamine have long been attributed to the fact that it is an NMDA receptor antagonist (Orser et al. 1997). However, a recent study in mice showed that blocking the NMDA receptor is only an intermediate step, with the subsequent activation of the AMPA receptor being crucial for ketamine's rapid antidepressant actions (Angelucci et al. 2012). It has recently been reported that sigma receptor–mediated neuronal remodeling may also contribute to the antidepressant effects of ketamine (Robson et al. 2012). Thus, a lot remains to be discovered about the antidepressant mechanisms of this drug.

2.4.5 Antiepilepsy

Glutamate overflow plays an important role during epileptic seizures and seizure-related brain damage. Antagonists of glutamate receptors, especially the ionotropic NMDA and AMPA receptors, have thus attracted much attention in treating epilepsy in recent years. As the only clinical drug of its type, the antiepileptic activity of ketamine has been evaluated in several animal models. For example, ketamine was shown to significantly prevent lidocaine- and transauricular electrode–induced seizures in mice (Guler et al. 2005; Manocha et al. 2001), and bicuculline-induced seizure in rats (Schneider and Rodríguez de Lores Arnaiz 2013). In humans, ketamine has also been shown to be a promising treatment for epilepsy, especially in status epilepticus. A ketamine–propofol combination was suggested to be an alternative strategy to enhance the seizure quality and clinical efficiency of electroconvulsive therapy (Erdogan Kayhan et al. 2012). In some neurological intensive care units, ketamine has also been used in cases of prolonged seizures. Some evidence indicates that the NMDA-blocking effect of the drug protects the neurons from glutamatergic damage during prolonged seizures (Fujikawa 1995). Most recently, a retrospective multicenter study showed that ketamine appears to be a relatively effective and safe drug for the treatment of refractory status epilepticus. In this study, permanent control of refractory status epilepticus was achieved in 57% (34 of 60) of the episodes. Ketamine was believed to have contributed to permanent control in 32% (19 of 60) of the episodes including 7 (12%) in which ketamine was the last drug to be administered and the most likely to cause immediate action (Gaspard et al. 2013).

2.4.6 Treatment for alcohol and heroin addiction

Ketamine-assisted psychotherapy (KPT) sessions have been used to treat heroin and alcohol addiction with encouraging results. A double-blind randomized clinical trial of KPT for heroin addiction showed that high-dose KPT (with ketamine at 2.0 mg/kg) elicited a full psychedelic experience in heroin addicts as assessed quantitatively by the Hallucinogen Rating Scale. On the other hand, low-dose KPT (with ketamine at 0.2 mg/kg) elicited "subpsychedelic" experiences with ketamine-facilitated guided imagery. High-dose KPT produced a significantly greater rate of abstinence in heroin addicts within the first 2 years of follow-up, a greater and longer-lasting reduction in craving for heroin, and a greater positive change in nonverbal unconscious emotional attitudes, than did low-dose KPT (Krupitsky et al. 2002). In another study, patients with heroin dependence were randomized into two treatment groups. One group received addiction counseling sessions followed by KPT injections while the other (control) group received addiction counseling sessions but no additional ketamine therapy sessions. The results showed that 50% of the KPT group remained abstinent, when compared with just 22.2% in the control group (Krupitsky et al. 2007). Similarly, Jovaisa et al. (2006) demonstrated attenuation of opiate withdrawal symptoms with ketamine, suggesting its potential use in treating drug addiction.

2.4.7 Others

Ketamine has also been proven to exert anti-inflammatory effects. As an NMDA receptor antagonist, ketamine acts at different levels of the inflammatory response, playing a role in inflammatory cell recruitment, cytokine production, and inflammatory mediator regulation. The resultant effect of these interactions confers to ketamine an anti-proinflammatory effect by limiting exacerbation of systemic inflammation without affecting local healing processes (Loix et al. 2011). A couple of recent studies also showed that ketamine could limit soman-induced neuroinflammation (Dhote et al. 2012; Dorandeu et al. 2013). Ketamine was also shown to attenuate postoperative cognitive dysfunction just 1 week after cardiac surgery and this effect may be related to its anti-inflammatory properties (Hudetz et al. 2009). Combinations of ketamine and atropine have been shown to be neuroprotective, reducing neuroinflammation after a toxic status epilepticus in mice (Dhote et al. 2012). Ketamine was also reported to have inhibitory effects on microglial activation, at least partially attributed to the inhibition of ERK1/2 phosphorylation (Chang et al. 2009). Furthermore, ketamine can inhibit proliferation of and induce apoptosis in pheochromocytoma cells in a concentration-dependent manner (Zuo et al. 2011). Ketamine was also observed to induce apoptosis in human lymphocytes and neuronal cells via the mitochondrial pathway (Braun et al. 2010). As more potential clinical applications of ketamine are reported, updated reviews on ketamine research become necessary.

2.5 Adverse effects

2.5.1 Acute adverse effects

Up to 40% of patients may experience side effects with continuous subcutaneous infusion of ketamine. These include dizziness, blurred vision, altered hearing, hypertension, nausea and vomiting, vivid dreams, and hallucinations (Quibell et al. 2011). However, ketamine is considered to be relatively safe and does not result in severe side effects when used at a low dose and within a short period.

As an NMDA receptor antagonist, ketamine can trigger excessive glutamate release and subsequent cortical excitation, which may induce psychosis-like behavior and cognitive anomalies (Yu et al. 2012). Acute effects of ketamine administration are primarily linked to aberrant activation of the prefrontal cortex and limbic structures with elevated glutamate and dopamine levels, which can provoke dose-dependent positive and negative schizophrenia-like symptoms. In rhesus macaques, at both fetal and neonatal stages of development, the brain is sensitive to the apoptogenic action of ketamine, and an exposure for 5 h is sufficient to induce a significant neuroapoptotic response (Brambrink et al. 2012). The pattern of neurodegeneration induced by ketamine was different in fetuses from that in neonates, and the loss of neurons attributable to ketamine exposure was 2.2 times greater in the fetal than in the neonatal brain. However, current research suggests that acute ketamine exposure does not cause significant neurotoxicity.

2.5.2 Chronic adverse effects

Ketamine is a common drug of abuse for youths worldwide (Lankenau and Sanders 2007; Lankenau et al. 2010; Pavarin 2006). The number of ketamine abusers has grown rapidly in recent years, because it is cheap and easily available, and can produce desirable short-term sensations of excitement, dream-like states, hallucinations, and vivid imagery (Gutkin et al. 2012). However, repeated use of ketamine can result in severe toxicity and cause health problems such as those outlined in the following.

2.5.2.1 Uropathy

It has been well documented that ketamine is excreted in the urine and can cause damage to the upper and lower urinary tracts (Chan et al. 2012). Patients may suffer from imperative urinary frequency, urgency, pollakisuria, dysuria, hematuria, and cystitis (Chen et al. 2011; Gutkin et al. 2012; Lieb et al. 2012). For example, a case of chronic cystitis associated with ureteral strictures in a young ketamine abuser has been reported (Huang et al. 2011).

Ketamine-induced ulcerative cystitis is a recently identified condition that can have a severe and potentially long-lasting impact on ketamine users. Shahani et al. (2007) first documented this condition in nine ketamine-dependent users. Computerized tomography scans of these individuals revealed a marked thickening of the bladder wall, a small bladder capacity, and perivesicular stranding, consistent with severe inflammation. All patients showed severe ulcerative cystitis at cystoscopy. Biopsies in four of these cases found denuded urothelial mucosa with thin layers of reactive and regenerating epithelial cells and ulcerations with vascular granulation tissue and scattered inflammatory cells. Urinary tract symptoms appear to be most common in those using ketamine on a daily basis over an extended period (Morgan and Curran 2012). Since ketamine is not administered chronically in a typical clinical setting, these symptoms have presented in just one medical case of ketamine treatment. However, following dose reduction, the symptoms were reduced (Morgan and Curran 2012). Management of these symptoms primarily involves keta-mine cessation, for which compliance is low. Other treatments that have been used include antibiotics, nonsteroidal anti-inflammatory drugs, ste-roids, anticholinergics, and cystodistension (Middela and Pearce 2011). Both hyaluronic acid instillation and combined pentosan polysulfate and ketamine cessation have been shown to provide relief in some patients. In the latter case, however, it is unclear whether relief resulted from keta-mine cessation, administration of pentosan polysulfate, or a combination of both. Further follow-up is required to fully assess the efficacy of these treatments (Middela and Pearce 2011).

2.5.2.2 Neurotoxicity

Animal research has shown that ketamine abuse can result in a perma-nent deficit in brain function, such as significant deterioration in neuro-muscular strength and nociception (Sun et al. 2011). Functional magnetic resonance imaging of adolescent cynomolgus monkey brains has also shown that repeated exposure to ketamine markedly reduced neural activity in the ventral tegmental area and substantia nigra in the midbrain, poste-rior cingulate cortex, and visual cortex (Yu et al. 2012). In addition, the mesolimbic, mesocortical, and entorhinostriatal systems were found to be functionally vulnerable to chronic ketamine administration, and dys-functions of these neural circuits have been implicated in several neuro-psychiatric disorders including depression, schizophrenia, and attention deficit disorder. Furthermore, the cerebella activity in humans and mice was down-regulated and the number of apoptotic cells significantly increased in ketamine users (Chan et al. 2012). Moreover, ketamine can induce neurotoxicity and changes in gene expression in the developing rat brain (Liu et al. 2011). In mice, chronic exposure to ketamine was found to impair working memory and up-regulate gene expression of GABAA5 in the prefrontal cortex (Tan et al. 2011). The neurodegenerative process induced by ketamine was also suggested to be similar to that during aging and Alzheimer's disease (Yeung et al. 2010). Short-term exposure of ketamine at high concentrations to cultures of GABAergic neurons led to a significant loss of differentiated cells in one study, and non-cell death–inducing concentrations of ketamine (10 µg/ml) may still initiate long-term alterations of dendritic arbors in differentiated neurons. The same study also demonstrated that chronic administration of ketamine at concentra-tions as low as 0.01 µg/ml interfered with the maintenance of dendritic arbor architecture. These results raise the possibility that chronic expo-sure to low, subanesthetic concentrations of ketamine, while not affecting cell survival, might still impair neuronal maintenance and development (Hargreaves et al. 1994; Vutskits et al. 2007).

2.5.2.3 Depression

Ketamine is generally acknowledged to possess antidepressant properties. However, increased depression (assessed via the Beck Depression
Inventory) in both daily users and ex-ketamine users was found over the
course of 1 year (Morgan et al. 2010) but not in current but infrequent (i.e.,
>1 per month/<3 times per week) users. This elevated level of depression
was at a subclinical level and the increase was not correlated with changes
in ketamine use. Evidence suggests that prolonged central nervous system depression probably results from the gabapentin and ketamine interaction during postoperative recovery following cervical laminoplasty
(Elyassi et al. 2011).

2.5.2.4 Cognitive impairment

The NMDA receptor mediates the form of synaptic plasticity known as
long-term potentiation, which is central for learning and memory. Given
that ketamine is an antagonist of the NMDA receptor, the consequences
of ketamine use on cognition have been widely investigated. In humans,
a single dose of ketamine induces a marked, dose-dependent impairment
in working and episodic memory, which can profoundly affect the users'
ability to function (Morgan et al. 2006).

Several studies have examined cognitive function in both infrequent
and frequent ketamine users (Curran and Monaghan 2001; Morgan et al.
2006; Morgan and Curran 2006; Narendran et al. 2005). Overall, infrequent
or recreational ketamine use does not cause cognitive deficits (Narendran
et al. 2005). The most robust findings are that frequent ketamine users
exhibit profound impairments in both short- and long-term memory
(Morgan et al. 2006). In a longitudinal study, frequent ketamine use caused
impairments in both visual recognition and spatial working memory
that correlated with changes in the level of ketamine use over 12 months
(Stewart 2001). Other impairments in planning and frontal functions have
also been observed but appear so far to be unrelated to the level of ketamine use (Morgan et al. 2009). Memory impairments may be reversible
when ketamine use is discontinued, as they were not found in a group of
30 ex-ketamine users who had been abstinent for at least a year. The cognitive consequences of repeated ketamine use in pediatric anesthesia would
also merit further investigation (Istaphanous and Loepke 2009).

2.5.2.5 Psychosis

Ketamine can induce transient positive and negative symptoms of schizophrenia in healthy volunteers (Krystal et al. 1994). In schizophrenic patients
who have been stabilized on antipsychotic medication, ketamine causes
a resurgence of psychotic symptoms (Lahti et al. 1995), which are similar
to those exhibited during the acute phase of their illness (Malhotra et al.
1997).

Preclinical studies demonstrated that a small number (i.e., ~5) of repeated doses of ketamine given to rats induced "schizophrenia-like" changes such as abnormal hippocampal neurogenesis, and the reduction in hippocampal parvalbumin–containing GABAergic interneurons (Keilhoff et al. 2004), as well as increased dopamine binding in the hippocampus and decreased glutamate binding in the prefrontal cortex (Becker et al. 2003). A study with infrequent (i.e., >1 per month/<3 times per week) and daily users of ketamine as well as polydrug users who do not use ketamine (controls) in which subclinical psychotic symptomatology were assessed found that scores on the measure of delusion, dissociation, and schizotypy are highest in daily ketamine users when compared with infrequent users and lowest in the polydrug controls (Curran and Morgan 2000; Morgan et al. 2010). Morgan et al. (2010) also found that daily ketamine users showed a similar pattern of "basic symptoms" to individuals prodromal for schizophrenia (Morgan et al. 2010). However, there is no evidence of clinical psychotic symptoms in infrequent ketamine users (Narendran et al. 2005). Despite the presence of anecdotal reports (Jansen 2001), there is little evidence of any link between chronic, heavy use of ketamine and diagnosis of a psychotic disorder.

2.5.2.6 Respiratory toxicity

According to a literature review, ketamine may also result in airway obstruction in 10% to 20% of users (Strayer and Nelson 2008), although this adverse event typically does not affect the beneficial outcomes of ketamine.

2.5.2.7 Cardiotoxicity

Long-term abuse of ketamine can cause significant ventricular myocardial apoptosis, fibrosis, and sympathetic sprouting, which alters the electrophysiological properties of the heart and increases its susceptibility to malignant arrhythmia leading to sudden cardiac death (Li et al. 2012).

2.5.2.8 Hepatotoxicity

The hepatotoxic effect of ketamine when it is used for chronic pain management has also been described (Noppers et al. 2011). In case reports of three patients treated with ketamine repeatedly to relieve chronic pain, liver enzyme abnormalities were observed, but they returned to the normal range on cessation of the drug. These findings suggest that liver enzymes must be monitored during such treatment (Sear 2011).

2.5.2.9 Kidney dysfunction

Another emerging physical health problem associated with the frequent use of high doses of ketamine appears to be hydronephrosis (water on the kidney), which is secondary to urinary tract problems. In a study of ketamine-induced ulcerative cystitis, Chu et al. (2008) reported that

30 (51%) patients presented with either unilateral (7%) or bilateral (44%) hydronephrosis. On initial assessment, four patients also showed papillary necrosis (destruction of kidney cells), and this led to renal failure in one patient who had complete obstruction of the urethra (Chu et al. 2008).

2.5.2.10 Other side effects

Other problems caused by a chronic administration of ketamine include severe abdominal pain in the gastrointestinal tract, biliary dilatation, and bilateral corneal edema (Gutkin et al. 2012; Starte et al. 2012).

2.6 Abuse

Ketamine is also used as a recreational drug worldwide at nightclubs, dance parties, and rave scenes where it is commonly known as "Special K," "Vitamin K," or "SuperK" (Wolff and Winstock 2006). Although ketamine is a controlled substance, in recent years, its illicit use has increased rapidly in many countries (United Nations Office on Drug Control 2010). Most commonly, ketamine comes as a powder, but it can also be obtained in liquid and tablet form. It can be insufflated or placed in beverages. At low doses, ketamine induces distortion of time and space, hallucinations, and mild dissociative effects. According to users, the most appealing aspects of ketamine use are the sensation of "melting into the surroundings," "visual hallucinations," "out-of body experiences," and "giggliness" (Stewart 2001). At high doses, ketamine induces a more severe state of dissociation commonly referred to as a "K-hole," wherein the user experiences intense detachment to the point that their perceptions appear completely divorced from their previous reality. Some users (i.e., astronauts of the psyche or "psychonauts") value these profoundly altered states of consciousness, whereas others see the resulting decreased sociability as a less appealing aspect of ketamine use.

In conclusion, ketamine, a noncompetitive antagonist of the NMDA receptor, is frequently used in human and veterinary medicine, especially in pediatric patients, as an analgesic and sedative agent under tightly controlled conditions. However, it is often illegally used as a recreational drug primarily by young adults, often at "rave" parties and nightclubs. The adverse effects caused by the repeated use of ketamine and the underlying mechanisms involved should be well studied in order to provide solutions to combat them.

Acknowledgments

Grant support: National Natural Science Foundation of China (No. 81171154), Collaborative Innovation Program of Shenzhen (GJHS20120621153317134), and Basic Research Program of Shenzhen (JCYJ20120613113228732).

References

Abrishamkar, S., Eshraghi, N., Feizi, A., Talakoub, R., Rafiei, A., and Rahmani, P. 2012. Analgesic effects of ketamine infusion on postoperative pain after fusion and instrumentation of the lumbar spine: A prospective randomized clinical trial. *Med. Arh.* 66:107–110.

Adams, H.A. 1997. S-(+)-ketamine. Circulatory interactions during total intravenous anesthesia and analgesia-sedation. *Anaesthesist* 46:1081–1087.

Ahmadi, A., Khalili, M., Hajikhani, R., Hosseini, H., Afshin, N., and Nahri-Niknafs, B. 2012. Synthesis and study the analgesic effects of new analogues of ketamine on female wistar rats. *Med. Chem.* 8:246–251.

Amr, Y.M. 2011. Effect of addition of epidural ketamine to steroid in lumbar radiculitis: One-year follow-up. *Pain Physician* 14:475–481.

Angelucci, F., Caltagirone, C., and Ricci, V. 2012. In response to: AMPA receptor potentially participates in the mediation of the increased brain-derived neurotrophic factor following chronic ketamine use by Yang, Jianjun; Zhou, Zhiqiang; Yang, Chun. *Psychopharmacology* 220:245.

Aroni, F., Iacovidou, N., Dontas, I., Pourzitaki, C., and Xanthos, T. 2009. Pharmacological aspects and potential new clinical applications of ketamine: Reevaluation of an old drug. *J. Clin. Pharmacol.* 49:957–964.

Autry, A.E., Adachi, M., Nosyreva, E., Na, E.S., Los, M.F., Cheng, P.F., Kavalali, E.T., and Monteggia, L.M. 2011. NMDA receptor blockade at rest triggers rapid behavioural antidepressant responses. *Nature* 475:91–95.

Azari, P., Lindsay, D.R., Briones, D., Clarke, C., Buchheit, T., and Pyati, S. 2012. Efficacy and safety of ketamine in patients with complex regional pain syndrome: A systematic review. *CNS Drugs* 26:215–228.

Bahetwar, S.K., Pandey, R.K., Saksena, A.K., and Chandra, G. 2011. A comparative evaluation of intranasal midazolam, ketamine and their combination for sedation of young uncooperative pediatric dental patients: A triple blind randomized crossover trial. *J. Clin. Pediatr. Dent.* 35:415–420.

Becker, A., Peters, B., Schroeder, H., Mann, T., Huether, G., and Grecksch, G. 2003. Ketamine-induced changes in rat behaviour: A possible animal model of schizophrenia. *Prog. Neuropsychopharmacol. Biol. Psychiatry* 27:687–700.

Bell, R.F. 2012. Ketamine for chronic noncancer pain: Concerns regarding toxicity. *Curr. Opin. Support Palliat. Care* 6:183–187.

Bergman, S.A. 1999. Ketamine: Review of its pharmacology and its use in pediatric anesthesia. *Anesth. Prog.* 46:10–20.

Berti, M., Baciarello, M., Troglio, R., and Fanelli, G. 2009. Clinical uses of low-dose ketamine in patients undergoing surgery. *Curr. Drug Targets* 10:707–715.

Bouwense, S.A., Buscher, H.C., vanGoor, H., and Wilder-Smith, O.H. 2011. S-ketamine modulates hyperalgesia in patients with chronic pancreatitis pain. *Reg. Anesth. Pain Med.* 36:303–307.

Brambrink, A.M., Evers, A.S., Avidan, M.S., Farber, N.B., Smith, D.J., Martin, L.D., Dissen, G.A., Vreeley, C.E., and Olney, J.W. 2012. Ketamine-induced neuroapoptosis in the fetal and neonatal rhesus macaque brain. *Anesthesiology* 116:372–384.

Braun, S., Gaza, N., Werdehausen, R., Hermanns, H., Bauer, I., Durieux, M.E., Hollmann, M.W., and Stevens, M.F. 2010. Ketamine induces apoptosis via the mitochondrial pathway in human lymphocytes and neuronal cells. *Br. J. Anaesth.* 105:347–354.

Brown, S.D., and Melton, T.C. 2011. Trends in bioanalytical methods for the determination and quantification of club drugs: 2000–2010. *Biomed. Chromatogr.* 25:300–321.

Cha, M.H., Eom, J.H., Lee, Y.S., Kim, W.Y., Park, Y.C., Min, S.H., and Kim, J.H. 2012. Beneficial effects of adding ketamine to intravenous patient-controlled analgesia with fentanyl after the Nuss procedure in pediatric patients. *Yonsei Med. J.* 53:427–432.

Chakraborty, K., Neogi, R., and Basu, D. 2011. Club drugs: Review of the "rave" with a note of concern for the Indian scenario. *Indian J. Med. Res.* 133:594–604.

Chan, W.M., Xu, J., Fan, M., Jiang, Y., Tsui, T.Y., Wai, M.S., Lam, W.P., and Yew, D.T. 2012. Downregulation in the human and mice cerebella after ketamine versus ketamine plus ethanol treatment. *Microsc. Res. Tech.* 75:258–264.

Chang, Y., Lee, J.J., Hsieh, C.Y., Hsiao, G., Chou, D.S., and Sheu, J.R. 2009. Inhibitory effects of ketamine on lipopolysaccharide-induced microglial activation. *Mediators Inflamm.* 2009:705379. doi: 10.1155/2009/705379.

Chen, C.H., Lee, M.H., Chen, Y.C., and Lin, M.F. 2011. Ketamine-snorting associated cystitis. *J. Formos. Med. Assoc.* 110:787–791.

Chu, P.S., Ma, W.K., Wong, S.C., Chu, R.W., Cheng, C.H., Wong, S., Tse, J.M., Lau, F.L., Yiu, M.K., and Man, C.W. 2008. The destruction of the lower urinary tract by ketamine abuse: A new syndrome? *BJU Int.* 102:1616–1622.

Cohen, S.P., Liao, W., Gupta, A., and Plunkett, A. 2011. Ketamine in pain management. *Adv. Psychosom. Med.* 30:139–161.

Corssen, G., and Domino, E.F. 1966. Dissociative anesthesia: Further pharmacologic studies and first clinical experience with the phencyclidine derivative CI-581. *Anesth. Analg.* 45:29–40.

Curran, H.V., and Monaghan, L. 2001. In and out of the K-hole: A comparison of the acute and residual effects of ketamine in frequent and infrequent ketamine users. *Addiction* 96:749–760.

Curran, H.V., and Morgan, C. 2000. Cognitive, dissociative and psychotogenic effects of ketamine in recreational users on the night of drug use and 3 days later. *Addiction* 95:575–590.

Damle, S.G., Gandhi, M., and Laheri, V. 2008. Comparison of oral ketamine and oral midazolam as sedative agents in pediatric dentistry. *J. Indian Soc. Pedod. Prev. Dent.* 26:97–101.

Dhote, F., Carpentier, P., Barbier, L., Peinnequin, A., Baille, V., Pernot, F., Testylier, G., Beaup, C., Foquin, A., and Dorandeu, F. 2012. Combinations of ketamine and atropine are neuroprotective and reduce neuroinflammation after a toxic status epilepticus in mice. *Toxicol. Appl. Pharmacol.* 259:195–209.

Diazgranados, N., Ibrahim, L., Brutsche, N.E., Newberg, A., Kronstein, P., Khalife, S., Kammerer, W.A., Quezado, Z., Luckenbaugh, D.A., Salvadore, G., Machado-Vieira, R., Manji, H.K., and Zarate, C.A. Jr. 2010. A randomized add-on trial of an N-methyl-D-aspartate antagonist in treatment-resistant bipolar depression. *Arch. Gen. Psychiatry* 67:793–802.

Domino, E.F. 2010. Taming the ketamine tiger. *Anesthesiology* 113:678–686.

Dorandeu, F., Barbier, L., Dhote, F., Testylier, G., and Carpentier, P. 2013. Ketamine combinations for the field treatment of soman-induced self-sustaining status epilepticus. Review of current data and perspectives. *Chem. Biol. Interact.* 203:154–159.

Elia, N., and Tramèr, M.R. 2005. Ketamine and postoperative pain—A quantitative systematic review of randomised trials. *Pain* 113:61–70.

Elyassi, A.R., Long, R.P., Bejnarowicz, R.P., and Schoneboom, B.A. 2011. Possible gabapentin and ketamine interaction causing prolonged central nervous system depression during post-operative recovery following cervical laminoplasty: A case report. *J. Med. Case Rep.* 28:167.

Engin, E., Treit, D., and Dickson, C.T. 2009. Anxiolytic- and antidepressant-like properties of ketamine in behavioral and neurophysiological animal models. *Neuroscience* 161:359–369.

Erdogan Kayhan, G., Yucel, A., Colak, Y.Z., Ozgul, U., Yologlu, S., Karlidag, R., and Ersoy, M.O. 2012. Ketofol (mixture of ketamine and propofol) administration in electroconvulsive therapy. *Anaesth. Intensive Care* 40:305–310.

Forero, M., Chan, P.S.L., and Restrepo-Garces, C.E. 2012. Successful reversal of hyperalgesia/myoclonus complex with low-dose ketamine infusion. *Pain Pract.* 12:154–158.

Fujikawa, D.G. 1995. Neuroprotective effect of ketamine administered after status epilepticus onset. *Epilepsia* 36:186–195.

Gaspard, N., Foreman, B., Judd, L.M., Brenton, J.N., Nathen, B.R., McCoy, B.M., Al-Otaibi, A., Kilbride, R., Fernández, I.S., Mendoza, L., Samuel, S., Zakaria, A., Kalamangalam, G.P., Legros, B., Szaflarski, J.P., Loddenkemper, T., Hahn, C.D., Goodkin, H.P., Claassen, J., Hirsch, L.J., and Laroche, S.M. 2013. Intravenous ketamine for the treatment of refractory status epilepticus: A retrospective multicenter study. *Epilepsia* 54:1498–1503.

Goforth, H.W., and Holsinger, T. 2007. Rapid relief of severe major depressive disorder by use of preoperative ketamine and electroconvulsive therapy. *J. ECT* 23:23–25.

Goldberg, M.E., Domsky, R., Scaringe, D., Hirsh, R., Dotson, J., Sharaf, I., Torjman, M.C., and Schwartzman, R.J. 2005. Multi-day low dose ketamine infusion for the treatment of complex regional pain syndrome. *Pain Physician* 8:175–179.

Goldberg, M.E., Torjman, M.C., Schwartzman, R.J., Mager, D.E., and Wainer, I.W. 2010. Pharmacodynamic profiles of ketamine (R-) and (S+) with five day inpatient infusion for the treatment of complex regional pain syndrome. *Pain Physician* 13:379–387.

Golpayegani, M.V., Dehghan, F., Ansari, G., and Shaveghi, S. 2012. Comparison of oral Midazolam-Ketamine and Midazolam-Promethazine as sedative agents in pediatric dentistry. *Dent. Res. J.* 9:36–40.

Guedes, A.G., Matthews, N.S., and Hood, D.M. 2012. Effect of ketamine hydrochloride on the analgesic effects of tramadol hydrochloride in horses with signs of chronic laminitis-associated pain. *Am. J. Vet. Res.* 73:610–619.

Guler, G., Erdogan, F., Golgeli, A., Akin, A., and Boyaci, A. 2005. Ketamine reduces lidocaine-induced seizures in mice. *Int. J. Neurosci.* 115:1239–1244.

Gutkin, E., Hussain, S.A., and Kim, S.H. 2012. Ketamine-induced biliary dilatation: From Hong Kong to New York. *J. Addict. Med.* 6:89–91.

Hargreaves, R.J., Hill, R.G., and Iversen, L.L. 1994. Neuroprotective NMDA antagonists: The controversy over their potential for adverse effects on cortical neuronal morphology. *Acta Neurochir. Suppl. (Wien)* 60:15–19.

Harrison, N.L., and Simmonds, M.A. 1985. Quantitative studies on some antagonists of N-methyl-D-aspartate in slices of rat cerebral cortex. *Br. J. Pharmacol.* 84:381–391.

Heshmati, F., Zeinali, M.B., Noroozinia, H., Abbacivash, R., and Mahoori, A. 2003. Use of ketamine in severe status asthmaticus in intensive care unit. *Iran. J. Allergy Asthma Immunol.* 2:175–180.

Hesselink, J.M., and Kopsky, D.J. 2012. Intractable neuropathic pain due to ulnar nerve entrapment treated with cannabis and ketamine 10%. *J. Clin. Anesth.* 24:78–79.

Honarmand, A., Safavi, M., and Karaky, H. 2012. Preincisional administration of intravenous or subcutaneous infiltration of low-dose ketamine suppresses postoperative pain after appendectomy. *J. Pain Res.* 5:1–6.

Huang, P.W., Meng, E., Cha, T.L., Sun, G.H., Yu, D.S., and Chang, S.Y. 2011. 'Walking-stick ureters' in ketamine abuse. *Kidney Int.* 80:895. doi: 10.1038/ki.2011.242.

Hudetz, J.A., Iqbal, Z., Gandhi, S.D., Patterson, K.M., Byrne, A.J., Hudetz, A.G., Pagel, P.S., and Warltier, D.C. 2009. Ketamine attenuates post-operative cognitive dysfunction after cardiac surgery. *Acta Anaesthesiol. Scand.* 53:864–872.

Istaphanous, G.K., and Loepke, A.W. 2009. General anesthetics and the developing brain. *Curr. Opin. Anaesthesiol.* 22:368–373.

Jansen, K. 2001. *Ketamine: Dreams and Realities.* Multidisciplinary Association for Psychedelic Studies, Sarasota, FL.

Jha, A.K., Bhardwaj, N., Yaddanapudi, S., Sharma, R.K., and Mahajan, J.K. 2013. A randomized study of surgical site infiltration with bupivacaine or ketamine for pain relief in children following cleft palate repair. *Paediatr. Anaesth.* 23:401–406.

Joe-Laidler, K., and Hunt, G. 2008. Sit down to float: The cultural meaning of ketamine use in Hong Kong. *Addict. Res. Theory* 16:259–271.

Jovaisa, T., Laurinenas, G., Vosylius, S., Sipylaite, J., Badaras, R., and Ivaskevicius, J. 2006. Effects of ketamine on precipitated opiate withdrawal. *Medicina* 42:625–634.

Kajiume, T., Sera, Y., Nakanuno, R., Ogura, T., Karakawa, S., Kobayakawa, M., Taguchi, S., Oshita, K., Kawaguchi, H., Sato, T., and Kobayashi, M. 2012. Continuous Intravenous infusion of ketamine and lidocaine as adjuvant analgesics in a 5-year-old patient with neuropathic cancer pain. *J. Palliat. Med.* 15:719–722.

Kaviani, N., Khademi, A., Ebtehaj, I., and Mohammadi, Z. 2011. The effect of orally administered ketamine on requirement for anesthetics and postoperative pain in mandibular molar teeth with irreversible pulpitis. *J. Oral Sci.* 53:461–465.

Keilhoff, G., Bernstein, H.G., Becker, A., Grecksch, G., and Wolf, G. 2004. Increased neurogenesis in a rat ketamine model of schizophrenia. *Biol. Psychiatry* 56:317–322.

Kranaster, L., Hoyer, C., Janke, C., and Sartorius, A. 2014. Preliminary evaluation of clinical outcome and safety of ketamine as an anesthetic for electroconvulsive therapy in schizophrenia. *World J. Biol. Psychiatry* 15:242–250.

Krupitsky, E.M., Burakov, A.M., Dunaevsky, I.V., Romanova, T.N., Slavina, T.Y., and Grinenko, A.Y. 2007. Single versus repeated sessions of ketamine-assisted psychotherapy for people with heroin dependence. *J. Psychoactive Drugs* 39:13–19.

Krupitsky, E., Burakov, A., Romanova, T., Dunaevsky, I., Strassman, R., and Grinenko, A. 2002. Ketamine psychotherapy for heroin addiction: Immediate effects and two-year follow-up. *J. Subst. Abuse Treat.* 23:273–283.

Krystal, J.H., Karper, L.P., Seibyl, J.P., Freeman, G.K., Delaney, R., Bremner, J.D., Heninger, G.R., Bowers, M.B. Jr., and Charney, D.S. 1994. Subanesthetic effects of the noncompetitive NMDA antagonist, ketamine, in humans. Psychotomimetic, perceptual, cognitive, and neuroendocrine responses. *Arch. Gen. Psychiatry* 51:199–214.

Kudoh, A., Takahira, Y., Katagai, H., and Takazawa, T. 2002. Small-dose ketamine improves the postoperative state of depressed patients. *Anesth. Analg.* 95:114–118.

Lahti, A.C., Koffel, B., LaPorte, D., and Tamminga, C.A. 1995. Subanesthetic doses of ketamine stimulate psychosis in schizophrenia. *Neuropsychopharmacology* 13:9–19.

Lankenau, S.E., Bloom, J.J., and Shin, C. 2010. Longitudinal trajectories of ketamine use among young injection drug users. *Int. J. Drug Policy* 21:306–314.

Lankenau, S.E., and Sanders, B. 2007. Patterns of ketamine use among young injection drug users. *J. Psychoactive Drugs* 39:21–29.

Le Cong, M., Gynther, B., Hunter, E., and Schuller, P. 2012. Ketamine sedation for patients with acute agitation and psychiatric illness requiring aeromedical retrieval. *Emerg. Med. J.* 29:335–337.

Li, Y., Shi, J., Yang, B.F., Liu, L., Han, C.L., Li, W.M., Dong, D.L., Pan, Z.W., Liu, G.Z., Geng, J.Q., Sheng, L., Tan, X.Y., Sun, D.H., Gong, Z.H., and Gong, Y.T. 2012. Ketamine-induced ventricular structural, sympathetic and electrophysiological remodelling: Pathological consequences and protective effects of metoprolol. *Br. J. Pharmacol.* 165:1748–1756.

Lieb, M., Bader, M., Palm, U., Stief, C.G., and Baghai, T.C. 2012. Ketamine-induced vesicopathy. *Psychiatr. Prax.* 39:43–45.

Liebrenz, M., Stohler, R., and Borgeat, A. 2009. Repeated intravenous ketamine therapy in a patient with treatment-resistant major depression. *World J. Biol. Psychiatry* 10:640–643.

Liu, F., Paule, M.G., Ali, S., and Wang, C. 2011. Ketamine-induced neurotoxicity and changes in gene expression in the developing rat brain. *Curr. Neuropharmacol.* 9:256–261.

Lizarraga, I., and Chambers, J. 2012. Use of analgesic drugs for pain management in sheep. *N. Z. Vet. J.* 60:87–94.

Loix, S., De Kock, M., and Henin, P. 2011. The anti-inflammatory effects of ketamine: State of the art. *Acta Anaesthesiol. Belg.* 62:47–58.

Lynch, M.E., Clark, A.J., Sawynok, J., and Sullivan, M.J. 2005. Topical amitriptyline and ketamine in neuropathic pain syndromes: An open-label study. *J. Pain* 6:644–649.

Malhotra, A.K., Pinals, D.A., Adler, C.M., Elman, I., Clifton, A., Pickar, D., and Breier, A. 1997. Ketamine-induced exacerbation of psychotic symptoms and cognitive impairment in neuroleptic-free schizophrenics. *Neuropsychopharmacology* 17:141–150.

Manocha, A., Sharma, K.K., and Mediratta, P.K. 2001. Possible mechanism of anticonvulsant effect of ketamine in mice. *Indian J. Exp. Biol.* 39:1002–1008.

Mathew, S.J., Shah, A., Lapidus, K., Clark, C., Jarun, N., Ostermeyer, B., and Murrough, J.W. 2012. Ketamine for treatment-resistant unipolar depression: Current evidence. *CNS Drugs* 26:189–204.

Mathisen, L.C., Skjelbred, P., Skoglund, L.A., and Ove, I. 1995. Effect of ketamine, an NMDA receptor inhibitor, in acute and chronic orofacial pain. *Pain* 61:215–220.

Middela, S., and Pearce, I. 2011. Ketamine-induced vesicopathy: A literature review. *Int. J. Clin. Pract.* 65:27–30.

Morgan, C.J., and Curran, H.V. 2006. Acute and chronic effects of ketamine upon human memory: A review. *Psychopharmacology* 188:408–424.

Morgan, C.J., and Curran, H.V. 2012. Ketamine use: A review. *Addiction* 107:27–38.

Morgan, C.J., Muetzelfeldt, L., and Curran, H.V. 2009. Ketamine use, cognition and psychological wellbeing: A comparison of frequent, infrequent and ex-users with polydrug and non-using controls. *Addiction* 104:77–87.

Morgan, C.J., Muetzelfeldt, L., and Curran, H.V. 2010. Consequences of chronic ketamine self-administration upon neurocognitive function and psychological wellbeing: A 1-year longitudinal study. *Addiction* 105:121–133.

Morgan, C.J., Rossell, S.L., Pepper, F., Smart, J., Blackburn, J., Brandner, B., and Curran, H.V. 2006. Semantic priming after ketamine acutely in healthy volunteers and following chronic self-administration in substance users. *Biol. Psychiatry* 59:265–272.

Mossetti, V., Vicchio, N., and Ivani, G. 2012. Local anesthetis and adjuvants in pediatric regional anesthesia. *Curr. Drug Targets* 13:952–960.

Motamed, F., Aminpour, Y., Hashemian, H., Soltani, A.E., Najafi, M., and Farahmand, F. 2012. Midazolam-ketamine combination for moderate sedation in upper GI endoscopy. *J. Pediatr. Gastroenterol. Nutr.* 54:422–426.

Muetzelfeldt, L., Kamboj, S.K., Rees, H., Taylor, J., Morgan, C.J., and Curran, H.V. 2008. Journey through the K-hole: Phenomenological aspects of ketamine use. *Drug Alcohol Depend.* 95:219–229.

Murrough, J.W. 2012. Ketamine as a novel antidepressant: From synapse to behavior. *Clin. Pharmacol. Ther.* 91:303–309.

Murrough, J.W., Iosifescu, D.V., Chang, L.C., Al Jurdi, R.K., Green, C.E., Perez, A.M., Iqbal, S., Pillemer, S., Foulkes, A., Shah, A., Charney, D.S., and Mathew, S.J. 2013. Antidepressant efficacy of ketamine in treatment-resistant major depression: A two-site randomized controlled trial. *Am. J. Psychiatry* 170:1134–1142.

Narendran, R., Frankle, W.G., Keefe, R., Gil, R., Martinez, D., Slifstein, M., Kegeles, L.S., Talbot, P.S., Huang, Y., Hwang, D.R., Khenissi, L., Cooper, T.B., Laruelle, M., and Abi-Dargham. 2005. Altered prefrontal dopaminergic function in chronic recreational ketamine users. *Am. J. Psychiatry* 162:2352–2359.

Noppers, I.M., Niesters, M., Aarts, L.P., Bauer, M.C., Drewes, A.M., Dahan, A., and Sarton, E.Y. 2011. Drug-induced liver injury following a repeated course of ketamine treatment for chronic pain in CRPS type 1 patients: A report of 3 cases. *Pain* 152:2173–2178.

Orser, B.A., Pennefather, P.S., and MacDonald, J.F. 1997. Multiple mechanisms of ketamine blockade of N-methyl-D-aspartate receptors. *Anesthesiology* 86:903–917.

Pabelick, C.M, Rehder, K., Jones, K.A., Shumway, R., Lindahl, S.G., and Warner, D.O. 1997. Stereospecific effects of ketamine enantiomers on canine tracheal smooth muscle. *Br. J. Pharmacol.* 121:1378–1382.

Pai, A., and Heining, M. 2007. Ketamine. *Contin. Educ. Anaesth. Crit. Care Pain* 7:59–63.

Patil, S., and Anitescu, M. 2012. Efficacy of outpatient ketamine infusions in refractory chronic pain syndromes: A 5-year retrospective analysis. *Pain Med.* 13:263–269.

Pavarin, R.M. 2006. Substance use and related problems: A study on the abuse of recreational and not recreational drugs in Northern Italy. *Ann. 1st Super Sanità* 42:477–484.

Peck, T.E., Hill, S.A., and Williams, M. 2008. *Pharmacology for Anaesthesia and Intensive Care*, 3rd edition. Cambridge University Press, Cambridge, p. 111.

Persson, J. 2010. Wherefore ketamine? *Curr. Opin. Anaesthesiol.* 23:455–460.

Pfenninger, E.G., Durieux, M.E., and Himmelseher, S. 2002. Cognitive impairment after small-dose ketamine isomers in comparison to equianalgesic racemic ketamine in human volunteers. *Anesthesiology* 96:357–366.

Pharmaceutical Society of Australia. 2011. General anaesthetics. In: *Australian Medicines Handbook*. Australian Medicines Handbook Pty Ltd, Adelaide, p. 13.

Polson, S., Taylor, P., and Yates, D. 2012. Analgesia after feline ovariohysterectomy under midazolam-medetomidine-ketamine anaesthesia with buprenorphine or butorphanol, and carprofen or meloxicam: A prospective, randomised clinical trial. *J. Feline Med. Surg.* 14:553–559.

Quibell, R., Prommer, E.E., Mihalyo, M., Twycross, R., and Wilcock, A. 2011. Ketamine. *J. Pain Symptom Manage.* 41:640–649.

Rakhman, E., Shmain, D., White, I., Ekstein, M.P., Kollender, Y., Chazan, S., Dadia, S., Bickels, J., Amar, E., and Weinbroum, A.A. 2011. Repeated and escalating preoperative subanesthetic doses of ketamine for postoperative pain control in patients undergoing tumor resection: A randomized, placebo-controlled, double-blind trial. *Clin. Ther.* 33:863–873.

Reynolds, B.S., Geffré, A., Bourgès-Abella, N.H., Vaucoret, S., Mourot, M., Braun, J.P., and Trumel, C. 2012. Effects of intravenous, low-dose ketamine-diazepam sedation on the results of hematologic, plasma biochemical, and coagulation analyses in cats. *J. Am. Vet. Med. Assoc.* 240:287–293.

Robson, M.J., Elliott, M., Seminerio, M.J., and Matsumoto, R.R. 2012. Evaluation of sigma (σ) receptors in the antidepressant-like effects of ketamine in vitro and in vivo. *Eur. Neuropsychopharmacol.* 22:308–317.

Safavi, M., Honarmand, A., Habibabady, M.R., Baraty, S., and Aghadavoudi, O. 2012. Assessing intravenous ketamine and intravenous dexamethasone separately and in combination for early oral intake, vomiting and postoperative pain relief in children following tonsillectomy. *Med. Arh.* 66:111–115.

Saito, O., Aoe, T., Kozikowski, A., Sarva, J., Neale, J.H., and Yamamoto, T. 2006. Ketamine and N-acetylaspartylglutamate peptidase inhibitor exert analgesia in bone cancer pain. *Can. J. Anaesth.* 53:891–898.

Schneider, P.G., and Rodríguez de Lores Arnaiz, G. 2013. Ketamine prevents seizures and reverses changes in muscarinic receptor induced by bicuculline in rats. *Neurochem. Int.* 62:258–264.

Schüttler, J. 1992. S-(+)-ketamine. The beginning of a new ketamine era? *Anaesthesist* 41:585–587.

Sear, J.W. 2011. Ketamine hepato-toxicity in chronic pain management: Another example of unexpected toxicity or a predicted result from previous clinical and pre-clinical data? *Pain* 152:1946–1947.

Shahani, R., Streutker, C., Dickson, B., and Stewart, R.J. 2007. Ketamine-associated ulcerative cystitis: A new clinical entity. *Urology* 69:810–812.

Starte, J.M., Fung, A.T., and Kerdraon, Y.A. 2012. Ketamine-associated corneal edema. *Cornea* 31:572–574.

Stewart, C.E. 2001. Ketamine as a street drug. *Emerg. Med. Serv.* 30:30, 32, 34 passim.

Strayer, R.J., and Nelson, L.S. 2008. Adverse events associated with ketamine for procedural sedation in adults. *Am. J. Emerg. Med.* 26:985–1028.

Sun, L., Lam, W.P., Wong, Y.W., Lam, L.H., Tang, H.C., Wai, M.S., Mak, Y.T., Pan, F., and Yew, D.T. 2011. Permanent deficits in brain functions caused by long-term ketamine treatment in mice. *Hum. Exp. Toxicol.* 30:1287–1296.

Suppa, E., Valente, A., Catarci, S., Zanfini, B.A., and Draisci, G. 2012. A study of low-dose S-ketamine infusion as "preventive" pain treatment for cesarean section with spinal anesthesia: Benefits and side effects. *Minerva Anestesiol.* 78:774–781.

Swartjes, M., Morariu, A., Niesters, M., Aarts, L., and Dahan, A. 2011. Nonselective and NR2B-selective N-methyl-D-aspartic acid receptor antagonists produce antinociception and long-term relief of allodynia in acute and neuropathic pain. *Anesthesiology* 115:165–174.

Tan, S., Rudd, J.A., and Yew, D.T. 2011. Gene expression changes in GABA$_A$ receptors and cognition following chronic ketamine administration in mice. *PLoS One* 6:e21328. doi: 10.1371/journal.pone.0021328.

Taylor, D.M., Bell, A., Holdgate, A., MacBean, C., Huynh, T., Thom, O., Augello, M., Millar, R., Day, R., Williams, A., Ritchie, P., and Pasco, J. 2011. Risk factors for sedation-related events during procedural sedation in the emergency department. *Emerg. Med. Australas.* 23:466–473.

Tokita, K., Yamaji, T., and Hashimoto, K. 2012. Roles of glutamate signaling in preclinical and/or mechanistic models of depression. *Pharmacol. Biochem. Behav.* 100:688–704.

United Nations Office on Drug Control. 2010. *World Drug Report 2010.* United Nations Publications, New York.

Vutskits, L., Gascon, E., Potter, G., Tassonyi, E., and Kiss, J.Z. 2007. Low concentrations of ketamine initiate dendritic atrophy of differentiated GABAergic neurons in culture. *Toxicology* 234:216–226.

Wagner, A.E., Mama, K.R., Steffey, E.P., and Hellyer, P.W. 2012. Evaluation of infusions of xylazine with ketamine or propofol to modulate recovery following sevoflurane anesthesia in horses. *Am. J. Vet. Res.* 73:346–352.

Weinbroum, A.A. 2012. Non-opioid IV adjuvants in the perioperative period: Pharmacological and clinical aspects of ketamine and gabapentinoids. *Pharmacol. Res.* 65:411–429.

White, P.F., Schüttler, J., Shafer, A., Stanski, D.R., Horai, Y., and Trevor, A.J. 1985. Comparative pharmacology of the ketamine isomers: Studies in volunteers. *Br. J. Anaesth.* 57:197–203.

Wiley, J.L., Evans, R.L., Grainger, D.B., and Nicholson, K.L. 2011. Locomotor activity changes in female adolescent and adult rats during repeated treatment with a cannabinoid or club drug. *Pharmacol. Rep.* 63:1085–1092.

Wolff, K., and Winstock, A.R. 2006. Ketamine: From medicine to misuse. *CNS Drugs* 20:199–218.

World Health Organization. 2011. WHO Model List of Essential Medicines.

Yeung, L.Y., Wai, M.S.M., Fan, M., Mak, Y.T., Lam, W.P., Li, Z., Lu, G., and Yew, D.T. 2010. Hyperphosphorylated tau in the brains of mice and monkeys with long-term administration of ketamine. *Toxicol. Lett.* 193:189–193.

Yu, H., Li, Q., Wang, D., Shi, L., Lu, G., Sun, L., Wang, L., Zhu, W., Mak, Y.T., Wong, N., Wang, Y., Pan, F., and Yew, D.T. 2012. Mapping the central effects of chronic ketamine administration in an adolescent primate model by functional magnetic resonance imaging (fMRI). *Neurotoxicology* 33:70–77.

Zarate, C.A. Jr., Singh, J.B., Carlson, P.J., Brutsche, N.E., Ameli, R., Luckenbaugh, D.A., Charney, D.S., and Manji, H.K. 2006. A randomized trial of an N-methyl-D-aspartate antagonist in treatment-resistant major depression. *Arch. Gen. Psychiatry* 63:856–864.

Zhou, Z.Q., Zhang, G.F., Li, X.M., Yang, C., and Yang, J.J. 2012. Fast-spiking interneurons and gamma oscillations may be involved in the antidepressant effects of ketamine. *Med. Hypotheses* 79:85–86.

Zuo, Y.Y., Zhao, Y.B., Jiang, X.G., Gu, Z.L., Guo, C.Y., and Bian, S.Z. 2011. Effects of ketamine on proliferation and apoptosis of pheochromocytoma cell. *Fa Yi Xue Za Zhi* 27:405–408, 412.

chapter three

Diverse pharmacological properties of ketamine

Ismail Laher, Xin Zhang, Ping Chung Leung, and Willmann Liang

Contents

3.1 Introduction

The dissociative anesthetic ketamine is best known as an uncompetitive *N*-methyl-D-aspartate (NMDA) receptor antagonist; it is known to be dissociative because it creates the sense of detachment whereby there is a sense that the mind and body are separated in users. There is much known about how ketamine interacts with NMDA receptors to elicit its effects and readers are directed to other review articles for detailed accounts of ketamine's interactions with NMDA receptors. The focus of this chapter is to provide an overview of the current knowledge related to the effects of ketamine that are unrelated to its interactions with the NMDA receptor. The majority of information presented here comes from reviewing the literature published since 2003 and focuses on findings of ketamine with a variety of non-NMDA targets that are receptors, ion channels, or enzyme ligands. These targets and their responses to ketamine are listed in Table 3.1.

3.2 Central neurotransmitter targets

3.2.1 Dopamine receptors

Most of the pharmacological actions of ketamine, in both clinical and illicit use, are attributed to actions that are mediated by the central nervous system (CNS). The neurotransmitter dopamine is implicated in disorders such as Parkinson's disease and psychosis. Phencyclidine, which is also an NMDA receptor antagonist much like ketamine, also binds to the ion channel site of NMDA receptors and possesses dopamine-like properties as well. For example, prolactin secretion is suppressed and rotational behavior is increased by phencyclidine in rodents (Lozovsky et al. 1983; Mele et al. 1998). The increased locomotor activity observed in rats is blocked more effectively by the D_1 receptor antagonist SCH23390 than the D_2 receptor antagonist raclopride in the nucleus accumbens (NAc) (Matulewicz et al. 2010). Ketamine also increases the association (or binding potential) of D_1 receptors in the rat striatum (Momosaki et al. 2004). The availability of D_1 receptors is higher in ketamine users in the human prefrontal cortex (Narendran et al. 2005). The binding potential of D_1 receptors in these human subjects positively correlated with the extent of ketamine use (Narendran et al. 2005). The interactions of ketamine with dopamine receptors may underlie dopamine-associated prefrontal impairments, notably working memory deficits that are often seen in chronic ketamine users (Brozoski et al. 1979). Monkeys treated with ketamine demonstrate cognitive and behavioral symptoms that resemble those of schizophrenia (Roberts et al. 2010). The atypical antipsychotic risperidone, which blocks D_2 receptors, did not alleviate ketamine-induced symptoms (Roberts et al. 2010). Instead, prior activation of D_1 receptors (by the full agonist A77646 or by the partial agonist SKF38393) improves spatial memory performance (Roberts et al. 2010). Gross motor and hallucinatory-like behaviors are reduced by SKF38393 but not by A77646, a finding the authors attributed to the more specific D_1 receptor–mediated actions on working memory (Roberts et al. 2010). The function of D_1 receptors in the

Table 3.1 Summary of Ketamine Effects on Non-NMDA Receptor Targets

Target	Effect
Dopamine receptors	Increases D_1 receptor binding potential and availability
	Binds to high-affinity form of D_2 receptors
5-HT receptors	Enhances 5-HT$_{1A}$ autoreceptor effects
	Increases binding to 5-HT$_{1B}$ receptors
	Decreases binding to 5-HT transporters
	Increases 5-HT release via 5-HT$_{2A}$ receptor activation
Gamma-aminobutyric acid (GABA) receptors	Enhances GABA$_A$ receptor activity
	Inhibits GABA$_B$ receptor downstream complex formation
Opioid receptors	Prevents quick desensitization of μ receptor and promotes its resensitization
	Binds to κ-receptor to induce psychotic symptoms
	Elicits antidepressant effect by σ-receptor activation
Endocannabinoid receptors	Is sensitive to CB$_1$ receptor blockade in eliciting its psychomimetic effects
ATP-sensitive K^+ channels	Blocks channel in antinociception
	May enhance adrenergic receptor-mediated contractile effects that are sensitive to channel blockade
Voltage-gated K^+ channels	Blocks channels
Ca^{2+}-activated K^+ channels	Activates channel during vasorelaxation
	Blocks channel in tactile allodynia
Ca^{2+} transport and Ca^{2+} sensitivity	Suppresses Ca^{2+} oscillations
	Blocks voltage-gated Ca^{2+} channels
	Decreases myofilament Ca^{2+} sensitivity in blood vessels
Hyperpolarization-activated cyclic nucleotide-gated channel and cGMP	Blocks channels (subtype 1)
	Increases cGMP production

(Continued)

Table 3.1 (Continued) Summary of Ketamine Effects on Non-NMDA Receptor Targets

Target	Effect
Na⁺ channels	Blocks channels
Nicotinic receptors	Inhibits channel-mediated currents
	Inhibits Ca^{2+} responses subsequent to nicotinic receptor activation
	Increases acetylcholine release associated with nicotinic receptor activation
Purinergic receptors	Increases A_{2A} receptor activity in antinociception
	Decreases A_{2A} receptor activity in locomotor activation
	Decreases P_{2X4} receptor expression
Histamine receptor	Increases histamine release and produces $H_{1/2}$ receptor–mediated hypotension (in the presence of sevoflurane)
Leukotriene receptors	Decreases LT_1 receptor expression
Immunomodulatory targets	Suppresses Toll-like receptor type 4 (TLR4) expression in intestines and astrocytes
	Suppresses lipopolysaccharide (LPS)-induced phosphorylation of proteins downstream of TLR4 signaling
	Decreases cytokine production induced by LPS and lipoteichoic acid
	Decreases expression of TLR2

prefrontal cortex and working memory is well established (Goldman-Rakic et al. 2000). Therefore, the observed spatial memory impairment and motor symptoms elicited by ketamine are likely to be D_1 receptor dependent, which may also be related to the ability of ketamine to modulate D_1 receptor binding.

While it is the D_1 receptor that is thought to be predominantly responsible for the low-frequency oscillations in the NAc in ketamine-associated locomotor symptoms (Matulewicz et al. 2010), other dopamine receptor subtypes that originate from the hippocampus also mediate the electrical activity of the NAc. The NAc is subject to inputs from various brain structures, and overstimulation of the NAc and dysregulated dopamine release could manifest as psychotic symptoms (Grace 2000). Hippocampal fimbria–induced impulses recorded in the NAc are suppressed in rats treated with ketamine; this effect is prevented by prior treatment with the typical antipsychotic haloperidol, which blocks D_2 and D_4 receptors (Hunt et al. 2005). The D_1 receptor antagonist SCH23390 did not alter these effects of ketamine (Hunt et al. 2005). Ketamine-induced psychotic symptoms positively correlated with D_2 and D_3 receptor availability in healthy humans as measured by using fallypride as a radioligand (Vernaleken et al. 2013). Fallypride binding is selective for the high-affinity form of the D_2 receptor, while raclopride binds to the low-affinity form of the receptor (Vernaleken et al. 2011). Ketamine binds more tightly (i.e., having a lower dissociative constant) to the high-affinity form of D_2 receptors than to NMDA receptors (Seeman et al. 2005), further supporting the view that ketamine-induced psychotic symptoms are D_2 receptor dependent.

3.2.2 5-HT receptors

A number of drugs indicated for depression have shared properties in enhancing 5-HT-mediated transmission, but the clinical antidepressant effect is often delayed, perhaps owing to desensitization of presynaptic 5-HT receptors (Yamanaka et al. 2014). The antidepressant effect of ketamine in laboratory studies (Autry et al. 2011; Maeng et al. 2008) has directed interest in the relationship between ketamine and 5-HT receptors. The basal firing rate of serotoninergic neurons in the rat dorsal raphe nucleus is reduced by ketamine treatment (McCardle and Gartside 2012). The suppressant effects on serotoninergic neuronal firing are enhanced by ketamine when 5-HT$_{1A}$ autoreceptors are activated (McCardle and Gartside 2012). In a novelty-suppressed feeding test study, the 5-HT$_{1A}$ receptor antagonist WAY100635 reduces the latency time in ketamine-pretreated mice (Fukumoto et al. 2014). It is possible that decreased serotoninergic neuronal activity prevents rapid 5-HT depletion, which in turn desensitizes presynaptic 5-HT release and effectively diminishes 5-HT neurotransmission. Indeed, prior application of the tryptophan hydroxylase inhibitor *para*-chlorophenylalanine abolishes ketamine-induced shortening of latency time (Fukumoto et al. 2014). Binding to 5-HT$_{1B}$ receptors was increased by ketamine as revealed by radioligand experiments (Yamanaka et al. 2014). In contrast, ketamine decreased binding of the 5-HT transporter, although the functional outcomes remain unclear (Yamanaka et al. 2014). Addition of the AMPA receptor antagonist NBQX abolished the increased 5-HT$_{1B}$ receptor binding induced by ketamine (Yamanaka et al. 2014), suggesting that the interaction between ketamine and 5-HT neurotransmission may also involve a glutamate component.

Unlike the 5-HT$_1$ receptor that is implicated in depression, 5-HT$_2$ receptor functions are associated with psychotic effects. Increased neuronal activity and 5-HT release in the prefrontal cortex is mediated by 5-HT$_2$ receptors (Jackson et al. 2004). Other studies show that psychotic symptoms are produced when 5-HT$_{2A/2C}$ receptors are activated (Gouzoulis-Mayfrank et al. 1998) or when NMDA receptors are blocked (Miyamoto et al. 2001). Ketamine-induced increases in 5-HT release are sensitive to clozapine and olanzapine, both acting via 5-HT$_{2A}$ antagonism (Amargós-Bosch et al. 2006). Although the NMDA receptor may be involved in producing the 5-HT-releasing effect, this may be sensitive to ketamine-induced 5-HT$_{2A}$ receptor binding (Waelbers et al. 2013). Binding of the 5-HT$_{2A}$-selective radiolabeled ligand 5-I-R91150 is decreased by ketamine, but it is unclear whether this is attributed to competition for binding or decreased binding affinity (Waelbers et al. 2013). Nevertheless, the psychotic symptoms associated with ketamine are better related to its interactions with 5-HT than dopamine receptors, since blockade of the latter (by sulpiride) did not affect the ketamine discriminative stimulus effect in trained rats (Yoshizawa et al. 2013).

3.2.3 Gamma-aminobutyric acid (GABA) receptors

The use of ketamine in anesthesia has led to suggestions that its action involve GABA$_A$ receptors (Irifune et al. 2000). The clinical importance of GABA neurotransmission is exemplified by the antiseizure effect of benzodiazepines, which facilitate GABA$_A$ channel opening. However, it is estimated that nearly half of the patients with status epilepticus are unresponsive to benzodiazepines (Alldredge et al. 2001), raising the possibility for a role for ketamine through its modulatory actions on benzodiazepine-binding or other sites on GABA$_A$ receptors. Support for ketamine enhancing GABA$_A$ receptor activity comes from experiments studying picrotoxin-sensitive apneustic breathing in cats (Budzińska 2005). Both bicuculline (a GABA$_A$ receptor antagonist that acts at a different site from picrotoxin) and picrotoxin (a GABA$_A$ receptor antagonist) failed to alter ketamine-induced apneustic breathing, an observation the authors attributed to ketamine actions on GABA$_C$ receptors (Budzińska 2005). The function of the GABA$_C$

receptor is poorly understood, and this receptor was not considered in other studies of ketamine and GABA receptors. Recombinant expression of different $GABA_A$ receptor subunit compositions suggests that ketamine induces GABA-mediated currents in receptors containing α6β2/3δ subunits (Hevers et al. 2008). A similar ketamine effect occurs in cerebellar granular neurons where the α6β2/3δ type of $GABA_A$ receptor is predominant (Hevers et al. 2008). Ketamine fails to enhance the GABA-mediated current if these neurons were from transgenic α6$^{-/-}$ and δ$^{-/-}$ mice, indicating the essential role of α6 and δ subunits in the ketamine-induced effect (Hevers et al. 2008). Other NMDA receptor antagonists such as phencyclidine and dizocilpine were unable to mimic ketamine, suggesting that ketamine acted via NMDA-independent mechanisms (Hevers et al. 2008). Either bicuculline or the diuretic furosemide (used as an α6-containing $GABA_A$ receptor antagonist here) blocked the ketamine-induced current, further demonstrating the $GABA_A$ receptor modulatory effect of ketamine (Hevers et al. 2008). On the basis of the functional relationship between benzodiazepine and $GABA_A$ receptors, studies were undertaken to determine if ketamine modulates benzodiazepine binding. An early study reported that the anesthetic effect of ketamine was increased by the benzodiazepine receptor antagonist flumazenil (Restall et al. 1990). In healthy human subjects, S-ketamine (but not the R-enantiomer) decreased binding of the radiolabeled iomazenil (a benzodiazepine receptor inverse agonist) to the benzodiazepine receptor at the prefrontal cortex (Heinzel et al. 2008). The authors postulated that ketamine altered the benzodiazepine-binding site so as to increase GABAergic neuronal activity (Heinzel et al. 2008), although it was unclear whether ketamine facilitated benzodiazepine receptor agonist binding as well, or simply inhibited binding of the inverse agonist. In addition to the effects of ketamine on benzodiazepine receptors, interactions with muscarinic receptors have also been implicated in the $GABA_A$ receptor effects of ketamine. In rats treated with pilocarpine to induce benzodiazepine-resistant status epilepticus, either the benzodiazepine diazepam or ketamine used in isolation was without effect within 5 h, but their combination reduced the number of animals with prolonged seizures during the same time frame (Martin and Kapur 2008). Whether ketamine also elicited an effect on muscarinic receptors in suppressing pilocarpine-induced seizures was not investigated in the study by Martin and Kapur (2008), but another study using bicuculline to induce seizures reported that ketamine altered muscarinic receptor affinity (Schneider et al. 2013). Thus, it is likely that the $GABA_A$ receptor activating or potentiating effect of ketamine may also be associated with activation of muscarinic receptors.

Activation of $GABA_B$ receptors underlies spasticity treatment, and the agonist baclofen is often used for this purpose (Ando et al. 2011). However, long-term baclofen treatment results in tolerance as $GABA_B$ receptors desensitize (Kanaide et al. 2007; Perroy et al. 2003). It is known that formation of the $GABA_B$ receptor/G protein–coupled receptor kinase (GRK) 4/5 complex contributes to receptor desensitization (Kanaide et al. 2007; Perroy et al. 2003). Experiments in oocytes coexpressing $GABA_B$ receptors, GRK 4/5, and G protein–activated inward-rectifying K^+ channel 1/2 indicated that S-ketamine inhibited the translocation of GRK 4/5 to the plasma membrane, suggesting that the $GABA_B$ receptor/GRK complex was not formed and the K^+ current was preserved (Ando et al. 2011).

3.2.4 Opioid receptors

Activation of µ opioid receptors is commonly associated with antinociception while some reports suggest that δ opioid receptors are also involved (Brandt et al. 2001). Since prolonged use of µ receptor agonists alone is also associated with adverse effects such as respiratory depression and constipation, pain management is improved from the adjuvant use of other analgesic agents. The combined use of µ and δ receptor agonists produces greater antinociceptive effects (Banks et al. 2010). The δ receptor agonist SNC162, when used together with the µ receptor agonist fentanyl, had better antinociceptive effects in monkeys compared to the combined use of fentanyl and ketamine (Banks et al. 2010). Experiments with coadministration of ketamine and SNC162 would have provided information on possible δ receptor effects of ketamine.

The antinociceptive effects mediated by µ receptors are enhanced in the presence of ketamine (Banks et al. 2010; Hama et al. 2006). In a rat model of hyperalgesia induced by formalin, ketamine and morphine cotreatment achieved higher levels of antinociception than with morphine alone (Hama et al. 2006). The ketamine effect was more pronounced in the later phases (after 15 min) of formalin administration (Hama et al. 2006). Another NMDA receptor antagonist, [Ser[1]]-histogranin, also enhanced the antinociceptive effect of morphine, but to a lesser extent (Hama et al. 2006). Earlier reports suggested that hyperalgesia developed after chronic opioid treatment as a result of NMDA receptor activation by opioids (Chu et al. 2008; Joly et al. 2005). Levels of NMDA receptor mRNA returned to normal (after an initial transient decrease) in mice treated chronically with morphine (Ohnesorge et al. 2013). Combined ketamine (or clonidine, an α_2 receptor agonist) and morphine treatment prevented the rebound of NMDA receptor mRNA as well as enhanced antinociception (Ohnesorge et al. 2013). The ketamine-enhanced antinociception was only partially related to α_2 receptor activation, as revealed by the small suppression of the ketamine effect by yohimbine (an α_2 receptor antagonist) (Campos et al. 2006). Instead, the opioid receptor antagonist naloxone fully abolished the enhanced antinociception (Campos et al. 2006), indicating a direct effect of ketamine on µ receptors.

Additional evidence of ketamine modulating μ receptor function comes from electrophysiological and protein biochemical studies. Ketamine suppressed Ca^{2+}-activated Cl^- currents after μ receptor activation (Minami et al. 2010), an effect attributed to actions on the receptor rather than on downstream Ca^{2+} signaling. It is unclear whether this inhibitory effect on μ receptors contributed to the added antinociception associated with down-regulated NMDA receptors mRNA reported elsewhere, since morphine alone increased NMDA receptor mRNA levels (Ohnesorge et al. 2013). An extended period of μ receptor stimulation also causes receptor desensitization as, for example, when phosphorylation of a mitogen-activated protein kinase ERK1/2 transiently increases and returns to baseline after a short period (Trapaidze et al. 2000). Ketamine increases the level and prolongs the duration of ERK1/2 phosphorylation (Gupta et al. 2011). Moreover, resensitization of μ receptors (achieved by the temporary removal of morphine) was facilitated by ketamine (Gupta et al. 2011). Although there was also evidence of increased μ receptor density in the rat hippocampus after ketamine treatment, nociception was not measured in that study to allow conclusions to be drawn on possible ketamine modulation on μ receptor function (Kekesi et al. 2011). Therefore, one can reasonably suggest that ketamine enhances antinociception by inhibiting μ receptors to prevent its rapid desensitization on the one hand, and shortens the receptor recovery time on the other hand.

Schizophrenia and depression are two examples of opioid receptor–associated affective disorders. Cognitive deficits in schizophrenia have been linked to κ receptor stimulation (Chavkin et al. 2004). The opioid receptor antagonist JDTic diminished attention deficit in rats induced by ketamine after selective κ receptor blockade (Nemeth et al. 2010). In humans, ketamine binds to κ receptors as an agonist with a binding affinity that is much weaker than at σ receptors (Nemeth et al. 2010). The σ receptor has been implicated in nerve growth factor (NGF)–mediated neuronal plasticity and antidepressant effects (Ishima et al. 2008). Ketamine promotes NGF-induced neurite growth in PC12 cells, an effect that is sensitive to NE-100 (a σ receptor antagonist) (Robson et al. 2012). This effect was attributed to σ receptor–mediated ERK signaling, a result similar to the inhibitory action of ketamine on σ receptors (Gupta et al. 2011). However, in an in vivo depression model of forced swimming, the shortened immobility time after ketamine treatment was insensitive to σ receptor blockade (Robson et al. 2012).

3.2.5 *Endocannabinoid receptors*

Endogenous cannabinoids are responsible for many actions in the CNS including analgesia and memory consolidation, and if consumed as exogenous agents, visual hallucination, perceptual disturbance, and psychotic symptoms (Pacher and Kunos 2013). In a study of discriminative stimuli in monkeys treated with Δ^9-tetrahydrocannabinol (a naturally occurring cannabinoid), only the cannabinoid-1 (CB_1) receptor antagonist SR141716A showed efficacy (McMahon 2006). Both ketamine and midazolam were without effect, suggesting that CB_1-mediated effects were independent of ketamine action. However, if rats were preconditioned with ketamine in a place preference system (Li et al. 2008), reapplication of ketamine after a period of abstinence reestablished place preference in a CB_1 receptor antagonist (rimonabant)–sensitive manner. This finding suggests a potential for CB_1 receptor blockade in preventing relapse of ketamine-associated psychotomimetic effects. In a trial of healthy humans treated with S-ketamine, two parameters showed contrasting outcomes after administration of cannabidiol (a CB_1 receptor antagonist and CB_2 inverse agonist) (Hallak et al. 2011). First, psychomotor activation was increased by cannabidiol, and second, depersonalization was marginally decreased (Hallak et al. 2011). It is unclear if the effects of ketamine were attributed to direct actions on CB receptors or indirect effects mediated by other receptors.

3.3 Ion channels

3.3.1 ATP-sensitive K^+ (K_{ATP}) channels

Activation of K_{ATP} channels promotes neuronal hyperpolarization in the CNS and thus elicits a neuroprotective effect (Fujimura et al. 1997). Opening of K_{ATP} channels is activated by low intracellular ATP levels, and ketamine is able to block channel opening in neurons from the rat substantia nigra (Ishiwa et al. 2004). The NMDA receptor antagonist AP-5 did not modulate K_{ATP}-mediated currents, suggesting a non-NMDA-dependent mechanism of ketamine (Ishiwa et al. 2004). Ketamine had no effect on K_{ATP} channels when intracellular ATP levels was kept at 2 mM (Ishiwa et al. 2004), but in reassociated K_{ATP} channels in the presence of Mg-ADP, the channel-inhibitory effect of ketamine was enhanced (Kawano et al. 2005). Reassociated K_{ATP} channels were made by coexpressing different inward-rectifying K^+ channel (Kir) subunits and sulfonylurea receptors (SUR) in COS-7 cells (Kawano et al. 2005). In a channel with the Kir6.2 subunit but lacking SUR, the K^+ current was inhibited by ketamine but did not reach full efficacy when compared with channels with SUR subunits (Kawano et al. 2005). The SUR was likely the ketamine-binding site for its action on K_{ATP} channels (Kawano et al. 2005). In both spontaneously open (Kawano et al. 2005) and nicorandil-stimulated K_{ATP} channels (Kawano et al. 2010), racemic ketamine was more potent than the S-enantiomer. However, Mg-ADP decreased the inhibitory effect of ketamine on K_{ATP} channel when nicorandil was present (Kawano et al. 2010). The disparity in findings between the two studies by Kawano et al. (2005, 2010) could be attributed to different states of the channel, channel subunits expressed, or other non-K_{ATP} channel properties (e.g., as a nitric oxide donor) of nicorandil.

The interrelationships between ketamine, K_{ATP} channels, and adrenergic receptors have also been studied. Although the attenuation of morphine-induced hyperalgesia by ketamine was not sensitive to the K_{ATP} channel blocker glibenclamide (Campos et al. 2006), a different conclusion was reached when prostaglandin E_2 was used to induce pain (Romero and Duarte 2013). Ketamine demonstrated antinociceptive effect in the hind paw of rats that had received a prior intraplantar injection of prostaglandin E_2 (Romero and Duarte 2013). The antinociceptive effect of ketamine was blocked by glibenclamide but not by other voltage-gated K^+ channel blockers, highlighting the role of K_{ATP} channels in this type of nociception (Romero and Duarte 2013). In a study of isolated human atrial appendage contractility, prior administration of ketamine enabled greater recovery of contractility in hypoxia-reoxygenated tissues (Hanouz et al. 2005). This effect of ketamine was blocked by antagonists of α (phentolamine) and β (propranolol) receptors (Hanouz et al. 2005). While ketamine effects on the isolated atrium could be attributed to actions on K_{ATP} channels and adrenergic receptors, it is unclear if the CNS could also have an effect or if combined treatment with various blockers could also augment the contractile recovery. It should also be noted that in the study by Hanouz et al. (2005), the atrial samples were taken from many patients who were also receiving treatment with benzodiazepine. Since ketamine alters benzodiazepine binding (Heinzel et al. 2008), it is possible that $GABA_A$ receptors may also have a role in the ketamine-mediated recovery of atrial contractile function in vivo.

3.3.2 *Voltage-gated K⁺ channels*

Ketamine has a stimulatory effect on the cardiovascular system to increase heart rate and blood pressure (Stanley et al. 1968). The voltage-gated K^+ (K_V) channel regulates the resting membrane potential and is down-regulated in hypertension (Cox 2002). The inhibitory effects of ketamine on K_V currents (via $K_V1.x$, $K_V2.x$, and $K_V3.x$ channel subfamilies) were similar for both racemic and S-ketamine and were independent of channel activation voltage (Kim et al. 2007). Ketamine also depolarized the membrane potential at rest (Kim et al. 2007), a finding the authors suggested correlated with ketamine-induced increases in blood pressure. Ketamine suppressed sympathetic activity in the isolated rat spinal cord (Ho and Su 2006). However, a variety of results have been reported for the effects of ketamine on the sympathetic nervous system in vivo (Akine et al. 2001; Cheng et al. 2004; Kienbaum et al. 2000; Sasao et al. 1996). It may therefore be premature to determine whether the K_V inhibitory effect of ketamine directly contributes to the increased depolarization of vascular smooth muscle and hypertension observed clinically.

The rapid-delayed rectifier K^+ current (I_{Kr}) (via the $K_V 11.x$ channel subfamily) is responsible for repolarization of the cardiac action potential. Contrasting findings have been reported for the effects of ketamine on I_{Kr}, possibly resulting from the different experimental models used. Transgenic rabbits with long QT syndrome type 2 (LQT2, which lacks I_{Kr}) failed to show any alteration in the QT interval after ketamine treatment, which also failed to affect the slow-delayed rectifier K^+ current (Odening et al. 2008). When I_{Kr} was measured in *Xenopus* oocytes expressing the human ether-a-go-go-related gene, the current was inhibited by ketamine, with increasing effect at more negative voltages and after longer depolarizations (Zhang et al. 2013). Plotting the normalized current versus voltage curves indicated that ketamine exerted its effect during channel inactivation, suggesting that binding occurred in the channel-open state (Zhang et al. 2013). The findings by Zhang et al. (2013) related to prolonged QT interval and decreased heart rate are seemingly contradictory to the clinical effects of ketamine. The authors attributed this disparity, first, to the in vitro conditions of their study and, second, to the more prominent participation of other channels in inducing cardiac dysrhythmia (Zhang et al. 2013).

3.3.3 *Ca²⁺-activated K⁺ channels*

In contracted vascular smooth muscle, the Ca^{2+}-activated K^+ (K_{Ca}) channel opens in response to cytosolic Ca^{2+} increases, allowing K^+ efflux and membrane hyperpolarization, and so contributes to the regulation of vascular tone. Under in vivo conditions, ketamine increases blood pressure, while under ex vivo conditions, ketamine induces relaxation of vascular smooth muscle (Jung and Jung 2012; Klockgether-Radke et al. 2005) by activation of large-conductance Ca^{2+}-activated K^+ (BK) channel. In pig coronary arteries, KCl- and prostaglandin $F_{2\alpha}$-precontracted vessel rings were relaxed by ketamine, with *S*-ketamine being more potent than either racemic mixture or *R*-ketamine (Klockgether-Radke et al. 2005). The ketamine-mediated vasorelaxation was diminished by tetraethylammonium (TEA), a nonselective inhibitor of K_{Ca} channels and not by glibenclamide, a K_{ATP} channel blocker (Klockgether-Radke et al. 2005). The suppression of contractions of rabbit renal arterial rings by ketamine is also inhibited by TEA and by iberiotoxin, suggesting a role for BK channels (Jung and Jung 2012). In contrast, an inhibitory effect of ketamine on BK channel was demonstrated in mouse microglia, which when hyperactivated in neural injury yields neuropathic pain (Scholz and Woolf 2007). The BK channel opener NS1619-induced current was suppressed by *S*-ketamine more potently than the racemic mixture (Hayashi et al. 2011). In summary, ketamine could, depending on tissue types, elicit different effects on BK channel activity.

3.3.4 Ca^{2+} transport and Ca^{2+} sensitization

It has been known for some time that ketamine blocks both L- and T-type Ca^{2+} channels (Sinner and Graf 2008). Ryanodine-mediated Ca^{2+} release from the endoplasmic reticulum (ER) and Ca^{2+} recycling via the ER Ca^{2+}-ATPase were not affected by ketamine treatment in renal arteries (Jung and Jung 2012). However, Ca^{2+} influx was smaller in the presence of ketamine after Ca^{2+} store depletion (Jung and Jung 2012). In rat hippocampal neurons, spontaneous Ca^{2+} oscillations were abolished by the NMDA receptor antagonist dizocilpine while oscillating amplitude was increased by the $GABA_A$ receptor antagonist bicuculline (Sinner et al. 2005). Although ketamine suppressed both oscillating frequency and amplitude, it could not be determined from the study by Sinner et al. (2005) whether this effect was due solely to NMDA receptor blockade since ketamine also enhances $GABA_A$ receptor activity to decrease oscillating amplitude (Hevers et al. 2008). Stereoselectivity of ketamine is important in suppressing Ca^{2+} oscillations, with the *S*-enantiomer being more potent than the *R*-enantiomer (Sinner et al. 2005).

Membrane depolarization and neurotransmitter release are two neuronal events that depend on voltage-gated Ca^{2+} entry. Drugs such as lithium and valproic acid that possess Ca^{2+} channel–blocking properties have considerable efficacy in seizures (Karydes et al. 2013; Yanagita et al. 2007). Seizure threshold (determined by the emergence of forelimb and full body clonus) in mice with chemical-induced clonic seizures was increased by ketamine and lithium (at noneffective concentrations when used individually) (Ghasemi et al. 2010). Seizure threshold was similarly increased when lithium was given together with L-type Ca^{2+} channel blockers including nifedipine (Ghasemi et al. 2010). The authors postulated a three-way interaction in the antiseizure effects of lithium, Ca^{2+} channel blockers, and ketamine because of their NMDA receptor antagonistic effects (Ghasemi et al. 2010). However, noting that ketamine also acts directly on Ca^{2+} channels, it may be useful to also investigate whether combined ketamine and nifedipine treatment further increases seizure threshold.

Besides inhibiting Ca^{2+} flux, ketamine also decreased myofilament Ca^{2+} sensitivity in dog pulmonary veins (Ding and Murray 2007). As in other blood vessel studies (Jung and Jung 2012; Klockgether-Radke et al. 2005), acetylcholine (ACh)-induced vasoconstriction was diminished by ketamine (Ding and Murray 2007). When venous tissue was permeabilized and exposed to increasing extracellular Ca^{2+} concentrations, ACh addition resulted in greater contractility that was associated with increased myofilament Ca^{2+} sensitivity (Ding and Murray 2007). Ketamine diminished the ACh-mediated effect by causing the translocation of protein kinase Cα to the plasma membrane (Ding and Murray 2007). While vasorelaxation induced by ketamine occurs in a number of tissues, its effect on myofilament Ca^{2+} sensitivity could be tissue specific since, for example, ketamine fails to alter myofilament Ca^{2+} sensitivity in dog tracheal smooth muscle (Hanazaki et al. 2000).

3.3.5 Hyperpolarization-activated cyclic nucleotide-gated (HCN) channels and cGMP

The hypnotic effect of general anesthetics is thought to be dependent on synchronized, slow rhythms in forebrain pyramidal neurons (Chen et al. 2009b; Zhou et al. 2013). Activity in these neurons is facilitated by the pacemaker current (I_h) via HCN channel opening, causing membrane depolarization owing to Na^+ entry (Pape 1996). Excessive synaptic inputs to the cortex are normally restricted as a result (Magee 2000); on the contrary, a decrease in I_h is present in animal epilepsy models (Shah et al. 2004). The synchronized firings under the influence of anesthetics are similar to those seen in deep sleep (Amzica and Steriade 1998) and are observed when I_h is inhibited (Carr et al. 2007). Ketamine, but not dizocilpine, blocks HCN subtype 1 (HCN1) channel–mediated I_h, and this blockade was greater with the S-enantiomer than with the racemic compound (Chen et al. 2009b). Conditionally knocking out HCN1 expression in the mouse forebrain (while preserving that in the cerebellum) rendered I_h insensitive to ketamine treatment (Zhou et al. 2013). The hypnotic effect was also diminished in knockout mice (Zhou et al. 2013), further suggesting HCN1–ketamine interactions in potentiating cortical synchronization.

There is no evidence supporting a link between HCN channel activity in the CNS and cGMP availability. However, HCN channel can be activated by cyclic nucleotides in other pacemaker cells such as SA nodal cells. In a rat model of PGE_2-induced hyperalgesia, the ketamine antinociceptive effect was blocked by the neuronal nitric oxide synthase (nNOS) inhibitor N^W-propyl-L-arginine but not by specific inhibitors of other NOS isoforms (Romero et al. 2011). Nitrite measurements indicated that more NO was released at the pain site after ketamine treatment (Romero et al. 2011). NO activates soluble guanylate cyclase (sGC) to generate cGMP (Moncada et al. 1991). The sGC inhibitor ODQ suppressed while the cGMP-selective phosphodiesterase inhibitor zaprinast enhanced ketamine antinociceptive effects (Romero et al. 2011). Given the role of cGMP in ketamine-mediated antinociception, it would be useful to also investigate whether a similar phenomenon is present in HCN function in the forebrain pyramidal neurons.

3.3.6 Na^+ channels

Several studies in the 1970s to 1980s demonstrated that ketamine had Na^+ channel–blocking properties resembling those of local anesthetics (Sinner and Graf 2008). The only other study to appear since then on ketamine and Na^+ channels examined alveolar Na^+ transport (Cui et al. 2011). Na^+ reabsorption via epithelial Na^+ channels prevents excessive fluid accumulation in the lungs and, thus, pulmonary edema. Ketamine use and abuse are associated with pulmonary edema (Pandey et al. 2000), thereby implicating Na^+ channel dysfunction. Ketamine treatment decreased alveolar fluid clearance in human lung lobes ex vivo, and this effect was absent when the epithelial Na^+ channel blocker amiloride was added (Cui et al. 2011). Ketamine also suppresses Na^+ currents in a human alveolar epithelial cell line (Cui et al. 2011). These findings, together with the inhibition of HCN channel (see above discussion) and of nicotinic receptor (see Section 3.4), collectively suggest an overall suppressant effect on Na^+ entry by ketamine.

3.4 Nicotinic receptors

Activation of nicotinic receptors results in the opening of Na^+/Ca^{2+} influx channels to cause membrane depolarization (Beker et al. 2003). In addition to blocking Na^+ entry via NMDA receptors, ketamine also blocks nicotinic receptors with similar potency (Harrison and Simmonds 1985). A variety of nicotinic receptor subunit compositions were investigated. The homomeric $\alpha7$ nicotinic receptor–mediated current was blocked by ketamine and its metabolite dehydronorketamine (Ho and Flood 2004; Moaddel et al. 2013). The ketamine-binding site is on transmembrane segment 2 in point mutated $\alpha7$ nicotinic receptors (Ho and Flood 2004). Other studies reported that ketamine binds near the intracellular end of the $\alpha7$ transmembrane domain (Bondarenko et al. 2013a). Both ketamine and its metabolite norketamine (Moaddel et al. 2013) inhibited the currents associated with heteromeric $\alpha3\beta4$ nicotinic receptors (Moaddel et al. 2013). More recently, multiple ketamine-binding sites were resolved by NMR at the heteromeric $\alpha4\beta2$ nicotinic receptor, one near the extracellular end of the $\beta2$ transmembrane domain and another near the intracellular end of the transmembrane domains between the $\alpha4$ and $\beta2$ subunits (Bondarenko et al. 2013b).

Ketamine suppresses Ca^{2+} transport and Ca^{2+} sensitivity (see Section 3.3.4). Nicotinic receptor–mediated Ca^{2+} responses are inhibited by ketamine in rat intracardiac ganglionic neurons (Weber et al. 2005). Unlike other cell types where voltage-gated Ca^{2+} channels are blocked by ketamine, depolarization-induced Ca^{2+} responses are unaffected by ketamine in intracardiac ganglionic neurons (Weber et al. 2005). Release of inositol

1,4,5-triphosphate–sensitive Ca^{2+} stores is also insensitive to ketamine treatment (Weber et al. 2005). However, since the ryanodine-sensitive Ca^{2+} store was not examined in this study, the precise Ca^{2+}-inhibitory mechanism(s) of ketamine is (are) unclear.

Functional interactions between nicotinic receptors and ketamine have been demonstrated in animal studies of schizophrenia and antinociception. The α7 nicotinic receptor agonist GTS-21, as well as the nicotine metabolite cotinine, reverses cognitive deficits associated with impaired strategy formation and revision in the prefrontal cortex in ketamine-treated monkeys (Buccafusco and Terry 2009; Cannon et al. 2013). Ketamine increases spinal ACh release in a rat spinal anesthesia study, where ACh release is associated with antinociception (Abelson et al. 2006; Kommalage and Höglund 2004). The nicotinic receptor agonist epibatidine potentiated the increase in ACh whereas this effect was diminished by the nicotinic receptor antagonist mecamylamine (Abelson et al. 2006).

3.5 Purinergic receptors

Adenosine activates the P1 (or A_x) family of receptors, also known as adenosine receptors. Ketamine increases adenosine release (by blocking NMDA receptors) to enhance its anti-inflammatory effects (Bong et al. 1996). The mortality rate after sepsis was reduced after adenosine treatment (McCallion et al. 2004). In mouse models of sepsis, ketamine increased plasma adenosine concentrations and animal survival (Mazar et al. 2005). Leukocyte recruitment, an indicator of immune activity, is suppressed by ketamine as well as the A2A receptor agonist CGS-21680 (Mazar et al. 2005). Conversely, ketamine effects are blocked by both the nonselective A2 receptor antagonist DMPX and the selective A2A receptor antagonist ZM241385 (Mazar et al. 2005). The release of the inflammatory cytokines TNFα and IL-6 was reduced after ketamine and CGS-21680 treatment, with this effect being sensitive to DMPX (Mazar et al. 2005). Neither A1 nor A3 receptor blockade altered this ketamine effect (Mazar et al. 2005).

Adenosine is also involved in locomotor activity, which is under regulation by GABAergic neuronal terminals expressing $A_{1/2}$ and $D_{1/2}$ receptors (Hettinger et al. 2001; Jin et al. 1993; Schiffman et al. 1991; Svenningsson et al. 1999). Motor symptoms associated with D_2 receptor antagonists were mimicked by A2A receptor agonists (Correa et al. 2004). Ketamine-treated mice display locomotor hyperactivity that is suppressed by CGS-21680 and the nonselective $A_{1/2}$ receptor antagonist NECA but not by selective A_1 activation (Mandryk et al. 2005). Locomotor activity was enhanced after DMPX and caffeine (used here as an $A_{1/2}$ receptor antagonist) whereas selective A_1 receptor blockade had no effect (Mandryk et al. 2005). Comparing the studies of Mandryk et al. (2005) and Mazar et al. (2005), it is interesting to note that although ketamine exerts an influence on A_2 receptor under different experimental settings, there appears to be contrasting effects with ketamine increasing A_2 receptor activity during antinociception but decreasing A_2 receptor activity when inducing motor symptoms.

ATP preferentially activates P2 receptors, of which the P2X1 and P2X4 subtypes have been the best-studied ketamine targets. The effects of ketamine on P2X1 receptors are discussed in detail in Chapter 10. Hyperactivation of microglia, via P2X4 receptor activation, increases the release of inflammatory mediators and, at the same time, decreases anti-nociceptive neurotransmission (Inoue and Tsuda 2009). Microglial P2X4 receptors are up-regulated after L4 nerve injury, and this is sensitive to inhibition by the BK-IK channel blocker charybdotoxin (Hayashi et al. 2011). The influence of ketamine on P2X4 receptors was only inferred from this study (Hayashi et al. 2011). Another study attempted but failed to show a direct effect of ketamine on P2X4 receptor–mediated currents (Hasaka et al. 2012). However, some of the shared characteristics of the anesthetics used in the study are worthy of discussion here. In addition to ketamine, thiopental also lacked an effect on P2X4 receptors (Hasaka et al. 2012) but enhanced P2X7 receptor–mediated currents in another study (Nakanishi et al. 2007). It may be useful to investigate whether ketamine also exerts similar effects on P2X7 receptors, the expression of which increases after status epilepticus (Henshall et al. 2013), and furthermore, to also investigate the relationship between ketamine, GABA$_A$ receptor, and Ca^{2+} transport in epilepsy as mentioned in Sections 3.2.3 and 3.3.4.

3.6 Histamine receptors

Ketamine stimulates histamine release from mast cells (Marone et al. 1993). A recent study examined ketamine actions related to histamine. Ketamine increased plasma histamine and lowered blood pressure in sevoflurane-anesthetized cats (Costa-Farré et al. 2005). This decrease in blood pressure was more pronounced after blockade of both H_1 receptors (by chlorphenamine and cyproheptadine) and H_2 receptors (by ranitidine) (Costa-Farré et al. 2005). Ketamine did not affect heart rate but decreased heart rate when both H_1 and H_2 receptors were blocked (Costa-Farré et al. 2005). The mechanisms for these effects of ketamine on heart rate and blood pressure are unclear, since other studies suggest that histamine should stimulate both (Stanley et al. 1968); one possibility could be related to the use of sevoflurane, which is known to decrease blood pressure (Degoute 2007). Since heart rate is usually increased when H_2 receptors are activated, it is surprising that ketamine did not alter heart rate by releasing additional histamine release in the study by Costa-Farré et al. (2005). Another factor to consider is that the decrease in blood pressure caused by ketamine should activate baroreceptors and also increase heart rate (and cardiac output). The individual contributions of H_1 and H_2

receptors to the heart rate and blood pressure responses were not determined in the study by Costa-Farré et al. (2005).

3.7 Leukotriene receptors

Leukotrienes are another group of autacoids that also interact with ketamine. In ischemia-induced neuronal injury, increased glutamate release and subsequent NMDA receptor stimulation activate 5-lipoxygenase to synthesize leukotrienes (Ge et al. 2006). Leukotriene (LT_1) receptor blockade produced neuroprotective effects in rodents with cerebral ischemia (Zeng et al. 2001). Up-regulation of NMDA receptors was also prevented by blocking LT_1 receptors (Zhang and Wei 2003). Both ketamine and the LT_1 receptor antagonist pranlukast reduced brain lesion volumes in mice injected with NMDA into the parietal cortex (Ding et al. 2006). The same study also reported that ketamine and pranlukast abolished the increases in LT_1 receptor mRNA and protein expression caused by injecting NMDA in the cortical area and in hippocampal CA1 neurons (Ding et al. 2006). This study did not examine whether a combined administration of ketamine and pranlukast had a greater effect; it is also unknown if a 5-lipoxygenase inhibitor could unveil a direct action of ketamine on leukotriene synthesis.

3.8 Immunomodulatory targets

Ketamine suppresses cytokine release in a mouse model of sepsis (Mazar et al. 2005). Most reports support the concept that ketamine affects Toll-like receptor (TLR)–mediated pathways. Lipopolysaccharides (LPS) derived from bacterial cell membranes form a complex with LPS-binding protein, which in turn activates TLR4 (Liu et al. 2012). The cascade of events that follows involves phosphorylation of Ras, Raf, mitogen-activated protein kinase kinases (MEKs), extracellular signal-regulated kinases (ERKs), inhibitor kappa-B kinase (IKK), and c-Jun N-terminal kinase (JNK), all of which ultimately trigger the activation of transcription factors such as nuclear factor kappa-B (NFκB) and activator protein-1 (AP-1) to produce cytokines (Chowdhury et al. 2006; Fan et al. 2004; Jones et al. 2001; Liu et al. 2012). Ketamine did not alter resting levels of TLR4 receptor expression but suppressed its up-regulation after LPS induction in rat intestines and astrocytes (Wu et al. 2012; Yu et al. 2006). However, LPS-induced TLR4 receptor expression was not affected by ketamine in mouse macrophages, suggesting cell-specific differences (Chen et al. 2009a). The expression of TLR4 mRNA in mouse macrophages is insensitive to ketamine, although the binding affinity of LPS was weakened (Chen et al. 2009b). Furthermore, introduction of TLR4 siRNA lowers TLR4 expression, which is potentiated by ketamine (Wu et al. 2008).

The signaling events mediated by TLR4 receptors are also altered by ketamine. LPS-induced phosphorylation of Ras, Raf, MEK1/2, ERK1/2, IKK, and JNK is suppressed by ketamine (Chen et al. 2009a; Wu et al. 2008). It should be noted that ketamine apparently has opposing effects on ERK1/2 phosphorylation under different stimulating conditions, as exemplified by the enhanced ERK1/2 response mediated by opioids (Gupta et al. 2011). The translocation and activity of NFκB mediated by IKK are suppressed by ketamine (Chen et al. 2009a; Wu et al. 2012; Yu et al. 2006). Ketamine reduces the translocation of c-Jun and c-Fos into the nucleus after the inhibition of JNK phosphorylation (Wu et al. 2008). The LPS-induced activity of AP-1 was lower after ketamine treatment (Wu et al. 2008). Finally, LPS-induced production of cytokines such as TNFα, IL-1β, and IL-6 is reduced by ketamine (Chen et al. 2009a; Wu et al. 2012).

Cecal ligation and puncture (CLP) and lipoteichoic acid (LTA) are other methods used to induce experimental sepsis, and ketamine has similar immunosuppressant effects in these experimental models. For example, ketamine reduces the expression of TLR4, NFκB, TNFα, and IL-6 in rats with CLP (Yu et al. 2007a). The TLR subtype TLR2 is associated with LTA-induced pathogenic responses (Yoshioka et al. 2007), and the up-regulation of TLR2 in rats with CLP is inhibited by ketamine (Yu et al. 2007b). The ketamine-induced reduction in the expression of TNFα and IL-6 was enhanced when combined with the ERK inhibitor PD98059 (Chang et al. 2010). A synergistic effect between ketamine and TLR2 siRNA has also been reported (Chang et al. 2010).

References

Abelson KS, Goldkuhl RR, Nylund A et al. 2006. The effect of ketamine on intra-spinal acetylcholine release: Involvement of spinal nicotinic receptors. *Eur J Pharmacol* 534(1–3):122–128.

Akine A, Suzuka H, Hayashida Y et al. 2001. Effects of ketamine and propofol on autonomic cardiovascular function in chronically instrumented rats. *Auton Neurosci* 87:201–208.

Alldredge BK, Gelb AM, Isaacs SM et al. 2001. A comparison of lorazepam, diazepam, and placebo for the treatment of out-of-hospital status epilepticus. *N Engl J Med* 345:631–637.

Amargós-Bosch M, López-Gil X, Artigas F et al. 2006. Clozapine and olanzapine, but not haloperidol, suppress serotonin efflux in the medial prefrontal cortex elicited by phencyclidine and ketamine. *Int J Neuropsychopharmacol* 9(5):565–573.

Amzica F, Steriade M. 1998. Electrophysiological correlates of sleep delta waves. *Electroencephalogr Clin Neurophysiol* 107:69–83.

Ando Y, Hojo M, Kanaide M et al. 2011. S(+)-ketamine suppresses desensitization of γ-aminobutyric acid type B receptor-mediated signaling by inhibition of the interaction of γ-aminobutyric acid type B receptors with G protein-coupled receptor kinase 4 or 5. *Anesthesiology* 114(2):401–411.

Autry AE, Adachi M, Nosyreva E et al. 2011. NMDA receptor blockade at rest triggers rapid behavioural antidepressant responses. *Nature* 475:91–95.

Banks ML, Folk JE, Rice KC et al. 2010. Selective enhancement of fentanyl-induced antinociception by the delta agonist SNC162 but not by ketamine in rhesus monkeys: Further evidence supportive of delta agonists as candidate adjuncts to mu opioid analgesics. *Pharmacol Biochem Behav* 97(2):205–212.

Beker F, Weber M, Fink RHA et al. 2003. Muscarinic and nicotinic ACh receptor activation differentially mobilize Ca^{2+} in rat intracardiac ganglionic neurons. *J Neurophysiol* 90:1956–1964.

Bondarenko V, Mowrey D, Liu LT et al. 2013a. NMR resolved multiple anesthetic binding sites in the TM domains of the α4β2 nAChR. *Biochim Biophys Acta* 1828(2):398–404.

Bondarenko V, Mowrey DD, Tillman TS et al. 2013b. NMR structures of the human α7 nAChR transmembrane domain and associated anesthetic binding sites. *Biochim Biophys Acta* 1838(5):1389–1395.

Bong GW, Rosengren S, Firestein GS. 1996. Spinal cord adenosine receptor stimulation in rats inhibits peripheral neutrophil accumulation: The role of N-methyl-D-aspartate receptors. *J Clin Invest* 98:2779–2785.

Brandt MR, Furness MS, Mello NK et al. 2001. Antinociceptive effects of (delta)-opioid agonists in rhesus monkeys: Effects on chemically induced thermal hypersensitivity. *J Pharmacol Exp Ther* 296:939–946.

Brozoski TJ, Brown RM, Rosvold HE et al. 1979. Cognitive deficit caused by regional depletion of dopamine in prefrontal cortex of rhesus monkey. *Science* 205:929–932.

Buccafusco JJ, Terry AV Jr. 2009. A reversible model of the cognitive impairment associated with schizophrenia in monkeys: Potential therapeutic effects of two nicotinic acetylcholine receptor agonists. *Biochem Pharmacol* 78(7):852–862.

Budzińska K. 2005. Divergent effects of bicuculline and picrotoxin on ketamine-induced apneustic breathing. *J Physiol Pharmacol* 56(4):39–46.

Campos AR, Santos FA, Rao VS. 2006. Ketamine-induced potentiation of morphine analgesia in rat tail-flick test: Role of opioid-, alpha2-adrenoceptors and ATP-sensitive potassium channels. *Biol Pharm Bull* 29(1):86–89.

Cannon CE, Puri V, Vivian JA et al. 2013. The nicotinic α7 receptor agonist GTS-21 improves cognitive performance in ketamine impaired rhesus monkeys. *Neuropharmacology* 64:191–196.

Carr DB, Andrews GD, Glen WB et al. 2007. α2-noradrenergic receptors activation enhances excitability and synaptic integration in rat prefrontal cortex pyramidal neurons via inhibition of HCN currents. *J Physiol* 584:437–450.

Chang HC, Lin KH, Tai YT et al. 2010. Lipoteichoic acid-induced TNF-α and IL-6 gene expressions and oxidative stress production in macrophages are suppressed by ketamine through downregulating Toll-like receptor 2-mediated activation of ERK1/2 and NFκB. *Shock* 33(5):485–492.

Chavkin C, Sud S, Jin W et al. 2004. Salvinorin A, an active component of the hallucinogenic sage Salvia divinorum is a highly efficacious kappa-opioid receptor agonist: Structural and functional considerations. *J Pharmacol Exp Ther* 308:1197–1203.

Chen TL, Chang CC, Lin YL et al. 2009a. Signal-transducing mechanisms of ketamine-caused inhibition of interleukin-1 beta gene expression in lipopolysaccharide-stimulated murine macrophage-like Raw 264.7 cells. *Toxicol Appl Pharmacol* 240(1):15–25.

Chen X, Shu S, Bayliss DA. 2009b. HCN1 channel subunits are a molecular substrate for hypnotic actions of ketamine. *J Neurosci* 29(3):600–609.

Cheng YJ, Chien CT, Chen CY et al. 2004. Differentiation of ketamine effects on renal nerve activity and renal blood flow in rats. *Acta Anaesthesiol Taiwan* 42:185–189.

Chowdhury P, Sacks SH, Sheerin NS. 2006. Toll-like receptors TLR2 and TLR4 initiate the innate immune response of the renal tubular epithelium to bacterial products. *Clin Exp Immunol* 145:346–356.

Chu LF, Angst MS, Clark DMDP. 2008. Opioid-induced hyperalgesia in humans: Molecular mechanisms and clinical considerations. *Clin J Pain* 24:479–496.

Correa M, Wisniecki A, Betz A et al. 2004. The adenosine A2A antagonist KF17837 reverses the locomotor suppression and tremulous jaw movements induced by haloperidol in rats: Possible relevance to parkinsonism. *Behav Brain Res* 148:47–54.

Costa-Farré C, García F, Andaluz A et al. 2005. Effect of H1- and H2-receptor antagonists on the hemodynamic changes induced by the intravenous administration of ketamine in sevoflurane-anesthetized cats. *Inflamm Res* 54(6):256–260.

Cox RH. 2002. Changes in the expression and function of arterial potassium channels during hypertension. *Vascul Pharmacol* 38:13–23.

Cui Y, Nie H, Ma H et al. 2011. Ketamine inhibits lung fluid clearance through reducing alveolar sodium transport. *J Biomed Biotechnol* 2011:460596.

Degoute CS. 2007. Controlled hypotension: A guide to drug choice. *Drugs* 67(7): 1053–1076.

Ding Q, Wei EQ, Zhang YJ et al. 2006. Cysteinyl leukotriene receptor 1 is involved in N-methyl-D-aspartate-mediated neuronal injury in mice. *Acta Pharmacol Sin* 27(12):1526–1536.

Ding X, Murray PA. 2007. The differential effects of intravenous anesthetics on myofilament Ca2+ sensitivity in pulmonary venous smooth muscle. *Anesth Analg* 105(5):1278–1286.

Fan H, Peck OM, Tempel GE et al. 2004. Toll-like receptor 4 coupled GI protein signalling pathways regulate extracellular signal-regulated kinase phosphorylation and AP-1 activation independent of NFkappaB activation. *Shock* 22:57–62.

Fujimura N, Tanaka E, Yamamoto S et al. 1997. Contribution of ATP-sensitive potassium channels to hypoxic hyperpolarization in rat hippocampal CA1 neurons in vitro. *J Neurophysiol* 77:378–385.

Fukumoto K, Iijima M, Chaki S. 2014. Serotonin-1A receptor stimulation mediates effects of a metabotropic glutamate 2/3 receptor antagonist, 2S-2-amino-2-(1S,2S-2-carboxycycloprop-1-yl)-3-(xanth-9-yl)propanoic acid (LY341495), and an N-methyl-D-aspartate receptor antagonist, ketamine, in the novelty-suppressed feeding test. *Psychopharmacology (Berl)* 231:2291–2298.

Ge QF, Wei EQ, Zhang WP et al. 2006. Activation of 5-lipoxygenase after oxygen-glucose deprivation is partly mediated via NMDA receptor in rat cortical neurons. *J Neurochem* 97:992–1004.

Ghasemi M, Shafaroodi H, Nazarbeiki S et al. 2010. Voltage-dependent calcium channel and NMDA receptor antagonists augment anticonvulsant effects of lithium chloride on pentylenetetrazole-induced clonic seizures in mice. *Epilepsy Behav* 18(3):171–178.

Goldman-Rakic PS, Muly EC 3rd, Williams GV. 2000. D(1) receptors in prefrontal cells and circuits. *Brain Res Brain Res Rev* 31:295–301.

Gouzoulis-Mayfrank E, Habermeyer E, Hermle L et al. 1998. Hallucinogenic drug-induced states resemble acute endogenous psychoses: Results of an empirical study. *Eur Psychiatry* 13:399–406.

Grace AA. 2000. Gating of information flow within the limbic system and the pathophysiology of schizophrenia. *Brain Res Brain Res Rev* 31:330–341.

Gupta A, Devi LA, Gomes I. 2011. Potentiation of μ-opioid receptor-mediated signaling by ketamine. *J Neurochem* 119(2):294–302.

Hallak JE, Dursun SM, Bosi DC et al. 2011. The interplay of cannabinoid and NMDA glutamate receptor systems in humans: Preliminary evidence of interactive effects of cannabidiol and ketamine in healthy human subjects. *Prog Neuropsychopharmacol Biol Psychiatry* 35(1):198–202.

Hama A, Basler A, Sagen J. 2006. Enhancement of morphine antinociception with the peptide N-methyl-D-aspartate receptor antagonist [Ser1]-histogranin in the rat formalin test. *Brain Res* 1095(1):59–64.

Hanazaki M, Jones KA, Warner DO. 2000. Effects of intravenous anesthetics on Ca^{2+} sensitivity in canine tracheal smooth muscle. *Anesthesiology* 92:133–139.

Hanouz JL, Zhu L, Persehaye E et al. 2005. Ketamine preconditions isolated human right atrial myocardium: Roles of adenosine triphosphate-sensitive potassium channels and adrenoceptors. *Anesthesiology* 102(6):1190–1196.

Harrison NL, Simmonds MA. 1985. Quantitative studies on some antagonists of N-methyl-D-aspartate in slices of rat cerebral cortex. *Br J Pharmacol* 84:381–391.

Hasaka M, Mori T, Matsuura T et al. 2012. Effects of general anesthetics on P2X4 receptors in a mouse microglial cell line. *Neuroreport* 23(10):601–605.

Hayashi Y, Kawaji K, Sun L et al. 2011. Microglial Ca(2+)-activated K(+) channels are possible molecular targets for the analgesic effects of S-ketamine on neuropathic pain. *J Neurosci* 31(48):17370–17382.

Heinzel A, Steinke R, Poeppel TD et al. 2008. S-ketamine and GABA-A-receptor interaction in humans: An exploratory study with I-123-iomazenil SPECT. *Hum Psychopharmacol* 23(7):549–554.

Henshall DC, Diaz-Hernandez M, Miras-Portugal MT et al. 2013. P2X receptors as targets for the treatment of status epilepticus. *Front Cell Neurosci* 7:237.

Hettinger BD, Lee A, Linden J et al. 2001. Ultrastructural localization of adenosine A2A receptors suggests multiple cellular sites for modulation of GABAergic neurons in rat striatum. *J Comp Neurol* 431:331–346.

Hevers W, Hadley SH, Lüddens H et al. 2008. Ketamine, but not phencyclidine, selectively modulates cerebellar GABA(A) receptors containing alpha6 and delta subunits. *J Neurosci* 28(20):5383–5393.

Ho CM, Su CK. 2006. Ketamine attenuates sympathetic activity through mechanisms not mediated by N-methyl-D-aspartate receptors in the isolated spinal cord of neonatal rats. *Anesth Analg* 102(3):806–810.

Ho KK, Flood P. 2004. Single amino acid residue in the extracellular portion of transmembrane segment 2 in the nicotinic alpha7 acetylcholine receptor modulates sensitivity to ketamine. *Anesthesiology* 100(3):657–662.

Hunt MJ, Kessal K, Garcia R. 2005. Ketamine induces dopamine-dependent depression of evoked hippocampal activity in the nucleus accumbens in freely moving rats. *J Neurosci* 25(2):524–531.

Inoue K, Tsuda M. 2009. Microglia and neuropathic pain. *Glia* 14:1469–1479.

Irifune M, Sato T, Kamata Y et al. 2000. Evidence for $GABA_A$ receptor agonistic properties of ketamine: Convulsive and anesthetic behavioural models in mice. *Anesth Analg* 91:230–236.

Ishima T, Nishimura T, Iyo M et al. 2008. Potentiation of nerve growth factor-induced neurite outgrowth in PC12 cells by donepezil: Role of sigma-1 receptors and IP3 receptors. *Prog Neuropsychopharmacol Biol Psychiatry* 32:1656–1659.

Ishiwa D, Kamiya Y, Itoh H et al. 2004. Effects of isoflurane and ketamine on ATP-sensitive K channels in rat substantia nigra. *Neuropharmacology* 46(8): 1201–1212.

Jackson ME, Homayoun H, Moghaddam B. 2004. NMDA receptor hypofunction produces concomitant firing rate potentiation and burst activity reduction in the prefrontal cortex. *Proc Nat Acad Sci U S A* 101:8467–8472.

Jin S, Johansson B, Fredholm BB. 1993. Effects of adenosine A1 and A2 receptor stimulation on electrically evoked dopamine and acetylcholine release from rat striatal slices. *J Pharmacol Ext Ther* 267:801–808.

Joly V, Richebe P, Guignard B et al. 2005. Remifentanil-induced postoperative hyperalgesia and its prevention with small-dose ketamine. *Anesthesiology* 103:147–155.

Jones BW, Means TK, Heldwein KA et al. 2001. Different Toll-like receptor agonists induce distinct macrophage responses. *J Leukoc Biol* 69:1036–1044.

Jung I, Jung SH. 2012. Vasorelaxant mechanisms of ketamine in rabbit renal artery. *Korean J Anesthesiol* 63(6):533–539.

Kanaide M, Uezono Y, Matsumoto M et al. 2007. Desensitization of GABA$_B$ receptor signalling by formation of protein complexes of GABA$_{B2}$ subunit with GRK4 or GRK5. *J Cell Physiol* 210:237–245.

Karydes H, Meehan TJ, Bryant SM. 2013. The toxic trio: Valproic acid, lithium, and carbamazepine. *Am J Ther.* 21:265–268.

Kawano T, Oshita S, Takahashi A et al. 2005. Molecular mechanisms underlying ketamine-mediated inhibition of sarcolemmal adenosine triphosphate-sensitive potassium channels. *Anesthesiology* 102(1):93–101.

Kawano T, Tanaka K, Yinhua et al. 2010. Effects of ketamine on nicorandil induced ATP-sensitive potassium channel activity in cell line derived from rat aortic smooth muscle. *J Med Invest* 57(3–4):237–244.

Kekesi O, Tuboly G, Szucs M et al. 2011. Long-lasting, distinct changes in central opioid receptor and urinary bladder functions in models of schizophrenia in rats. *Eur J Pharmacol* 661(1–3):35–41.

Kienbaum P, Heuter T, Michel MC et al. 2000. Racemic ketamine decreases muscle sympathetic activity but maintains the neural response to hypotensive challenges in humans. *Anesthesiology* 92:94–101.

Kim SH, Bae YM, Sung DJ et al. 2007. Ketamine blocks voltage-gated K(+) channels and causes membrane depolarization in rat mesenteric artery myocytes. *Pflugers Arch* 454(6):891–902.

Klockgether-Radke AP, Huneck S, Meyberg S et al. 2005. Ketamine enantiomers differentially relax isolated coronary artery rings. *Eur J Anaesthesiol* 22(3):215–221.

Kommalage M, Höglund AU. 2004. (+/−) Epibatidine increases acetylcholine release partly through an action on muscarinic receptors. *Basic Clin Pharmacol Toxicol* 94:238–244.

Li F, Fang Q, Liu Y et al. 2008. Cannabinoid CB(1) receptor antagonist rimonabant attenuates reinstatement of ketamine conditioned place preference in rats. *Eur J Pharmacol* 589(1–3):122–126.

Liu FL, Chen TL, Chen RM. 2012. Mechanisms of ketamine-induced immunosuppression. *Acta Anaesthesiol Taiwan* 50(4):172–177.

Lozovsky D, Saller CF, Bayorh MA et al. 1983. Effects of phencyclidine on rat prolactin, dopamine receptor and locomotor activity. *Life Sci* 32:2725–2731.

Maeng S, Zarate CA Jr, Du J et al. 2008. Cellular mechanisms underlying the antidepressant effects of ketamine: Role of alpha-amino-3-hydroxy-5-methylisoxazole-4-propionic acid receptors. *Biol Psychiatry* 63:349–352.

Magee JC. 2000. Dendritic integration of excitatory synaptic input. *Nat Rev Neurosci* 1:181–190.

Mandryk M, Fidecka S, Poleszak E et al. 2005. Participation of adenosine system in the ketamine-induced motor activity in mice. *Pharmacol Rep* 57(1):55–60.

Marone G, Stellato C, Mastronardi P et al. 1993. Mechanisms of activation of human mast cells and basophils by general anesthetic drugs. *Ann Fr Anesth Reanim* 12:116–125.

Martin BS, Kapur J. 2008. A combination of ketamine and diazepam synergistically controls refractory status epilepticus induced by cholinergic stimulation. *Epilepsia* 49(2):248–255.

Matulewicz P, Kasicki S, Hunt MJ. 2010. The effect of dopamine receptor blockade in the rodent nucleus accumbens on local field potential oscillations and motor activity in response to ketamine. *Brain Res* 1366:226–232.

Mazar J, Rogachev B, Shaked G et al. 2005. Involvement of adenosine in the antiinflammatory action of ketamine. *Anesthesiology* 102(6):1174–1181.

McCallion K, Harkin DW Gardiner KR. 2004. Role of adenosine in immunomodulation: Review of the literature. *Crit Care Med* 32:273–277.

McCardle CE, Gartside SE. 2012. Effects of general anaesthetics on 5-HT neuronal activity in the dorsal raphe nucleus. *Neuropharmacology* 62(4):1787–1796.

McMahon LR. 2006. Discriminative stimulus effects of the cannabinoid CB1 antagonist SR 141716A in rhesus monkeys pretreated with Delta9-tetrahydrocannabinol. *Psychopharmacology (Berl)* 188(3):306–314.

Mele A, Wozniak KM, Hall FS et al. 1998. The role of striatal dopaminergic mechanisms in rotational behaviour induced by phencyclidine and phencyclidine-like drugs. *Psychopharmacology* 135:107–118.

Minami K, Sudo Y, Shiraishi S et al. 2010. Analysis of the effects of anesthetics and ethanol on mu-opioid receptor. *J Pharmacol Sci* 112(4):424–431.

Miyamoto Y, Yamada K, Noda Y et al. 2001. Hyperfunction of dopaminergic and serotonergic neuronal systems in mice lacking the NMDA receptor $\varepsilon1$ subunit. *J Neurosci* 21:750–757.

Moaddel R, Abdrakhmanova G, Kozak J et al. 2013. Sub-anesthetic concentrations of (R,S)-ketamine metabolites inhibit acetylcholine-evoked currents in $\alpha7$ nicotinic acetylcholine receptors. *Eur J Pharmacol* 698(1–3):228–234.

Momosaki S, Hatano K, Kawasumi Y et al. 2004. Rat-PET study without anesthesia: Anesthetics modify the dopamine D1 receptor binding in rat brain. *Synapse* 54(4):207–213.

Moncada S, Palmer RMJ, Higgs EA. 1991. Nitric oxide: Physiology, pathophysiology and pharmacology. *Pharmacol Rev* 43:109–142.

Nakanishi M, Mori T, Nishikawa K et al. 2007. The effects of general anesthetics on P2X7 and P2Y receptors in a rat microglial cell line. *Anesth Analg* 104:1136–1144.

Narendran R, Frankle WG, Keefe R et al. 2005. Altered prefrontal dopaminergic function in chronic recreational ketamine users. *Am J Psychiatry* 162(12):2352–2359.

Nemeth CL, Paine TA, Rittiner JE et al. 2010. Role of kappa-opioid receptors in the effects of salvinorin A and ketamine on attention in rats. *Psychopharmacology (Berl)* 210(2):263–274.

Odening KE, Hyder O, Chaves L et al. 2008. Pharmacogenomics of anesthetic drugs in transgenic LQT1 and LQT2 rabbits reveal genotype-specific differential effects on cardiac repolarization. *Am J Physiol Heart Circ Physiol* 295(6):H2264–H2272.

Ohnesorge H, Feng Z, Zitta K et al. 2013. Influence of clonidine and ketamine on m-RNA expression in a model of opioid-induced hyperalgesia in mice. *PLoS One* 8(11):e79567.

Pacher P, Kunos G. 2013. Modulating the endocannabinoid system in human health and disease—Successes and failures. *FEBS J* 280(9):1918–1943.

Pandey CK, Mathur N, Singh N et al. 2000. Fulminant pulmonary edema after intramuscular ketamine. *Can J Anaesth* 47:894–896.

Pape HC. 1996. Queer current and pacemaker: The hyperpolarization-activated cation current in neurons. *Annu Rev Physiol* 58:299–327.

Perroy J, Adam L, Qanbar R et al. 2003. Phosphorylation-independent desensitization of GABA$_B$ receptor by GRK4. *EMBO J* 22:3816–3824.

Restall J, Johnston IG, Robinson DN. 1990. Flumazenil in ketamine and midazolam anaesthesia. *Anaesthesia* 45:938–940.

Roberts BM, Seymour PA, Schmidt CJ et al. 2010. Amelioration of ketamine-induced working memory deficits by dopamine D1 receptor agonists. *Psychopharmacology (Berl)* 210(3):407–418.

Robson MJ, Elliott M, Seminerio MJ et al. 2012. Evaluation of sigma (σ) receptors in the antidepressant-like effects of ketamine in vitro and in vivo. *Eur Neuropsychopharmacol* 22(4):308–317.

Romero TR, Duarte ID. 2013. Involvement of ATP-sensitive K(+) channels in the peripheral antinociceptive effect induced by ketamine. *Vet Anaesth Analg* 40(4):419–424.

Romero TR, Galdino GS, Silva GC et al. 2011. Ketamine activates the L-arginine/ Nitric oxide/cyclic guanosine monophosphate pathway to induce peripheral antinociception in rats. *Anesth Analg* 113(5):1254–1259.

Sasao J, Taneyama C, Kohno N et al. 1996. The effects of ketamine on renal sympathetic nerve activity and phrenic nerve activity in rabbits (with vagotomy) with and without afferent inputs from peripheral receptors. *Anesth Analg* 82:362–367.

Schiffman SN, Jacobs O, Vanderhaeghen JJ. 1991. The striatal restricted adenosine A2 receptor (RDC8) is expressed by enkephaline but not by substance P neurons: An in-situ hybridization study. *J Neurochem* 57:1062–1067.

Schneider PG, Rodríguez de Lores Arnaiz G. 2013. Ketamine prevents seizures and reverses changes in muscarinic receptor induced by bicuculline in rats. *Neurochem Int* 62(3):258–264.

Scholz J, Woolf CJ. 2007. The neuropathic pain triad: Neurons, immune cells and glia. *Nat Neurosci* 10:1361–1368.

Seeman P, Ko F, Tallerico T. 2005. Dopamine receptor contribution to the action of PCP, LSD and ketamine psychotomimetics. *Mol Psychiatry* 10(9):877–883.

Shah MM, Anderson AE, Leung V et al. 2004. Seizure-induced plasticity of h channels in entorhinal cortical layer III pyramidal neurons. *Neuron* 44:495–508.

Sinner B, Friedrich O, Zink W et al. 2005. Ketamine stereoselectively inhibits spontaneous Ca2+-oscillations in cultured hippocampal neurons. *Anesth Analg* 100(6):1660–1666.

Sinner B, Graf BM. 2008. Ketamine. In *Handbook of Experimental Pharmacology*, eds. J. Schüttler and H. Schwilden, 313–333. Berlin: Springer-Verlag.

Stanley V, Hunt J, Willis KW et al. 1968. Cardiovascular and respiratory function with CI-581. *Anesth Analg* 47:760–768.

Svenningsson P, Le Moine C, Fisone G et al. 1999. Distribution, biochemistry and function of striatal adenosine A2A receptors. *Prog Neurobiol* 59:335–396.

Trapaidze N, Gomes I, Cvejic S et al. 2000. Opioid receptor endocytosis and activation of MAP kinase pathway. *Mol Brain Res* 76:220–228.

Vernaleken I, Klomp M, Moeller O et al. 2013. Vulnerability to psychotogenic effects of ketamine is associated with elevated D2/3-receptor availability. *Int J Neuropsychopharmacol* 16(4):745–754.

Vernaleken I, Peters L, Raptis M et al. 2011. The applicability of SRTM in [(18)F] fallypride PET investigations: Impact of scan durations. *J Cereb Blood Flow Metab* 31:1958–1966.

Waelbers T, Polis I, Vermeire S et al. 2013. 5-HT2A receptors in the feline brain: 123I-5-I-R91150 kinetics and the influence of ketamine measured with micro-SPECT. *J Nucl Med* 54(8):1428–1433.

Weber M, Motin L, Gaul S et al. 2005. Intravenous anaesthetics inhibit nicotinic acetylcholine receptor-mediated currents and Ca^{2+} transients in rat intracardiac ganglion neurons. *Br J Pharmacol* 144(1):98–107.

Wu GJ, Chen TL, Ueng YF et al. 2008. Ketamine inhibits tumor necrosis factor-alpha and interleukin-6 gene expressions in lipopolysaccharide-stimulated macrophages through suppression of toll-like receptor 4-mediated c-Jun N-terminal kinase phosphorylation and activator protein-1 activation. *Toxicol Appl Pharmacol* 228(1):105–113.

Wu Y, Li W, Zhou C et al. 2012. Ketamine inhibits lipopolysaccharide-induced astrocytes activation by suppressing TLR4/NF-κB pathway. *Cell Physiol Biochem* 30(3):609–617.

Yamanaka H, Yokoyama C, Mizuma H et al. 2014. A possible mechanism of the nucleus accumbens and ventral pallidum 5-HT1B receptors underlying the antidepressant action of ketamine: A PET study with macaques. *Transl Psychiatry* 4:e342. doi: 10.1038/tp.2013.112.

Yanagita T, Maruta T, Uezono Y et al. 2007. Lithium inhibits function of voltage-dependent sodium channels and catecholamine secretion independent of glycogen synthase kinase-3 in adrenal chromaffin cells. *Neuropharmacology* 53:881–889.

Yoshioka M, Fukuishi N, Iriguchi S et al. 2007. Lipoteichoic acid downregulates Fc(epsilon)RI expression on human mast cells through Toll-like receptor 2. *J Allergy Clin Immunol* 120:452–461.

Yoshizawa K, Mori T, Ueno T et al. 2013. Involvement of serotonin receptor mechanisms in the discriminative stimulus effects of ketamine in rats. *J Pharmacol Sci* 121(3):237–241.

Yu M, Shao D, Feng X et al. 2007a. Effects of ketamine on pulmonary TLR4 expression and NF-kappa-B activation during endotoxemia in rats. *Methods Find Exp Clin Pharmacol* 29(6):395–399.

Yu M, Shao D, Liu J et al. 2007b. Effects of ketamine on levels of cytokines, NF-kappaB and TLRs in rat intestine during CLP-induced sepsis. *Int Immunopharmacol* 7(8):1076–1082.

Yu M, Shao D, Yang J et al. 2006. Ketamine suppresses intestinal TLR4 expression and NF-kappaB activity in lipopolysaccharide-treated rats. *Croat Med J* 47(6):825–831.

Zeng LH, Zhang WP, Wang RD et al. 2001. Protective effect of ONO-1078, a leu-kotriene antagonist, on focal cerebral ischemia in mice. *Yao Xue Xue Bao* 36:148–150.

Zhang LH, Wei EQ. 2003. Neuroprotective effect of ONO-1078, a leukotriene recep-tor antagonist, on transient global cerebral ischemia in rats. *Acta Pharmacol Sin* 24:1241–1247.

Zhang P, Xing J, Luo A et al. 2013. Blockade of the human ether-a-go-go-related gene potassium channel by ketamine. *J Pharm Pharmacol* 65(9):1321–1328.

Zhou C, Douglas JE, Kumar NN et al. 2013. Forebrain HCN1 channels contribute to hypnotic actions of ketamine. *Anesthesiology* 118(4):785–795.

chapter four

Developmental neurotoxicity of ketamine in the developing brain

Chaoxuan Dong

Contents

4.1 Introduction

Pharmacological agents and biomedical devices are the most frequently used and most important tools used to facilitate the recovery of the human body from diseases or injuries. A characteristic paradox exists in nature—anything can have both beneficial and harmful effects. Similarly, any artificial or natural substance introduced into the complex human body may have valuable health-promoting effects but may also have detrimental side effects. Commonly used drugs are no exception. When drugs are used to cure diseases, their side effects may bring additional severe problems to patients. Among numerous drugs, anesthetics are very special because they usually act on the normal nervous system in order to block the pain and stress during surgical operations. As an exogenous chemical, these drugs surely produce side effects on the normal nervous system. Therefore, a neurotoxicological study, a systematic study of the toxic effects of anesthetic drugs, is necessary to avoid potential side effects in the normal and abnormal brain.

The developing nervous system is susceptible to all kinds of external stimuli and environmental influences. Some gentle external stimuli are beneficial to stimulating the normal development of the nervous system like light, sound, touch, and so on. However, in some emergent situations like surgery or critical care, it is evitable for infants and children to be exposed to potentially toxic external stimuli such as pediatric anesthetics, analgesics, sedatives, or other psychotropic agents. To minimize or avoid severe neurotoxic effects from the use of these drugs, developmental neurotoxicology was developed to evaluate the risk of neuroactive drugs in the developing nervous system.

Ketamine has been widely used as a pediatric anesthetic, analgesic, sedative, and antidepressant agent in pediatric and obstetric clinical practice and is also consumed as an illicit abuse drug by teenagers and young adults. Physicians and scientists are paying more attention and research interests on the potential side effects of ketamine on the developing brain. This chapter introduces the potential neurotoxic effects of ketamine on the developing brain.

4.2 Ketamine

Ketamine was synthesized as a substitute for phencyclidine (PCP) in the 1960s. Dr. Harold Maddox synthesized PCP in 1956 and introduced it for clinical study as a safe anesthetic agent in humans at the time. However, schizophrenic findings and the incidence of severe delirium relating to PCP were not acceptable for clinical use for human anesthesia. Given this situation, a series of short-acting derivative compounds of PCP were synthesized by Dr. Calvin Lee Stevens at Wayne State University in 1962 (Domino 2010). Finally, one of the compounds was screened to produce excellent anesthesia with a rapid-acting onset and short-term effects and was selected for human trials as CI-581 (the clinical investigation number). This compound (cyclohexylamine, CI-581) is well known as ketamine.

On August 3, 1964, ketamine was first used in a human study and was found to produce remarkable anesthesia with minimal side effects. After that, physicians extended applications of ketamine to more clinical practices: ophthalmic practice (Harris et al. 1968), surgery (Del Prete et al. 1968), neurosurgical diagnostic procedures (Corssen et al. 1969), pediatric procedures (Ginsberg and Gerber 1969; Szappanyos et al. 1969; Wilson et al. 1969), and so on under the following trade names: Ketalar, Ketaset, Ketajet, Ketavet, Vetamine, Vetaket, and Ketamine Hydrochloride (HCl) Injection. Widespread use of ketamine in humans occurred during the Vietnam War in the 1970s. Because of a big margin of safety, ketamine was given to injured American soldiers as an anesthetic, analgesic, and sedative (Mercer 2009). Within the past half-century, despite recent concerns on developmental neurotoxicity and drug abuse, ketamine has been developed as an anesthetic for premedication, sedation, induction, and maintenance of general anesthesia especially for trauma victims, for patients with hypovolemic and septic shock, and for patients with pulmonary diseases; as an analgesic for acute and chronic pain management including postoperative analgesia; and as a sedative for pain relief to intensive care patients, especially during ventilator management (Sinner and Graf 2008).

Unexpectedly, during popular uses of ketamine in clinics and hospitals, the feeling after the intake of ketamine attracted more people to consume it as a recreational drug, even as a date rape drug (Graeme 2000; Jansen 2000). Meanwhile, ketamine was given new street names like "Bump, CatValium, K, Ket, Kit Kat, Kizzo, Special K, Super Acid, and Vitamin K." Considering the increasing abuse of ketamine and more crimes relating to ketamine, the US DEA (Drug Enforcement Administration of the United States) listed ketamine as a Schedule III nonnarcotic substance under the Controlled Substances Act in 1999.

In the same year, one study about ketamine inducing neurodegeneration in the developing brain (Ikonomidou et al. 1999) ignited heated discussions on the neurotoxicity of ketamine use in infants and children. Subsequent studies indicated that a high dose of ketamine induces neuronal cell death, especially apoptosis, in many in vivo and in vitro models from mice, rats, and monkeys (Ikonomidou et al. 1999; Rudin et al. 2005;

Scallet et al. 2004; Slikker et al. 2007; Walker et al. 2010; Young et al. 2005). Additionally, ketamine has been reported to disturb normal neurogenesis of the developing brain (Dong et al. 2012). More recent preclinical studies indicate that the early exposure to ketamine can result in long-lasting cognitive deficits in rhesus monkeys (Paule et al. 2011). These findings forced scientists and physicians to reconsider the safety and toxic effects of ketamine in pediatric use (Mellon et al. 2007).

4.3 The developing brain exposed to ketamine

In clinical practice, ketamine has been used widely in four major fields: anesthesia, analgesia, sedation in the intensive care unit, and antidepressant effects (Domino 2010).

4.3.1 Anesthesia

Ketamine-induced anesthesia is described as dissociative anesthesia, characterized by profound analgesia and amnesia, with retention of protective airway reflexes, spontaneous respirations, and cardiopulmonary stability (Green et al. 2011). Under ketamine anesthesia, blood pressure is well maintained even in the case of hypovolemia, and spontaneous breathing and laryngeal reflexes are preserved. This has made ketamine become the "first choice" anesthetic agent for prehospital anesthesia/analgesia. In children, ketamine combined with other anesthetics like propofol or midazolam is utilized in pediatric plastic surgery (Zook et al. 1971), oral surgery (Birkhan et al. 1971), neurosurgery (Chadduck and Manheim 1973), cardiac anesthesia (Koruk et al. 2010; Radnay et al. 1976), ophthalmic surgery (Raju 1980), gastrointestinal procedures (Shemesh et al. 1982), and pediatric interventional cardiac procedures (Singh et al. 2000). A propofol–ketamine mixture is routinely used for anesthesia in pediatric patients undergoing cardiac catheterization (Kogan et al. 2003).

4.3.2 Analgesia and pain management

A subanesthetic dose of ketamine can produce analgesic effects as an "antihyperalgesic," "antiallodynic," or "tolerance-protective" agent (Visser and Schug 2006). Ketamine is used not only in acute pain management but also in the chronic pain setting (Amr 2010; Blonk et al. 2010; Noppers et al. 2010; Visser and Schug 2006). Systemic or regional administration of ketamine is used for acute pain management (Chazan et al. 2008). Ketamine produces potent analgesia in children (da Conceicao et al. 2006; Dal et al. 2007). Although ketamine was under reevaluation for its potential as a neurotoxicant in the developing brain, it is being used increasingly to supplement opioids for the pain after major surgery in pediatric clinical settings (Anderson and Palmer 2006). Therefore, it is necessary to be cautious when using high doses or long-term ketamine in this population.

4.3.3 Sedation

In the emergency department, intensive care unit, and in some medical examination procedures, a sedative state is necessary for further treatment, especially in pediatric settings. A large number of clinical studies indicate that the combinations of ketamine and midazolam (McGlone 2009; Sener et al. 2011), ketamine and dexmedetomidine (McVey and Tobias 2010), or ketamine and propofol (Weatherall and Venclovas 2010) can be very useful and safe for sedation and pain relief in pediatric intensive care patients, especially during ventilator management.

4.3.4 Recreational use

Ketamine has been listed as an illicit club drug with many street names as "Bump," "CatValium," "K," "Ket," "Kit Kat," "Kizzo," "Special K," "Super Acid," and "Vitamin K." It is mainly abused among teenagers and young adults at bars, nightclubs, concerts, and parties. Usually, ketamine is snorted, or injected intramuscularly, but sometimes injected intravenously by drug abusers. Studies show that low-dose intoxication impairs attention, learning ability, and memory (Morgan and Curran 2006; Morgan et al. 2009, 2010). At high doses, ketamine can cause dreamlike states and hallucinations, and at an extremely high dose, ketamine can result in delirium and amnesia.

Ketamine is used recreationally as a club drug because it decreases social inhibitions. Ketamine has been used to heighten sexual experience (Parks and Kennedy 2004), which increases the incidence of accidental pregnancy, and even leads to the increased risk of infectious disease transmission including human immunodeficiency virus (Romanelli et al. 2003; Semple et al. 2009). Ketamine-abusing pregnant mothers may have opportunities to take large doses of ketamine for long periods during pregnancy. Ketamine can easily cross the placenta, enter the body of a fetus (Craven 2007), and rapidly distribute to the embryonic brain. This situation results in greater chances for ketamine to manifest its neurotoxic effects in the embryonic brain, thus defining one of the risks of consuming ketamine as an abuse drug.

4.4 Pharmacological mechanisms of ketamine

4.4.1 An NMDA (N-methyl-D-aspartate) receptor antagonist

The major pharmacological effects of ketamine are related to the antagonism of NMDA receptors. Ketamine, like PCP, dizocilpine (MK-801), and memantine, noncompetitively binds to the PCP site inside the NMDA receptor, a calcium ion channel, and blocks the influx of calcium (Bolger et al. 1986; Ffrench-Mullen and Rogawski 1992; O'Shaughnessy and Lodge 1988) (Figure 4.1). Ketamine-produced antagonism of NMDA receptors depends on the opening state of the calcium ion channel.

The NMDA receptor is one type of ionotropic glutamate receptors (iGluRs) mediating excitatory synaptic transmission in the central nervous system (CNS) (Collingridge and Lester 1989). It is a tetrameric transmembrane channel receptor (Hirota et al. 1999a; Kubota et al. 1999; Ulbrich and Isacoff 2007) composed of four subunits: two obligatory NR1 (NMDA receptor 1) subunits and two regulatory subunits NR2 (NR2A–D) and NR3 (NR3A and B) (Furukawa et al. 2005). Varying composition of NMDA receptor subunits determines temporal and regional specificity and unique functional properties (Monyer et al. 1992; Paoletti and Neyton 2007). NR2A and NR2B subunits mainly constitute NMDA receptors in the forebrain. These two NMDA receptor subunits undergo a well-characterized developmental shift in the cortex. NR2B subunits are abundant in the early postnatal brain, and NR2A levels increase progressively with the brain development (Flint et al. 1997; Mierau et al. 2004; Quinlan et al. 1999; Roberts and Ramoa 1999; Sheng et al. 1994) (Figure 4.1).

NMDA-activated iGluRs are permeable to Ca^{++} (Mayer and Westbrook 1987) and are blocked by Mg^{++} (Mayer et al. 1984) at hyperpolarized membrane potentials, whereas non-NMDA receptor–activated iGluRs (AMPA [α-amino-3-hydroxy-5-methyl-4-isoxazolepropionic acid] and kainate) are also permeable to Ca^{++} (Iino et al. 1990) but cannot be blocked by Mg^{++}. This important difference is determined by different amino acid residues at the key action site in M2 regions of these ion channels. In NMDA

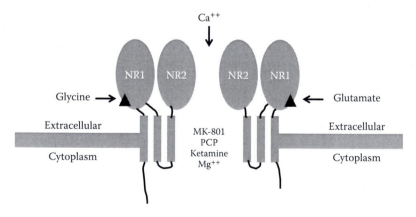

Figure 4.1 NMDA receptor model showing potential sites for ketamine. This figure shows the action site of noncompetitive NMDA receptor antagonists (ketamine, MK-801, and PCP) inside NMDA receptors.

receptors, the site is occupied by asparagines and substituted by glutamine or arginine in non-NMDA receptors (Kutsuwada et al. 1992; Mori et al. 1992; Moriyoshi et al. 1991), which is the action site of PCP (Ferrer-Montiel et al. 1996). Moreover, the site of the NMDA receptor M2 region is also related to the Mg^{++} blockade NMDA receptors (Burnashev et al. 1992). The conserved asparagine residue in segment M2 constitutes the Mg^{++} block site of the NMDA receptor channel and the MK-801/ketamine action site, indicating that the ketamine-induced antagonism of NMDA receptors depends on the Mg^{++} evacuation of the M2 segment site in NMDA receptors, namely, an open channel–dependent mechanism (Figure 4.1).

4.4.2 Other receptors binding

Besides the antagonism of excitatory amino acids on NMDA receptors, ketamine has also several mechanisms of action on other receptors (Table 4.1). Ketamine, especially $S(+)$-ketamine, can reduce opioid consumption after surgeries (Lahtinen et al. 2004) and reverse opioid tolerance in pain management (Mercadante et al. 2003), indicating that ketamine may interact with opioid receptors to some extent. Recent studies have testified that $S(+)$-ketamine is two to three times more potent than $R(-)$-ketamine at μ, κ, and δ opioid receptors (Hirota et al. 1999b,c) (Table 4.1). Ketamine at 100 μM reverses the enkephalin (μ) and spiradoline (κ) inhibition of cyclic adenosine monophosphate. Additionally, ketamine potentiates the anti-nociceptive effects of μ but not κ or δ agonists in a mouse acute pain model (Baker et al. 2002a,b).

Recent publications indicate that PCP stimulates dopaminergic receptors (D2) in vitro (Seeman and Guan 2008; Seeman et al. 2005) with even higher affinities than NMDA receptors (Kapur and Seeman 2002; Seeman and Guan 2009; Seeman and Lasaga 2005; Seeman et al. 2009) (Table 4.1).

Table 4.1 Summary of NMDA versus Non-NMDA Receptor Pharmacology of Ketamine

	NMDA[a]	Opioid receptor MOR DOR KOR	Dopamine D2	5HT 5HT-2	Ion channels
Ki (μ)	0.42	26.8 101 85	0.5	15	>50 or >100
References	Komhuber et al. (1989)	MOR DOR KOR Smith et al. (1987)	Dopamine Kapur and Seeman (2002)	5HT	Eide and Stabhaug (1997)

Note: This table presents the most probable ketamine-binding receptors. Among these data, the NMDA receptor and the D2 receptor have low Ki values, 0.42 and 0.5, respectively, suggesting that ketamine has high affinity to bind to these two receptors. MOR, μ-opioid receptor; DOR, δ-opioid receptor; KOR, κ-opioid receptor; 5HT, serotonin transporter binding.

[a] Ki for inhibition of [3H](+) MK-801 binding to cortical membranes.

The effects of ketamine on dopaminergic receptors support the theory for animal models of schizophrenia: ketamine and PCP can directly inter-act with dopaminergic receptors or indirectly block NMDA receptors to increase the release of dopamine (Javitt 2010). Dopamine dysfunction in the brain is associated with signs and symptoms and the incidence of schizophrenia (Seeman et al. 2005).

4.5 Expression of NMDA receptors in the early stage of brain development

4.5.1 Expression of NMDA receptors in the developing brain

The expression of NMDA receptor subunits is temporal and brain regional dependent in the developing brain. In the human dorsolateral prefron-tal cortex, NR1 expression is low prenatally, peaks in adolescence, and remains high throughout life, suggesting the lifelong importance of NMDA receptor function (Henson et al. 2008). Additionally, NR1 is an essential subunit for the formation of a functional NMDA receptor, ini-tially expressed in restricted areas such as the temporal region of the cere-bral cortex and the hippocampus in the fetal brain at embryonic day 18 (E18) and E20. In neonates, the expression of NR1 spreads widely through-out the whole brain. Using the NR1(N)-antiserum staining, the strongest signals are detected in the hippocampus, followed by the cortex, striatum, and thalamus, and weaker staining is observed in the brainstem and cer-ebellum of adult brain (Benke et al. 1995). For the cerebral cortex of the developing rat, the relative distribution profile of NMDA receptor sub-units is as follows: NR2B > NR1 > NR2A > NR2C. In the cerebellum, it is NR2C = NR1 > NR2A = NR2B (Sircar et al. 1996).

Compared to the wide distribution of the NR1 expression throughout the lifespan, the expression profiles of NR2 in the brain are developmen-tally and regionally regulated (Ishii et al. 1993). The distributions of NR2A and NR2C showed temporal and spatial similarities to that of NR1, but the expression of NR2B shows differences in the intensity and distribu-tion (Takai et al. 2003). In the rat, NR2B and NR2D subunits predomi-nate in the neonatal brain, and they are replaced by the NR2A and NR2C subunits in some brain regions (Akazawa et al. 1994; Monyer et al. 1994). Furthermore, NR2A expression occurs mainly at the cerebral cortex and hippocampus, while NR2B predominates in the forebrain, NR2C in the cerebellum, and diencephalon and NR2D in the lower brainstem regions (Ishii et al. 1993; Monyer et al. 1992; Watanabe et al. 1994a,b).

The developmental changes of expression patterns of NMDA recep-tor subunits can be altered by NMDA receptor antagonists such as PCP, MK-801, and ketamine. Acute PCP exposure to the rat on postnatal day (PND) 7 increases membrane levels of both NR1 and NR2B proteins in

the frontal cortex but decreases NR1 and NR2B levels in the endoplasmic reticulum fraction (Anastasio and Johnson 2008). Zou et al. (2009b)
reported remarkable increases in NR1 mRNA signals in the frontal cortex under the exposure of ketamine, using in situ hybridization. NMDA
receptor subunit mRNAs in the developing brain are rapidly altered after
MK-801 exposure. The increase in NR2A mRNA is larger than that in
NR1, NR2B, or NR2D (Wilson et al. 1998). Chronic PCP administration
in postnatal rats produced significant reduction in NR2B subunits in the
cerebral cortex, whereas the expression of other NMDA receptor subunits
was not altered in the cerebral cortex after the drug treatment (Sircar et al.
1996). The above evidence indicates that the expression of NMDA receptor subunits may be involved in neurotoxic effects in the developing brain
exposed to NMDA receptor antagonist.

4.5.2 *Expression of NMDA receptors in neural stem progenitor cells of the developing brain*

Stem cells are defined as undifferentiated cells that have the potency to
self-renew, thus maintaining an undifferentiated state, and to differentiate into a diverse range of specialized cell types (Maric and Barker 2005;
Mitalipov and Wolf 2009; Ulloa-Montoya et al. 2005). Neural stem cells
(NSCs) are multipotent stem cells that only differentiate into a closely
related family of cells, neurons, and glia (astrocytes and oligodendrocytes)
(Gage 2000). NPCs (neuronal and glial progenitors) have very limited differentiation abilities, but still have self-renewal properties, which distinguish
them from mature cells. In current studies, neural stem progenitor cells
(NSPCs) isolated for cultures are composed of NSCs and NPCs (Lovell-
Badge 2001). Presently, no effective cell sorting method can accurately separate one type of cells from the culture mixture (Maric and Barker 2005).

Being the precursors of neurons and glia, NSCs or NPCs are considered to have no or little expression of NMDA receptor until differentiated
completely. However, the fact is that NMDA receptor agonists (NMDA
or glutamate) can induce Ca^{++} currents in NSPCs, which can be blocked
by NMDA receptor antagonists in NPCs (Wang et al. 1996). NMDA can
also increase cytosolic Ca^{++} and inward currents in differentiating neuronal progenitors with $BrdU^+$ (bromodeoxyuridine, a marker of proliferating cells) and $Tuj-1^+$ (beta-tubulin III, a biomarker of newborn neurons)
staining. In contrast, proliferating ($BrdU^+$ and $Tuj-1^-$) NSCs fail to respond
to any iGluR agonists (Maric et al. 2000). After 3 weeks of in vitro differentiation of human NPCs, glutamate and NMDA elicited currents in
93% of NPCs, which can be markedly inhibited by memantine, an NMDA
receptor antagonist (Wegner et al. 2009). These data indicate that NPCs
(but probably not NSCs) have active NMDA receptors, which play a major
role in proliferation, differentiation, synaptogenesis, and neural plasticity.

NMDA receptor subunits have been detected to be expressed in NSPCs. Immunocytochemical analysis showed expression of NR1, NR2A, and NR2B subunits in cultured neurospheres (Joo et al. 2007; Kitayama et al. 2004; Mochizuki et al. 2007; Ramirez and Lamas 2009). The expression of NMDA receptors containing NR1 and NR2B was found in human NPCs (Suzuki et al. 2006; Wegner et al. 2009; Zhang et al. 2004). Plus, NR2B is the predominant NR2 subunit, in rat or human NSCs (Dong et al. 2012; Hu et al. 2008). The expression patterns of NMDA receptor subunits are also temporal dependent in NSPCs. NR2A, NR2B, and NR2D subunit transcription are present in both nondifferentiated and neuronally differentiated cultures, while NR2C subunits are expressed only transiently, during the early period of neural differentiation (Yoneyama et al. 2008).

Functional NMDA receptors are absent in NSPCs until a certain stage of neuronal commitment or the appearance of synaptic communication, despite the expression of their subunits (Jelitai et al. 2002; Muth-Kohne et al. 2010; Varju et al. 2001). However, some studies showed that stimulation of NPCs leads to either proliferation or neuronal differentiation, which depends on the level of NMDA receptor activation, suggesting that functional NMDA receptors have a role in the regulation of neurogenesis (Joo et al. 2007; Maric et al. 2000; Wang et al. 1996). The functions of the NMDA receptors in NSPCs remain to be determined in future studies.

4.6 Ketamine induces neuronal cell death in the developing brain

4.6.1 Neurons

During the brain growth spurt period, neuronal apoptosis can be triggered by the blockade of NMDA receptors. In clinical settings, many anesthetics produce the antagonism of NMDA receptors, such as ketamine and isoflurane. After the Ikonomidou et al. study, Scallet et al. (2004) published that a single dose of 20 mg/kg of ketamine (the blood ketamine level is close to anesthetic level in humans) failed to produce neurodegeneration in neonatal rats at PND7, whereas repeated 20-mg/kg doses did increase the number of silver-positive (degenerating) neurons. In neonatal mice, 50 mg/kg S.C. ketamine induced a severe degeneration of cells in the parietal cortex, which resulted in apparent deficits in habituation, acquisition learning, and retention memory at the age of 2 months (Fredriksson and Archer 2004; Fredriksson et al. 2004). This study indicated that ketamine-triggered neuroapoptosis in the developing brain was dose dependent; exposure of less than 20 mg/kg for 5 h does not induce a significant increase of caspase-3-positive neurons (Young et al. 2005). However, a clinically relevant single dose of ketamine can produce long-lasting neuronal apoptosis in certain brain areas of neonatal mice at PND7 (Rudin et al. 2005). In monkey models, immature monkeys at earlier developmental stages are more sensitive to ketamine-induced cell death, although low doses of ketamine at short-term exposures do not result in cell death in the monkey at PND5 (Slikker et al. 2007). A long-term exposure (9 or 24 h) of ketamine (20 mg/kg I.M. followed by I.V.) produced extensive increases in the number of caspase-3-positive neurons in infant monkey brains (Zou et al. 2009a). In addition, intrathecal ketamine can also increase apoptosis in spinal neurons and leads to impaired long-term functional outcomes in neonatal rats (Walker et al. 2010). All these in vivo data suggest that ketamine can induce apoptosis in the developing brain in a dose- and time-dependent manner.

Ketamine-induced apoptosis in the developing brains from in vivo studies has also been confirmed in in vitro studies with neuronal cultures. In rat forebrain cultures, 10 and 20 μM ketamine exposure for 12 h induces a substantial increase of TUNEL-positive cells relating to the up-regulation of NR1 after ketamine administration (Takadera et al. 2006; Wang et al. 2005). The same doses of ketamine also elevate apoptotic changes in cultured frontal cortical neurons isolated from a monkey at PND3 (Wang et al. 2006). Ketamine can induce apoptosis of immature GABAergic (gamma-aminobutyric acid) neurons (Desfeux et al. 2010). Although low subanesthetic concentrations of ketamine do not affect cell survival, it can impair neuronal morphology and dendritic arbor development in immature GABAergic neurons. This indicates that even low doses of ketamine can disturb the buildup of neural networks in the developing brain (Vutskits et al. 2006, 2007).

Presently, plenty of experimental evidence has confirmed the fact that high-dose ketamine triggers cell death in the developing brain. The mechanism of ketamine-induced cell death in neurons has been investigated in some studies. Most previous studies testified that ketamine-induced cell death occurs via apoptosis (Braun et al. 2010; Fredriksson et al. 2004; Rudin et al. 2005; Scallet et al. 2004; Takadera et al. 2006; Wang et al. 2008b; Young et al. 2005); however, a few studies indicate that necrotic cell death can also take place in high-dose ketamine exposure (Slikker et al. 2007). For the mechanism on how ketamine induces apoptosis in neurons, Drs. Wang and Slikker first reported that prolonged or large-dose ketamine exposure produces up-regulation of NMDA receptors and subsequent overstimulation of the glutamatergic system by endogenous glutamate, triggering apoptosis in developing neurons (Shi et al. 2010; Wang et al. 2005, 2006). Furthermore, ketamine-induced apoptosis may be mediated by the PI3K (phosphoinositide-3-kinase)/Akt-glycogen synthase kinase 3beta (GSK-3beta) signaling pathway (Shang et al. 2007; Takadera et al. 2006), via the Bax/Bcl-2 ratio pathway and caspase-3 in the differentiated neuronal cells (Mak et al. 2010). Elevated nitric oxide (NO) levels induced by ketamine in vitro (Wang et al. 2008a) can also lead to apoptosis in

newborn rat forebrain cultures. High-dose ketamine can induce apoptosis via the mitochondrial pathway, independent of death receptor signaling (Braun et al. 2010). Another possible apoptotic mechanism is related
to hyperphosphorylated tau in the brains of mice and monkeys with
long-term administration of ketamine (Yeung et al. 2010). More recently,
Soriano et al. (2010) report that ketamine can induce aberrant cell cycle
reentry, leading to apoptotic cell death in the developing rat brain.

4.6.2 Neural stem progenitor cells

Ketamine did not induce significant increases in the apoptosis or necrosis of cultured NSPCs. Hsieh et al. (2003) report that neural progenitor
cells (NPCs) can block the apoptosis of ischemia-induced excitatory neurotoxicity and may resist neurotoxic factors by themselves. The study also
addresses the potential mechanism that the resistance of NPCs to neurotoxicity may be partially attributed to a lack of response to glutamate,
because they found that the stimulation of glutamate receptors could not
induce a significant influx of Ca^{++} in the progenitor cultured as neurospheres, which is consistent with our recent findings in rat cortical NSPCs
(Dong et al. 2012). Neurospheres may contain more NSCs, while NSCs at
the earlier undifferentiated stage may not express NR1 receptors, indicating lack of functional NMDA receptors in neurospheres. Although
NMDA receptor expression is detected in the adherent monolayer NSPC
cultures, no direct evidence confirmed that functional NMDA receptors
are built up using these NMDA receptor subunits in the cultured NSPCs.

Ketamine-induced neuronal apoptosis in the developing brain is
not directly caused by the antagonism of NMDA receptor but indirectly
caused by ketamine-induced up-regulation of NMDA receptor in neurons
(Slikker et al. 2007; Wang et al. 2005; Zou et al. 2009b). Thus, endogenous
glutamate release can result in excitatory neurotoxic effects in neurons by
these up-regulated NMDA receptors in the developing brain. Ketamine
also induces similar up-regulation of NR1 in cultured NSPCs, but we do
not know that these NMDA receptor subunits can build up functional
NMDA receptors. In contrast, quite a few studies support a lack of functional NMDA receptors in NSPCs (Wang et al. 1996). Moreover, other corresponding studies also support that even high levels of glutamate, up to
1 mM, fail to induce toxic effects on neural precursor cultures until the
late differentiation stage of the neuron (Brazel et al. 2005; Buzanska et al.
2009; Hsieh et al. 2003). NSPCs appear to have resistance to the neurotoxicity induced by ketamine, compared to primary cultured neurons.

A high dosage of ketamine induced a significant increase of caspase-
3-positive cells in cultured NSPCs (Dong et al. 2012). This seems contradictory to the resistance of NSPCs to ketamine-induced neurotoxic effects,
but ketamine critically enhanced neuronal differentiation of NSPCs in a

dose-dependent manner within 24 h. Under high-dose ketamine exposure, NSPCs are induced to further differentiate into neuronal progenitor cells or neurons at the more differentiated state. According to previous evidence demonstrating that NMDA receptors mediate ketamine-induced neuronal cell death, maturation of these neurons and NPCs at the late differentiation phase would express functional NMDA receptors, which perhaps make them more susceptible to high concentrations of ketamine. This may explain why significant increases in apoptosis appeared in the NSPC cultures exposed to a high dose of ketamine.

Interestingly, recent experiments report that NSPCs may protect neurons from the ischemia-induced apoptotic damage by releasing progenitor-generated neurotrophic factors (Hsieh et al. 2003; Ono et al. 2010). For instance, chronic ketamine use can increase serum levels of brain-derived neurotrophic factor (BDNF) (Ricci et al. 2010). BDNF prevents PCP- or MK-801-induced apoptosis in the developing brain by the parallel activation of both ERK and PI3K/Akt pathways (Hansen et al. 2004; Xia et al. 2010b). Decreased BDNF signaling might play a part in NMDA receptor antagonist–induced neurotoxicity (Fumagalli et al. 2003; Semba et al. 2006). One recent study also showed that the differentiated NSCs lentivitally transduced with BDNF express NR2 and are resistant to neurotoxicity mediated by PI3K/Akt signaling pathways (Casalbore et al. 2010). IGF-1 (insulin-like growth factor 1) also plays an important role in the cellular survival of NSPCs mediated via IGF-1 receptors. Recently, it was reported that glutamate can block the effect of IGF-1 by decreasing IGF-1 receptor signaling and responsiveness, hence attenuating the survival properties of IGF-1 in neuronal cells. Conversely, this suggests that ketamine may block glutamate-induced inhibition of IGF-1 receptors and release the protective effects of IGF-1 (Zheng and Quirion 2009). Although detailed mechanisms on the changes of these growth factors in NSPCs exposed to ketamine are not clear, these findings may provide another potential explanation on the resistance of NSPCs to the ketamine-induced neurotoxic effects.

4.7 Ketamine alters neurogenesis in early developing brains

Neurogenesis is defined as the process by which new nerve cells are generated, including proliferation and differentiation. During neurogenesis, undifferentiated NSPCs determine the quantity of the cells in the developing brain and the formation of neurons, astrocytes, glia, and other neural lineages. Meanwhile, NSPCs maintain the undifferentiated cell pool via proliferation. In developmental neurotoxicological studies, in vitro cultured NSPCs can serve as a high-throughput screening tool to clarify and evaluate the potential risks of those toxic factors to the developing brain (Breier et al. 2010). To determine the potential neurotoxic effects of ketamine on the developing brain, cultured NSPCs will be employed as an in vitro developmental model to investigate neurogenesis in the developing brain.

NMDA receptors play important roles in synaptogenesis, apoptosis, and excitotoxicity during brain development (Choi 1994; Constantine-Paton and Cline 1998). In the developing fetal neocortex, functional NMDA receptors are observed in neurons of the cortical plate (CP) and in NPCs in the subventricular zone (SVZ) (LoTurco et al. 1991), suggesting the involvement of NMDA receptors in synaptic plasticity, neuronal development/survival, and the regulation of NPCs through activation of various intracellular signaling pathways (Adams and Sweatt 2002; Orban et al. 1999; Sweatt 2001). Recent publications indicate that NMDA receptor antagonists like PCP, MK-801, and ketamine have negative effects on the increased neurogenesis occurring in some pathological states (ischemia, stroke) (Arvidsson et al. 2001; Bernabeu and Sharp 2000; Wang et al. 2008b). Systematic injection of MK-801 completely blocks the increasing proliferation in the SVZ and the subgranular zone (SGZ) of the dentate gyrus (DG) (Chen et al. 2005; Maeda et al. 2007; Shen and Zhang 2007). Moreover, cortical spreading depression (SD), an epiphenomenon of neurological disorders, such as stroke or traumatic brain injury, can significantly increase proliferative newborn cells in the SGZ within 2–4 days. However, the mitogenic action of SD can be suppressed by systemic administration of MK-801 (Urbach et al. 2008). Winkelheide et al. (2009) reported that S(+)-ketamine can inhibit postischemic increase in newly generated neurons in a dose-dependent manner. Pretreatment with ketamine led to significant attenuation of mossy fiber sprouting, indicating that ketamine may interfere with synaptogenesis and cognitive recovery (Lamont et al. 2005). Direct evidence showed that MK-801 reduces the proliferation of NPCs of neurospheres (Mochizuki et al. 2007). These independent lines of evidence suggest that the antagonism of NMDA receptors can damage the neurogenic repair in the brain.

However, previous studies with a contrary conclusion do not support the hypothesis that NMDA receptor antagonists inhibit neurogenesis in the developing brain. Tung et al. (2008) found that prolonged ketamine exposure only induced a slight suppressive effect on the proliferation in older rats. Glutamate receptor antagonists may trigger neurogenesis and synaptogenesis in the DG of the normal animal (Cameron et al. 1995; Gould 1994; Petrus et al. 2009; Sharp et al. 2000). Application of MK-801 during lesion induction significantly enhances neurogenesis in the DG (Kluska et al. 2005). MK-801 also prevents the corticosterone-induced decrease in proliferating cells in the adult rat DG (Cameron et al. 1998). Guidi et al. (2005) reported that administration of MK-801 to PND1–PND5 guinea pigs caused an increase in cell proliferation restricted to the dorsal

DG. Furthermore, even subanesthetic doses of ketamine enhance neurogenesis in the rat SGZ (Keilhoff et al. 2004), at least in part, relating to the reestablishment of disturbed cell proliferation (Keilhoff et al. 2010). It seems that most of these ketamine pro-neurogenesis effects were found in the DG of hippocampus. Therefore, it is likely that the effects of ketamine on the neurogenesis of the brain may be brain regional specific.

Ketamine also affects important proteins involved in normal neurogenesis of the brain and induces functional deficits in the adult individual (Viberg et al. 2008). Keilhoff et al. (2004) report that ketamine per se has no effect on cell proliferation. Its withdrawal, however, significantly induced cellular proliferation and survival in the hippocampus. Ketamine withdrawal activates the expression of BDNF, whereas chronic ketamine use can also increase serum level of BDNF but not nerve growth factor levels (Ricci et al. 2010). These data indicate that growth factors induced by ketamine may play a promotional role in neurogenesis.

Furthermore, the NMDA receptor expression changes in the developing brain exposed to ketamine may alter the neurogenesis of NSPCs. For example, NR2B-containing NMDA receptor subtypes negatively regulate neurogenesis in the adult hippocampus by activating neuronal NO synthase (nNOS) activity (Hu et al. 2008). Exercise-induced increases in neurogenesis are blocked in a mouse mutant lacking the NR1 subunit (Kitamura et al. 2003). This indicates that the neurogenesis may be associated with the expression of NMDA receptors and the composition of NMDA receptor subunits in the developing brain.

In humans, the brain growth spurt period extends from the sixth month of gestation to several years after birth. During this critical period, immature neurons in the CNS are prone to die from the exposure of intoxicating concentrations of NMDA receptor antagonists, suggesting that human infants are more susceptible to the treatments that may be considered safe in older patients. Early possible abnormalities of neuronal death and neurogenesis in the developing CNS induced by the exposure to ketamine would likely lead to later cognitive and motor impairment. Therefore, it is necessary to reevaluate dose and time windows of ketamine use in neonates and children and to investigate the potential intracellular mechanisms of ketamine-induced cellular damage.

4.8 Ketamine inhibits proliferation of NSPCs in the developing brain

A high dose of ketamine can inhibit the proliferation of NPSCs in a dose-dependent manner (Dong et al. 2012). Overdoses of ketamine significantly reduced the number of BrdU- and Ki-67-positive cells in cultured NSPCs, indicating that ketamine can reduce proliferation capacity. When one proliferating cell exits the cell cycle, it would become a BrdU-negative cell and then stay at intermitotic times between the undifferentiated and the differentiated state, at which the cell would be Ki-67 positive and BrdU negative. When one proliferative NPC (BrdU−|Ki-67+) completely loses proliferative ability and becomes differentiated cells, it would appear BrdU negative and Ki-67 negative (BrdU−|Ki-67−). A high dose of ketamine reduced the number of BrdU- and Ki-67-positive cells in cultured NSPCs, suggesting that a high dose of ketamine not only can inhibit the activity of the proliferation but also can reduce the capacity of the proliferation and enhance significantly the differentiation of NSPCs. This hypothesis was also demonstrated by the following differentiation study showing that ketamine promotes neuronal differentiation in NSPCs. We found that the percentage of BrdU-positive cells in Ki-67-positive cells had no significant change in cultured NSPCs exposed to different concentrations of ketamine, suggesting that ketamine inhibits the proliferation of NSPCs while also promoting differentiation.

Ketamine-induced inhibition of the proliferation of cultured NSPCs is time dependent. For the time course of ketamine effects on the proliferation of NSPCs, the long-term exposure to 10 µM of ketamine (24 or 48 h) reduced the numbers of BrdU- and Ki-67-positive cells and the ratio of BrdU-positive cells in Ki-67-positive cells had no change. The short-term exposure to a small dose of ketamine is insufficient to inhibit the proliferation of NSPCs. Only ≥10 h exposure to a regular dose of ketamine induced the significant promotion of the proliferation in cultured NSPCs (Dong et al. 2012). Therefore, it may be safe for the small dose of ketamine to be administrated to pediatric patients for a short term in clinical settings, as far as the neurogenesis of NSPCs is concerned.

An in vivo study demonstrated that ketamine dose-dependently reduced BrdU-positive cells in the ventricular zone (VZ) and the SVZ of the fetal brains exposed to high doses of ketamine (20, 40, and 100 mg/kg). This in vivo result was verified by the results obtained from the in vitro studies described previously and was also consistent with the results of several previous studies (Arvidsson et al. 2001; Bernabeu and Sharp 2000; Chen et al. 2005; Maeda et al. 2007; Shen and Zhang 2007; Urbach et al. 2008; Wang et al. 2008b; Winkelheide et al. 2009). Similar results were obtained although most previous studies explored this issue by observing different neurogenic regions (the SVZ or the SGZ) and in varying models (normal or pathological, such as ischemia or psychiatric diseases). Therefore, the in vivo and in vitro experiments confirm that ketamine can inhibit the proliferation of NSPCs in the cortex of the fetal rat brain at E17/E19.

Proliferating cells in the VZ migrate from the apical side to the basal side when the cells are experiencing the cell cycle from the G1, S, G2, to the M phase. Mitotic cells (at the M phase) lie close to the lateral ventricles and newly generated cells migrate toward the SVZ. Within this process, the cells enter the G1 phase. When the cells reach the connecting line of the VZ and the SVZ, the cell cycle enters the S phase (Figure 4.2). The arrangement of proliferating cells at different cell cycle phases was altered by exposure to high doses of ketamine. Under exposure to 100 mg/kg of ketamine, BrdU-positive cells with large cell bodies at the S-phase region (close to the SVZ) significantly decrease in the VZ and the orientation of BrdU-positive cells in the VZ appears disordered. This pattern indicates that a number of NSPCs exposed to high-dose ketamine exit the cell cycle and migrate to the SVZ, possibly because ketamine inhibits the proliferation of NSPCs at the G1/S checkpoint of the cell cycle.

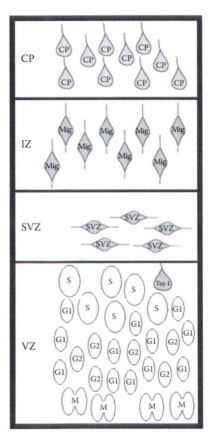

Figure 4.2 Diagram depicting the composition of the cell cycle of NSPCs in the cortex. The sketch graph shows that proliferating cells migrate within the VZ from the apical side to the basal side when the cells are experiencing their cell cycles from the G1, S, and G2, to the M phase. CP, cortical plate; Mig, migrating cells; SVZ, subventricular zone; RG, radial glia cells; VZ, ventricular zone; IZ, intermediate zone; Tuj-1, newborn neuron.

Greater reduction of BrdU-positive cells is found in the SVZ after keta-mine exposure than in the VZ. NSCs are completely undifferentiated cells and stay in the VZ, which is like a "proliferation engine" that produces high numbers of NSCs and NPCs to maintain further proliferation and dif-ferentiation during brain development. In the VZ, newly generated NPCs would move toward the SVZ. As a consequence, there are more NPCs in the SVZ, which are more differentiated than NSCs in the VZ. Neuronal differentiation of NSPCs can be enhanced by ketamine. When the fetal brain is exposed to ketamine, NPCs in the SVZ are induced to differentiate into more differentiated NPCs or newborn neurons, which have limited or no proliferative potency, respectively, whereas NSCs in the VZ may be induced to become NPCs while keeping their proliferative capacity.

4.9 Ketamine enhances the neuronal differentiation of NSPCs in the developing brain

NSPCs can multipotently differentiate into neurons, astrocytes, and oli-godendrocytes. Neuronal and glial differentiation is critical to neurogen-esis during normal brain development. Any endogenous and exogenous factor that can damage or interfere with normal neuronal differentiation could result in abnormalities in the quantity of neurons, neuronal migra-tion, synaptic formation and plasticity, and even brain structure abnormal-ity. Normally, neuronal differentiation is regulated by growth factors and neurotrophins like EGF (epidermal growth factor), bFGF (basic fibroblast growth factor), insulin, BDNF, NT-3 (neurotrophin), Shh (sonic hedgehog), BMP2 (bone morphogenic protein 2), LIF (leukemia inhibiting factor), VEGF (vascular endothelial growth factor) (Babu et al. 2007), and DISC1 (disrupted in schizophrenia 1 protein), mediating a molecular switch from neuronal proliferation to differentiation (Ishizuka et al. 2011). Recent stud-ies show that the repression of calcium signaling mediated by glutamate receptors is associated with the elevation of neuronal differentiation in human precursor cells (Feliciano and Edelman 2009; Whitney et al. 2008).

Several independent lines of evidence indicate that NMDA receptors are involved in regulating the neuronal differentiation of NSPCs (Aruffo et al. 1987; Bading et al. 1995). Glutamate application on postnatal rat whole-brain dissociated cell cultures promotes neuronal growth and differen-tiation (Aruffo et al. 1987). The activation of NMDA receptors transiently expressing in undifferentiated NPCs in the fetal rat neocortex promotes neuronal differentiation (Yoneyama et al. 2008). Excitatory stimuli act directly on adult hippocampal NPCs to increase neuron production medi-ated via L-type Ca^{++} channels and NMDA receptors. The excitation on the proliferating precursors can inhibit expression of the glial fate genes Hes1 and Id2 and increase expression of NeuroD, a positive regulator of neuronal differentiation (Deisseroth et al. 2004). In fact, there is conflicting evidence on whether the activation of glutamate receptors promotes or inhibits the neuronal differentiation of NSPCs.

Ketamine inhibits the proliferation of NSPCs by enhancing cell cycle arrest. As we know, when an NPC loses its proliferative capacity during neurogenesis, it enters the process of differentiation into mature differentiated cells. Ketamine has been reported to promote neuronal differentiation of NSPCs in a dose-dependent manner, which mutually supports that ketamine inhibits the proliferation in NSPCs.

The activation of NMDA receptors would promote neuronal differentiation of NPCs via NMDA receptors (Aruffo et al. 1987; Deisseroth et al. 2004; Joo et al. 2007; Pearce et al. 1987; Yoneyama et al. 2008). The finding that an NMDA receptor antagonist (ketamine) promotes the neuronal differentiation of NSPCs appears to be inconsistent with these studies. However, in fact, the two results are consistent if we analyze the potential effects of ketamine on the expression of NMDA receptors. Since calcium influx may play a role in the neuronal differentiation of NPSCs, how does ketamine influence the calcium influx? On examining the expression of NMDA receptors in NSPCs, it is evident that there are few, if any, functional NMDA receptors on NSPCs according to the above description, which explains the resistance of NSPCs to ketamine-induced neurotoxic effects. Previous studies show that ketamine can induce the expression of NR1 in mature neurons (Wang et al. 2005) and in NSPCs. When NSPCs were exposed to different concentrations of ketamine for 24 h, ketamine increased the expression of NMDA receptor subunits in these NSPCs. Ketamine was washed off after 24 h and fresh NSPC differentiation medium was added to the cultures to allow NSPCs a 3-week period of spontaneous differentiation. In the process of spontaneous differentiation, the ketamine-induced expression of NMDA receptor subunits would be expected to build up more functional NMDA receptors as the differentiation proceeds. These newly induced NMDA receptors can trigger calcium influx to promote neuronal differentiation. The findings suggest that ketamine can induce functional NMDA receptors in NSPCs, which mediates the differentiation of NSCs and NPCs. Therefore, the finding that ketamine promotes the neuronal differentiation of NSPCs is consistent with those of previous studies.

Additionally, enhanced neuronal differentiation of cultured NSPCs exposed to ketamine may also result from the change in differentiation direction of NSCs. One of two specific properties of NSCs is multipotent differentiation. An NSC can differentiate into a neuronal progenitor cell or a glial progenitor cell. NPCs have limited differentiation potency. Normally, neuronal progenitor cells only can differentiate into neurons, while glial progenitor cells are strictly committed to glial differentiation. This means that one NSC cannot become a glia, when its fate is determined to neuronal differentiation. Recent studies indicate that NSCs exposed to ketamine prefer neuronal differentiation, namely, differentiation into neuronal progenitor cells, but this does not mean that glial progenitor cells can be changed into neuronal progenitor cells or neurons under exposure to ketamine. Those glial progenitor cells in NSPC cultures exposed to ketamine still differentiate into glia.

Previous studies also support this point of view. For example, excitatory stimuli via NMDA receptors act directly on adult hippocampal NPCs to favor neuron production. It is likely that excitation of NMDA receptors on NSPCs inhibits the expression of the glial fate genes Hes1 and Id2 and increases expression of NeuroD, a positive regulator of neuronal differentiation (Deisseroth et al. 2004). Although functional NMDA receptor subunits are also expressed in glia (Schipke et al. 2001), the biophysical and pharmacological properties of glial NMDA receptors are very different to those of neuronal NMDA receptors (Palygin et al. 2011). All these lines of evidence combined with our results indicate that ketamine exposure may promote the neuronal differentiation of NSPCs, rather than leading to glial precursors.

In detecting the time course response of ketamine on the neuronal differentiation of cultured NSPCs, longer than 8-h exposure to regular ketamine resulted in a significant increase of neuronal differentiation in NSPCs. The percentage of Tuj-1$^+$ in DAPI$^+$ cells significantly increased after exposure to ketamine for more than 8 h. This indicates that a time threshold may exist for the ketamine-induced neuronal differentiation of cultured NSPCs. Ketamine can up-regulate the expression of NMDA receptors in NSPC cultures. Ketamine-induced expression of NMDA receptors is likely cumulative over the first 8 h, but the amount of NMDA receptor subunits may not be sufficient to build up functional NMDA receptors or to trigger enough Ca^{++} influx to activate pro-neuronal differentiation mechanisms. After 10 h of exposure to ketamine, the cumulative NMDA receptor activity in NSPCs or NPCs may be sufficient to promote an increase in the neuronal differentiation of NSPCs. Thus, ketamine can increase the neuronal differentiation of cultured NSPCs in a dose- and time-dependent manner.

4.10 Potential mechanisms of ketamine-induced neurogenic changes in the developing brain

4.10.1 Up-regulation of NMDA receptors in the developing brain exposed to ketamine

Ketamine can dose-dependently up-regulate the expression of NR1, whereas high expression of NR2B in NSPCs was not changed by ketamine exposure.

The up-regulated NMDA receptors in NSPCs can be supported by the finding that high-dose ketamine inhibits proliferation and promotes neuronal differentiation in NSPCs (Dong et al. 2012). Ketamine-induced neuronal differentiation of NSPCs produces more neurons or NPCs in the cultured NSPCs. Previous studies have reported that high doses of ketamine can up-regulate NR1 expression in differentiated neurons and significantly induce cell death of these neurons in the developing brain (Slikker et al. 2007; Wang et al. 2005; Zou et al. 2009b). Thus, it is likely that the increased NR1 expression of cultured NSPCs exposed to ketamine results from an increase of NR1 expression in newly generated neurons or NPCs.

High-dose ketamine can significantly increase the apoptosis of cultured NSPCs. High-dose ketamine induces more newly differentiated neurons or NPCs in the cultured NSPCs and up-regulates the expression of NMDA receptors in these neurons or NPCs (Dong et al. 2012). According to previous studies, up-regulated NMDA receptors may result in the apoptosis of these newly differentiated neurons or NPCs. Therefore, the up-regulation of NMDA receptors and promotion of neuronal differentiation associated with ketamine exposure may offer a plausible explanation for why cultured NSPCs were resistant to the neurotoxic effects of low-dose ketamine but not resistant to high doses of ketamine. Moreover, despite up-regulated NR1 expression and the existence of NR2B in NSPCs, functional NMDA receptors composed of NR1 and NR2B to initiate calcium influx remain undetected in NSPCs. However, it still cannot be confirmed that functional NMDA receptors are involved in ketamine-induced changes of cell death and neurogenesis of NSPCs.

4.10.2 PI3K/Akt signaling inhibited by ketamine exposure

The expression of Akt in neurospheres and adherent monolayer NSPC cultures (Paling et al. 2004; Sinor and Lillien 2004; Watanabe et al. 2006), the key factor of the PI3K/Akt signaling pathway, suggests that this signaling pathway plays a potential role in antiapoptosis, prosurvival, and cell cycle regulation in adherent cultured NSPCs (Figure 4.3). It is well known that NSPCs are regulated by growth factors, neurotrophic factors, insulin, and other molecules. All these factors are reported to regulate cellular activity such as the proliferation and the differentiation of NSPCs directly or indirectly, mediated by the PI3K/Akt signaling pathway.

Present studies demonstrated that ketamine dose-dependently reduced the phosphorylation of Akt in NSPCs. Recent studies have indicated that the activation of NMDA receptors has a positive effect on the PI3K signaling pathway (Hisatsune et al. 1999; Zhong et al. 2010). For instance, NMDA receptor agonists protect neurons in acute brain injury and chronic neurodegenerative disorders associated with excessive activation of NMDA receptors by activating the PI3K/Akt pathway (Zhong et al.

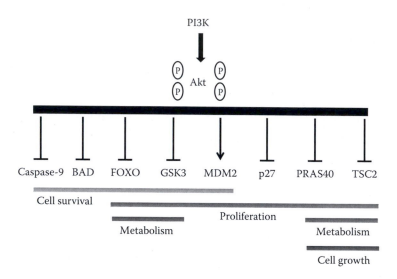

Figure 4.3 PI3K/Akt signaling pathway, downstream factors, and cellular functions. Akt-mediated phosphorylation of these downstream proteins leads to their activation (arrows) or inhibition (blocking arrows). Regulation of these substrates by Akt contributes to activation of the various cellular processes shown (i.e., survival, proliferation, metabolism, and growth). As illustrated by these eight targets, a high degree of functional versatility and overlap exists among Akt substrates.

2010). Habas et al. (2006) showed that NMDA acting via NR2B can protect against LY294002-induced apoptosis mediated by a proapoptotic kinase, GSK3beta on the PI3K signaling pathway. In cortical neuronal cultures, PI3K inhibitors wortmannin and MK-801 significantly reduce the phosphorylation of Akt at Serine 473, indicating that activity-dependent glutamate release maintains Akt activation through an NMDA receptor/PI3K pathway (Gines et al. 2003; Sutton and Chandler 2002).

In cultured NSPCs, p27 is the key downstream factor of Akt proteins. The PI3K/Akt signaling pathway inhibitor LY-294002 can dephosphorylate Akt proteins and up-regulate the expression of p27. In dose–response assays, ketamine dose-dependently up-regulates the expression of p27 coupled with the inhibition of Akt phosphorylation. Up-regulated p27 may increase the inhibition of cdk2, which would reduce the proliferation of NSPCs. This finding supports the finding from in vivo and vitro studies that ketamine can inhibit proliferation of cultured NSPCs in the developing brain.

4.10.3 BDNF changes are involved in neurogenic changes in NSPCs

Ketamine can dose/time dependently inhibit the proliferation of cultured NSPCs isolated from the cortex of the developing brain. However, little is known about the cellular mechanisms regulating the transition of NPCs from proliferation to differentiation. Presently, some signaling molecules are beginning to be identified, such as several growth factors, including bFGF, EGF, BDNF, and notch ligands (Benraiss et al. 2001; Caldwell et al. 2001; Vescovi et al. 1993; Zigova et al. 1998), modulating the proliferation and the differentiation of NPCs. BDNF can be induced by NMDA receptor antagonists (Ricci et al. 2010) and is also associated with the effects induced by NMDA receptor antagonists in neurons or NPCs (Casalbore et al. 2010; Fumagalli et al. 2003; Hansen et al. 2004; Semba et al. 2006; Xia et al. 2010a). Cheng et al. (2003) report that BDNF can reduce NPC proliferation and increase the expression of nNOS in differentiating neurons, and another publication indicates that ketamine can up-regulate the expression of nNOS mediating ketamine-induced neurotoxicity (Wang et al. 2008a). We propose a potential mechanism that the exposure to ketamine may regulate the secretion of BDNF and induce NO in cultured NSPCs, resulting in the inhibition of the proliferation and the promotion of the differentiation. This hypothesis will be investigated in future studies.

BDNF changes in NSPCs exposed to ketamine may help explain the result that ketamine promotes neuronal differentiation of NSPCs. Neurogenesis (proliferation and differentiation) of NSPCs is regulated by BDNF (Horne et al. 2010). NMDA receptor blockade can up-regulate the expression of BDNF and long-term use of NMDA receptor antagonist drugs can increase the serum level of BDNF (Casalbore et al. 2010; Fumagalli et al. 2003; Hansen et al. 2004; Ricci et al. 2010; Semba et al. 2006; Xia et al. 2010b). On one hand, increased BDNF itself can regulate neuronal differentiation of precursor cells (Babu et al. 2007). For instance, 10 days after a one-time exposure to BDNF, single EGF-generated neurospheres showed a twofold increase in neuron number and a marked enhancement in neurite outgrowth (Ahmed et al. 1995). On the other hand, BDNF can increase the expression of nNOS in differentiating neurons (Cheng et al. 2003). Ketamine can also up-regulate the expression of nNOS mediating ketamine-induced neurotoxicity (Wang et al. 2008a). We propose that ketamine induces NSPCs to secrete BDNF, and BDNF promotes neighboring NSPCs to increase the expression of nNOS and release NO; the released NO diffuses into neighboring cells to inhibit proliferation and promote neuronal differentiation (Ahmed et al. 1995; Cheng et al. 2003). This hypothesis will be testified in our future studies.

4.11 *Ketamine damages the cognitive and learning abilities of the developing brain*

Long-term abuse of ketamine has been reported to induce cognitive impairment, especially within the domains of episodic and semantic memory

(Fletcher and Honey 2006; Morgan and Curran 2006; Morgan et al. 2009, 2010). Direct experimental evidence with nonhuman primate models has also shown that a single 24-h ketamine exposure during a sensitive period of brain development can result in long-lasting deficits in cognition leading to subsequent functional deficits (Paule et al. 2011). The ketamine-induced cognition and learning impairment are widely believed to reflect the relevant damage in the hippocampus and cerebral cortex. Cell death has been reported in frontal cortex after ketamine exposure, and chronic ketamine administration can induce neurodegeneration in brain areas crucial for cognition (Ikonomidou et al. 1999; Scallet et al. 2004; Wang et al. 2005; Young et al. 2005; Zou et al. 2009b). Additionally, ketamine has been reported to decrease proliferation, promote neuronal differentiation, and inhibit neuronal migration in the developing cortex (Dong et al. 2012). Taken together, ketamine disturbs the normal brain development and induces brain structure changes, which would be reflected as disorders in brain functions including cognition, learning, and memory abilities.

4.12 Conclusions

Numerous studies have confirmed that high-dose ketamine significantly induces cell death of neurons in the developing brain (Slikker et al. 2007; Wang et al. 2005; Zou et al. 2009b). However, major differences exist between mature neurons and neural progenitors in cellular activities and the mechanisms regulating the activities. Therefore, the neurotoxic effects of ketamine exposure in mature neurons with the same effects in NSPCs cannot be extrapolated. In fact, ketamine does not increase cell death (necrosis or apoptosis) in NSPCs except at extremely high concentrations, indicating the resistance of NSPCs to ketamine-induced cell death. This is consistent with several independent lines of evidence (Brazel et al. 2005; Buzanska et al. 2009; Hsieh et al. 2003). Hsieh et al. (2003) found that stimulating glutamate receptors could not induce significant Ca^{++} influx into their progenitors cultured as neurospheres, suggesting a lack of functional NMDA receptor activity in NSPCs. Other studies also suggest that functional NMDA receptors may not be built up, despite the expression of NMDA receptor subunits in NPCs. Therefore, the resistance of NSPCs to ketamine neurotoxicity may not result from the activation of functional NMDA receptors in NSPCs.

A thin layer of neuroectodermal cells (neural precursors) generates the billions of neurons and glia well organized in the mature brain. The neurogenesis of neural stem and progenitor cells plays a major role in the brain-forming process. Genetic regulation of brain development appears to precisely control every developmental event at a restricted time point. Therefore, any factor disturbing the neurogenesis at a specific time point during the brain development would result in abnormalities or disorders in the structure and functions of the brain. Ketamine, an anesthetic, analgesic, and sedative widely used in pediatric patients, can potentially disturb the neurogenesis of NSPCs. Exposure to high-dose ketamine reduced the number of BrdU-positive cells in the VZ and SVZ of the fetal cortex, which is consistent with several lines of independent evidence (Arvidsson et al. 2001; Bernabeu and Sharp 2000; Brazel et al. 2005; Chen et al. 2005; Maeda et al. 2007; Mochizuki et al. 2007; Shen and Zhang 2007; Urbach et al. 2008; Wang et al. 2008b). Additionally, in vitro experiments revealed that ketamine can dose- and time-dependently inhibit the proliferation of cultured NSPCs. The inhibition of proliferation means that NSPCs will exit the cell cycle and enter a differentiation state. An increased neuronal differentiation of NSPCs is exposed to ketamine in a dose-/time-dependent manner. Taken together, ketamine exposure does alter normal neurogenesis of NSPCs in a dose- and time-dependent manner.

The disturbance of ketamine on NSPCs can speculate what will be the consequences of ketamine-induced disturbances in the neurogenesis of NSPCs in the developing brain. First, inhibition of the proliferation of NSPCs in the VZ could lead to a reduction of the number of neural cells in the developing brain, thus leading to a paucity of neuronal/glial cells for building up normal neural circuits. Second, the differentiation of NSPCs may be completed at the wrong time or at an erroneous position in the cortex. Normally, as newly generated NSPCs in the VZ/SVZ migrate to the CP, the migrating cells differentiate toward a mature cellular phenotype. If the entire process is disturbed, it is possible that more differentiated neurons would appear in the SVZ or IZ (intermediate zone) of the cortex. Thus, ketamine could disturb the internal structure of the cortex and the formation of neuronal circuits and may lead to seizure disorders or other altered brain functions.

4.13 Clinical implications

Presently, ketamine is used widely as an anesthetic, analgesic, and sedative in pediatric settings and is also an illicit club drug consumed by pregnant drug abusers. Therefore, current findings on ketamine's developmental neurotoxicity may provide some important clinical implications. As an anesthetic for obstetric and pediatric patients, ketamine should not be used in large doses and for prolonged periods. Preclinical studies showed that clinical doses of ketamine are unlikely to induce the cell death of NSPCs in infants, but this may lead to ketamine-induced changes in the neurogenesis of NSPCs. Actually, there is no accurate relationship between the experimental dose of ketamine in rats and its clinical dose in patients. Therefore, we cannot directly apply these experimental data to clinical settings. To obtain further clinical implications, we should further test these conclusions using primate animal models or human NSCs.

Additionally, considering that ketamine is a club drug consumed by pregnant drug abusers and that this population may repeatedly consume large doses of ketamine in the long term, the conclusions of this study suggest that the DEA should strictly control ketamine in the illicit drug market. Meanwhile, physicians could persuade drug-abusing pregnant patients to stop ketamine consumption through detoxification programs.

References

Adams, J.P., and J.D. Sweatt. 2002. Molecular psychology: Roles for the ERK MAP kinase cascade in memory. *Annu Rev Pharmacol Toxicol* 42:135–163.

Ahmed, S., B.A. Reynolds, and S. Weiss. 1995. BDNF enhances the differentiation but not the survival of CNS stem cell-derived neuronal precursors. *J Neurosci* 15:5765–5778.

Akazawa, C., R. Shigemoto, Y. Bessho, S. Nakanishi, and N. Mizuno. 1994. Differential expression of five N-methyl-D-aspartate receptor subunit mRNAs in the cerebellum of developing and adult rats. *J Comp Neurol* 347:150–160.

Amr, Y.M. 2010. Multi-day low dose ketamine infusion as adjuvant to oral gabapentin in spinal cord injury related chronic pain: A prospective, randomized, double blind trial. *Pain Physician* 13:245–249.

Anastasio, N.C., and K.M. Johnson. 2008. Differential regulation of the NMDA receptor by acute and sub-chronic phencyclidine administration in the developing rat. *J Neurochem* 104:1210–1218.

Anderson, B.J., and G.M. Palmer. 2006. Recent developments in the pharmacological management of pain in children. *Curr Opin Anaesthesiol* 19:285–292.

Aruffo, C., R. Ferszt, A.G. Hildebrandt, and J. Cervos-Navarro. 1987. Low doses of L-monosodium glutamate promote neuronal growth and differentiation in vitro. *Dev Neurosci* 9:228–239.

Arvidsson, A., Z. Kokaia, and O. Lindvall. 2001. N-methyl-D-aspartate receptor-mediated increase of neurogenesis in adult rat dentate gyrus following stroke. *Eur J Neurosci* 14:10–18.

Babu, H., G. Cheung, H. Kettenmann, T.D. Palmer, and G. Kempermann. 2007. Enriched monolayer precursor cell cultures from micro-dissected adult mouse dentate gyrus yield functional granule cell-like neurons. *PLoS One* 2:e388.

Bading, H., M.M. Segal, N.J. Sucher, H. Dudek, S.A. Lipton, and M.E. Greenberg. 1995. N-methyl-D-aspartate receptors are critical for mediating the effects of glutamate on intracellular calcium concentration and immediate early gene expression in cultured hippocampal neurons. *Neuroscience* 64:653–664.

Baker, A.K., V.L. Hoffmann, and T.F. Meert. 2002a. Dextromethorphan and ketamine potentiate the antinociceptive effects of mu- but not delta- or kappa-opioid agonists in a mouse model of acute pain. *Pharmacol Biochem Behav* 74:73–86.

Baker, A.K., V.L. Hoffmann, and T.F. Meert. 2002b. Interactions of NMDA antagonists and an alpha 2 agonist with mu, delta and kappa opioids in an acute nociception assay. *Acta Anaesthesiol Belg* 53:203–212.

Benke, D., A. Wenzel, L. Scheuer, J.M. Fritschy, and H. Mohler. 1995. Immuno-biochemical characterization of the NMDA-receptor subunit NR1 in the developing and adult rat brain. *J Recept Signal Transduct Res* 15:393–411.

Benraiss, A., E. Chmielnicki, K. Lerner, D. Roh, and S.A. Goldman. 2001. Adenoviral brain-derived neurotrophic factor induces both neostriatal and olfactory neuronal recruitment from endogenous progenitor cells in the adult forebrain. *J Neurosci* 21:6718–6731.

Bernabeu, R., and F.R. Sharp. 2000. NMDA and AMPA/kainate glutamate receptors modulate dentate neurogenesis and CA3 synapsin-I in normal and ischemic hippocampus. *J Cereb Blood Flow Metab* 20:1669–1680.

Birkhan, J., R. Shamash, and D. Gutman. 1971. Ketamine-dissociative anesthesia in pediatric oral surgery. *J Oral Surg* 29:853–857.

Blonk, M.I., B.G. Koder, P.M. van den Bemt, and F.J. Huygen. 2010. Use of oral ketamine in chronic pain management: A review. *Eur J Pain* 14:466–472.

Bolger, G.T., M.F. Rafferty, and P. Skolnick. 1986. Enhancement of brain calcium antagonist binding by phencyclidine and related compounds. *Pharmacol Biochem Behav* 24:417–423.

Braun, S., N. Gaza, R. Werdehausen, H. Hermanns, I. Bauer, M.E. Durieux, M.W. Hollmann, and M.F. Stevens. 2010. Ketamine induces apoptosis via the mitochondrial pathway in human lymphocytes and neuronal cells. *Br J Anaesth* 105:347–354.

Brazel, C.Y., J.L. Nunez, Z. Yang, and S.W. Levison. 2005. Glutamate enhances survival and proliferation of neural progenitors derived from the subventricular zone. *Neuroscience* 131:55–65.

Breier, J.M., K. Gassmann, R. Kayser, H. Stegeman, D. De Groot, E. Fritsche, and T.J. Shafer. 2010. Neural progenitor cells as models for high-throughput screens of developmental neurotoxicity: State of the science. *Neurotoxicol Teratol* 32:4–15.

Burnashev, N., R. Schoepfer, H. Monyer, J.P. Ruppersberg, W. Gunther, P.H. Seeburg, and B. Sakmann. 1992. Control by asparagine residues of calcium permeability and magnesium blockade in the NMDA receptor. *Science* 257:1415–1419.

Buzanska, L., J. Sypecka, S. Nerini-Molteni, A. Compagnoni, H.T. Hogberg, R. del Torchio, K. Domanska-Janik, J. Zimmer, and S. Coecke. 2009. A human stem cell-based model for identifying adverse effects of organic and inorganic chemicals on the developing nervous system. *Stem Cells* 27:2591–2601.

Caldwell, M.A., X. He, N. Wilkie, S. Pollack, G. Marshall, K.A. Wafford, and C.N. Svendsen. 2001. Growth factors regulate the survival and fate of cells derived from human neurospheres. *Nat Biotechnol* 19:475–479.

Cameron, H.A., B.S. McEwen, and E. Gould. 1995. Regulation of adult neurogenesis by excitatory input and NMDA receptor activation in the dentate gyrus. *J Neurosci* 15:4687–4692.

Cameron, H.A., P. Tanapat, and E. Gould. 1998. Adrenal steroids and N-methyl-D-aspartate receptor activation regulate neurogenesis in the dentate gyrus of adult rats through a common pathway. *Neuroscience* 82:349–354.

Casalbore, P., I. Barone, A. Felsani, I. D'Agnano, F. Michetti, G. Maira, and C. Cenciarelli. 2010. Neural stem cells modified to express BDNF antagonize trimethyltin-induced neurotoxicity through PI3K/Akt and MAP kinase pathways. *J Cell Physiol* 224:710–721.

Chadduck, W.M., and A. Manheim. 1973. Ketamine anesthesia for pediatric neurosurgical procedures. *Va Med Mon (1918)* 100:333–335.

Chazan, S., M.P. Ekstein, N. Marouani, and A.A. Weinbroum. 2008. Ketamine for acute and subacute pain in opioid-tolerant patients. *J Opioid Manag* 4:173–180.

Chen, J., C.T. Lee, S. Errico, X. Deng, J.L. Cadet, and W.J. Freed. 2005. Protective effects of Delta(9)-tetrahydrocannabinol against N-methyl-D-aspartate-induced AF5 cell death. *Brain Res Mol Brain Res* 134:215–225.

Cheng, A., S. Wang, J. Cai, M.S. Rao, and M.P. Mattson. 2003. Nitric oxide acts in a positive feedback loop with BDNF to regulate neural progenitor cell proliferation and differentiation in the mammalian brain. *Dev Biol* 258: 319–333.

Choi, D.W. 1994. Glutamate receptors and the induction of excitotoxic neuronal death. *Prog Brain Res* 100:47–51.

Collingridge, G.L., and R.A. Lester. 1989. Excitatory amino acid receptors in the vertebrate central nervous system. *Pharmacol Rev* 41:143–210.

Constantine-Paton, M., and H.T. Cline. 1998. LTP and activity-dependent synaptogenesis: The more alike they are, the more different they become. *Curr Opin Neurobiol* 8:139–148.

Corssen, G., E.H. Groves, S. Gomez, and R.J. Allen. 1969. Ketamine: Its place for neurosurgical diagnostic procedures. *Anesth Analg* 48:181–188.

Craven, R. 2007. Ketamine. *Anaesthesia* 62 Suppl 1:48–53.

da Conceicao, M.J., D.B. da Conceicao, and C.C. Leao. 2006. Effect of an intravenous single dose of ketamine on postoperative pain in tonsillectomy patients. *Paediatr Anaesth* 16:962–967.

Dal, D., N. Celebi, E.G. Elvan, V. Celiker, and U. Aypar. 2007. The efficacy of intravenous or peritonsillar infiltration of ketamine for postoperative pain relief in children following adenotonsillectomy. *Paediatr Anaesth* 17:263–269.

Deisseroth, K., S. Singla, H. Toda, M. Monje, T.D. Palmer, and R.C. Malenka. 2004. Excitation-neurogenesis coupling in adult neural stem/progenitor cells. *Neuron* 42:535–552.

Del Prete, S., E. Margaria, and U. Bosio. 1968. [CI-581 in pediatric surgery]. *Minerva Anestesiol* 34:1376–1378.

Desfeux, A., F. El Ghazi, S. Jegou, H. Legros, S. Marret, V. Laudenbach, and B.J. Gonzalez. 2010. Dual effect of glutamate on GABAergic interneuron survival during cerebral cortex development in mice neonates. *Cereb Cortex* 20:1092–1108.

Domino, E.F. 2010. Taming the ketamine tiger. 1965. *Anesthesiology* 113:678–684.

Dong, C., C.R. Rovnaghi, and K.J. Anand. 2012. Ketamine alters the neurogenesis of rat cortical neural stem progenitor cells. *Crit Care Med* 40:2407–2416.

Eide, P.K., and A. Stabhaug. 1997. Relief of glossopharyngeal neuroalgia by ketamine-induced N-methyl-aspartate receptor blockade. *Neurosurgery* 41(2):505–508.

Feliciano, D.M., and A.M. Edelman. 2009. Repression of Ca2+/calmodulin-dependent protein kinase IV signaling accelerates retinoic acid-induced differentiation of human neuroblastoma cells. *J Biol Chem* 284:26466–26481.

Ferrer-Montiel, A.V., W. Sun, and M. Montal. 1996. A single tryptophan on M2 of glutamate receptor channels confers high permeability to divalent cations. *Biophys J* 71:749–758.

Ffrench-Mullen, J.M., and M.A. Rogawski. 1992. Phencyclidine block of calcium current in isolated guinea-pig hippocampal neurones. *J Physiol* 456:85–105.

Fletcher, P.C., and G.D. Honey. 2006. Schizophrenia, ketamine and cannabis: Evidence of overlapping memory deficits. *Trends Cogn Sci* 10:167–174.

Flint, A.C., U.S. Maisch, J.H. Weishaupt, A.R. Kriegstein, and H. Monyer. 1997. NR2A subunit expression shortens NMDA receptor synaptic currents in developing neocortex. *J Neurosci* 17:2469–2476.

Fredriksson, A., and T. Archer. 2004. Neurobehavioural deficits associated with apoptotic neurodegeneration and vulnerability for ADHD. *Neurotox Res* 6: 435–456.

Fredriksson, A., T. Archer, H. Alm, T. Gordh, and P. Eriksson. 2004. Neurofunctional deficits and potentiated apoptosis by neonatal NMDA antagonist administration. *Behav Brain Res* 153:367–376.

Fumagalli, F., R. Molteni, M. Roceri, F. Bedogni, R. Santero, C. Fossati, M. Gennarelli, G. Racagni, and M.A. Riva. 2003. Effect of antipsychotic drugs on brain-derived neurotrophic factor expression under reduced N-methyl-D-aspartate receptor activity. *J Neurosci Res* 72:622–628.

Furukawa, H., S.K. Singh, R. Mancusso, and E. Gouaux. 2005. Subunit arrangement and function in NMDA receptors. *Nature* 438:185–192.

Gage, F.H. 2000. Mammalian neural stem cells. *Science* 287:1433–1438.

Gines, S., E. Ivanova, I.S. Seong, C.A. Saura, and M.E. MacDonald. 2003. Enhanced Akt signaling is an early pro-survival response that reflects N-methyl-D-aspartate receptor activation in Huntington's disease knock-in striatal cells. *J Biol Chem* 278:50514–50522.

Ginsberg, H., and J.A. Gerber. 1969. Ketamine hydrochloride: A clinical investigation in 60 children. *S Afr Med J* 43:627–628.

Gould, E. 1994. The effects of adrenal steroids and excitatory input on neuronal birth and survival. *Ann N Y Acad Sci* 743:73–92; discussion 92–93.

Graeme, K.A. 2000. New drugs of abuse. *Emerg Med Clin North Am* 18:625–636.

Green, S.M., M.G. Roback, R.M. Kennedy, and B. Krauss. 2011. Clinical practice guideline for emergency department ketamine dissociative sedation: 2011 update. *Ann Emerg Med.* 57:449–461.

Guidi, S., E. Ciani, S. Severi, A. Contestabile, and R. Bartesaghi. 2005. Postnatal neurogenesis in the dentate gyrus of the guinea pig. *Hippocampus* 15:285–301.

Habas, A., G. Kharebava, E. Szatmari, and M. Hetman. 2006. NMDA neuroprotection against a phosphatidylinositol-3 kinase inhibitor, LY294002 by NR2B-mediated suppression of glycogen synthase kinase-3beta-induced apoptosis. *J Neurochem* 96:335–348.

Hansen, H.H., T. Briem, M. Dzietko, M. Sifringer, A. Voss, W. Rzeski, B. Zdzisinska, F. Thor, R. Heumann, A. Stepulak, P. Bittigau, and C. Ikonomidou. 2004. Mechanisms leading to disseminated apoptosis following NMDA receptor blockade in the developing rat brain. *Neurobiol Dis* 16:440–453.

Harris, J.E., R.D. Letson, and J.J. Buckley. 1968. The use of CI-581, a new parenteral anesthetic, in ophthalmic practice. *Trans Am Ophthalmol Soc* 66:206–213.

Henson, M.A., A.C. Roberts, K. Salimi, S. Vadlamudi, R.M. Hamer, J.H. Gilmore, L.F. Jarskog, and B.D. Philpot. 2008. Developmental regulation of the NMDA receptor subunits, NR3A and NR1, in human prefrontal cortex. *Cereb Cortex* 18:2560–2573.

Hirota, K., T. Kubota, H. Ishihara, and A. Matsuki. 1999a. The effects of nitrous oxide and ketamine on the bispectral index and 95% spectral edge frequency during propofol-fentanyl anaesthesia. *Eur J Anaesthesiol* 16:779–783.

Hirota, K., H. Okawa, B.L. Appadu, D.K. Grandy, L.A. Devi, and D.G. Lambert. 1999b. Stereoselective interaction of ketamine with recombinant mu, kappa, and delta opioid receptors expressed in Chinese hamster ovary cells. *Anesthesiology* 90:174–182.

Hirota, K., K.S. Sikand, and D.G. Lambert. 1999c. Interaction of ketamine with mu2 opioid receptors in SH-SY5Y human neuroblastoma cells. *J Anesth* 13:107–109.

Hisatsune, C., H. Umemori, M. Mishina, and T. Yamamoto. 1999. Phosphorylation-dependent interaction of the N-methyl-D-aspartate receptor epsilon 2 sub-unit with phosphatidylinositol 3-kinase. *Genes Cells* 4:657–666.

Horne, M.K., D.R. Nisbet, J.S. Forsythe, and C.L. Parish. 2010. Three-dimensional nanofibrous scaffolds incorporating immobilized BDNF promote prolifera-tion and differentiation of cortical neural stem cells. *Stem Cells Dev* 19:843–852.

Hsieh, W.Y., Y.L. Hsieh, D.D. Liu, S.N. Yang, and J.N. Wu. 2003. Neural progeni-tor cells resist excitatory amino acid-induced neurotoxicity. *J Neurosci Res* 71:272–278.

Hu, M., Y.J. Sun, Q.G. Zhou, L. Chen, Y. Hu, C.X. Luo, J.Y. Wu, J.S. Xu, L.X. Li, and D.Y. Zhu. 2008. Negative regulation of neurogenesis and spatial memory by NR2B-containing NMDA receptors. *J Neurochem* 106:1900–1913.

Iino, M., S. Ozawa, and K. Tsuzuki. 1990. Permeation of calcium through excit-atory amino acid receptor channels in cultured rat hippocampal neurones. *J Physiol* 424:151–165.

Ikonomidou, C., F. Bosch, M. Miksa, P. Bittigau, J. Vockler, K. Dikranian, T.I. Tenkova, V. Stefovska, L. Turski, and J.W. Olney. 1999. Blockade of NMDA receptors and apoptotic neurodegeneration in the developing brain. *Science* 283:70–74.

Ishii, T., K. Moriyoshi, H. Sugihara, K. Sakurada, H. Kadotani, M. Yokoi, C. Akazawa, R. Shigemoto, N. Mizuno, M. Masu et al. 1993. Molecular charac-terization of the family of the N-methyl-D-aspartate receptor subunits. *J Biol Chem* 268:2836–2843.

Ishizuka, K., A. Kamiya, E.C. Oh, H. Kanki, S. Seshadri, J.F. Robinson, H. Murdoch, A.J. Dunlop, K.I. Kubo, K. Furukori, B. Huang, M. Zeledon, A. Hayashi-Takagi, H. Okano, K. Nakajima, M.D. Houslay, N. Katsanis, and A. Sawa. 2011. DISC1-dependent switch from progenitor proliferation to migration in the developing cortex. *Nature* 473(7345):92–96.

Jansen, K.L. 2000. A review of the nonmedical use of ketamine: Use, users and consequences. *J Psychoactive Drugs* 32:419–433.

Javitt, D.C. 2010. Glutamatergic theories of schizophrenia. *Isr J Psychiatry Relat Sci* 47:4–16.

Jelitai, M., K. Schlett, P. Varju, U. Eisel, and E. Madarasz. 2002. Regulated appear-ance of NMDA receptor subunits and channel functions during in vitro neu-ronal differentiation. *J Neurobiol* 51:54–65.

Joo, J.Y., B.W. Kim, J.S. Lee, J.Y. Park, S. Kim, Y.J. Yun, S.H. Lee, H. Rhim, and H. Son. 2007. Activation of NMDA receptors increases proliferation and differ-entiation of hippocampal neural progenitor cells. *J Cell Sci* 120:1358–1370.

Kapur, S., and P. Seeman. 2002. NMDA receptor antagonists ketamine and PCP have direct effects on the dopamine D(2) and serotonin 5-HT(2)receptors-implications for models of schizophrenia. *Mol Psychiatry* 7:837–844.

Keilhoff, G., H.G. Bernstein, A. Becker, G. Grecksch, and G. Wolf. 2004. Increased neurogenesis in a rat ketamine model of schizophrenia. *Biol Psychiatry* 56:317–322.

Keilhoff, G., G. Grecksch, and A. Becker. 2010. Haloperidol normalized prenatal vitamin D depletion-induced reduction of hippocampal cell proliferation in adult rats. *Neurosci Lett* 476:94–98.

Kitamura, T., M. Mishina, and H. Sugiyama. 2003. Enhancement of neurogenesis by running wheel exercises is suppressed in mice lacking NMDA receptor epsilon 1 subunit. *Neurosci Res* 47:55–63.

Kitayama, T., M. Yoneyama, K. Tamaki, and Y. Yoneda. 2004. Regulation of neuronal differentiation by N-methyl-D-aspartate receptors expressed in neural progenitor cells isolated from adult mouse hippocampus. *J Neurosci Res* 76:599–612.

Kluska, M.M., O.W. Witte, J. Bolz, and C. Redecker. 2005. Neurogenesis in the adult dentate gyrus after cortical infarcts: Effects of infarct location, N-methyl-D-aspartate receptor blockade and anti-inflammatory treatment. *Neuroscience* 135:723–735.

Kogan, A., R. Efrat, J. Katz, and B.A. Vidne. 2003. Propofol-ketamine mixture for anesthesia in pediatric patients undergoing cardiac catheterization. *J Cardiothorac Vasc Anesth* 17:691–693.

Komhuber, J., F. Mack-Burkhardt, M.E. Komhuber, and P. Riederer. 1989. [3H] MK-801 binding sites in post-mortem human frontal cortex. *Eur J Pharmacol* 162(3):483–490.

Koruk, S., A. Mizrak, B. Kaya Ugur, O. Ilhan, O. Baspinar, and U. Oner. 2010. Propofol/dexmedetomidine and propofol/ketamine combinations for anesthesia in pediatric patients undergoing transcatheter atrial septal defect closure: A prospective randomized study. *Clin Ther* 32:701–709.

Kubota, T., A. Miyata, A. Maeda, K. Hirota, H. Ishihara, and A. Matsuki. 1999. A hemodynamic evaluation of propofol/fentanyl compared with isoflurane/fentanyl anesthesia in coronary artery bypass grafting. *J Anesth* 13:44–47.

Kutsuwada, T., N. Kashiwabuchi, H. Mori, K. Sakimura, E. Kushiya, K. Araki, H. Meguro, H. Masaki, T. Kumanishi, M. Arakawa et al. 1992. Molecular diversity of the NMDA receptor channel. *Nature* 358:36–41.

Lahtinen, P., H. Kokki, T. Hakala, and M. Hynynen. 2004. S(+)-ketamine as an analgesic adjunct reduces opioid consumption after cardiac surgery. *Anesth Analg* 99:1295–1301; table of contents.

Lamont, S.R., B.J. Stanwell, R. Hill, I.C. Reid, and C.A. Stewart. 2005. Ketamine pre-treatment dissociates the effects of electroconvulsive stimulation on mossy fibre sprouting and cellular proliferation in the dentate gyrus. *Brain Res* 1053:27–32.

LoTurco, J.J., M.G. Blanton, and A.R. Kriegstein. 1991. Initial expression and endogenous activation of NMDA channels in early neocortical development. *J Neurosci* 11:792–799.

Lovell-Badge, R. 2001. The future for stem cell research. *Nature* 414:88–91.

Maeda, K., H. Sugino, T. Hirose, H. Kitagawa, T. Nagai, H. Mizoguchi, K. Takuma, and K. Yamada. 2007. Clozapine prevents a decrease in neurogenesis in mice repeatedly treated with phencyclidine. *J Pharmacol Sci* 103:299–308.

Mak, Y.T., W.P. Lam, L. Lu, Y.W. Wong, and D.T. Yew. 2010. The toxic effect of ketamine on SH-SY5Y neuroblastoma cell line and human neuron. *Microsc Res Tech* 73:195–201.

Maric, D., and J.L. Barker. 2005. Fluorescence-based sorting of neural stem cells and progenitors. *Curr Protoc Neurosci* Chapter 3:Unit 3.18.

Maric, D., Q.Y. Liu, G.M. Grant, J.D. Andreadis, Q. Hu, Y.H. Chang, J.L. Barker, J. Joseph, D.A. Stenger, and W. Ma. 2000. Functional ionotropic glutamate receptors emerge during terminal cell division and early neuronal differentiation of rat neuroepithelial cells. *J Neurosci Res* 61:652–662.

Mayer, M.L., and G.L. Westbrook. 1987. Permeation and block of N-methyl-D-aspartic acid receptor channels by divalent cations in mouse cultured central neurones. *J Physiol* 394:501–527.

Mayer, M.L., G.L. Westbrook, and P.B. Guthrie. 1984. Voltage-dependent block by Mg2+ of NMDA responses in spinal cord neurones. *Nature* 309:261–263.

McGlone, R. 2009. Emergency sedation in children. Utility of low dose ketamine. *BMJ* 339:b5575.

McVey, J.D., and J.D. Tobias. 2010. Dexmedetomidine and ketamine for sedation during spinal anesthesia in children. *J Clin Anesth* 22:538–545.

Mellon, R.D., A.F. Simone, and B.A. Rappaport. 2007. Use of anesthetic agents in neonates and young children. *Anesth Analg* 104:509–520.

Mercadante, S., P. Villari, and P. Ferrera. 2003. Burst ketamine to reverse opioid tolerance in cancer pain. *J Pain Symptom Manage* 25:302–305.

Mercer, S.J. 2009. 'The Drug of War'—A historical review of the use of Ketamine in military conflicts. *J R Nav Med Serv* 95:145–150.

Mierau, S.B., R.M. Meredith, A.L. Upton, and O. Paulsen. 2004. Dissociation of experience-dependent and -independent changes in excitatory synaptic transmission during development of barrel cortex. *Proc Natl Acad Sci U S A* 101:15518–15523.

Mitalipov, S., and D. Wolf. 2009. Totipotency, pluripotency and nuclear reprogramming. *Adv Biochem Eng Biotechnol* 114:185–199.

Mochizuki, N., N. Takagi, K. Kurokawa, T. Kawai, S. Besshoh, K. Tanonaka, and S. Takeo. 2007. Effect of NMDA receptor antagonist on proliferation of neurospheres from embryonic brain. *Neurosci Lett* 417:143–148.

Monyer, H., N. Burnashev, D.J. Laurie, B. Sakmann, and P.H. Seeburg. 1994. Developmental and regional expression in the rat brain and functional properties of four NMDA receptors. *Neuron* 12:529–540.

Monyer, H., R. Sprengel, R. Schoepfer, A. Herb, M. Higuchi, H. Lomeli, N. Burnashev, B. Sakmann, and P.H. Seeburg. 1992. Heteromeric NMDA receptors: Molecular and functional distinction of subtypes. *Science* 256:1217–1221.

Morgan, C.J., and H.V. Curran. 2006. Acute and chronic effects of ketamine upon human memory: A review. *Psychopharmacology (Berl)* 188:408–424.

Morgan, C.J., L. Muetzelfeldt, and H.V. Curran. 2009. Ketamine use, cognition and psychological wellbeing: A comparison of frequent, infrequent and ex-users with polydrug and non-using controls. *Addiction* 104:77–87.

Morgan, C.J., L. Muetzelfeldt, and H.V. Curran. 2010. Consequences of chronic ketamine self-administration upon neurocognitive function and psychological wellbeing: A 1-year longitudinal study. *Addiction* 105:121–133.

Mori, H., H. Masaki, T. Yamakura, and M. Mishina. 1992. Identification by mutagenesis of a Mg(2+)-block site of the NMDA receptor channel. *Nature* 358:673–675.

Moriyoshi, K., M. Masu, T. Ishii, R. Shigemoto, N. Mizuno, and S. Nakanishi. 1991. Molecular cloning and characterization of the rat NMDA receptor. *Nature* 354:31–37.

Muth-Kohne, E., J. Terhag, S. Pahl, M. Werner, I. Joshi, and M. Hollmann. 2010. Functional excitatory GABAA receptors precede ionotropic glutamate receptors in radial glia-like neural stem cells. *Mol Cell Neurosci* 43:209–221.

Noppers, I., M. Niesters, L. Aarts, T. Smith, E. Sarton, and A. Dahan. 2010. Ketamine for the treatment of chronic non-cancer pain. *Expert Opin Pharmacother* 11: 2417–2429.

Ono, T., E. Hashimoto, W. Ukai, T. Ishii, and T. Saito. 2010. The role of neural stem cells for in vitro models of schizophrenia: Neuroprotection via Akt/ERK signal regulation. *Schizophr Res* 122:239–247.

Orban, P.C., P.F. Chapman, and R. Brambilla. 1999. Is the Ras-MAPK signalling pathway necessary for long-term memory formation? *Trends Neurosci* 22:38–44.

O'Shaughnessy, C.T., and D. Lodge. 1988. N-methyl-D-aspartate receptor-mediated increase in intracellular calcium is reduced by ketamine and phencyclidine. *Eur J Pharmacol* 153:201–209.

Paling, N.R., H. Wheadon, H.K. Bone, and M.J. Welham. 2004. Regulation of embryonic stem cell self-renewal by phosphoinositide 3-kinase-dependent signaling. *J Biol Chem* 279:48063–48070.

Palygin, O., U. Lalo, and Y. Pankratov. 2011. Distinct pharmacological and functional properties of NMDA receptors in mouse cortical astrocytes. *Br J Pharmacol.* 163:1755–1766.

Paoletti, P., and J. Neyton. 2007. NMDA receptor subunits: function and pharmacology. *Curr Opin Pharmacol* 7:39–47.

Parks, K.A., and C.L. Kennedy. 2004. Club drugs: Reasons for and consequences of use. *J Psychoactive Drugs* 36:295–302.

Paule, M.G., M. Li, R.R. Allen, F. Liu, X. Zou, C. Hotchkiss, J.P. Hanig, T.A. Patterson, W. Slikker, Jr., and C. Wang. 2011. Ketamine anesthesia during the first week of life can cause long-lasting cognitive deficits in rhesus monkeys. *Neurotoxicol Teratol.* 33:220–230.

Pearce, I.A., M.A. Cambray-Deakin, and R.D. Burgoyne. 1987. Glutamate acting on NMDA receptors stimulates neurite outgrowth from cerebellar granule cells. *FEBS Lett* 223:143–147.

Petrus, D.S., K. Fabel, G. Kronenberg, C. Winter, B. Steiner, and G. Kempermann. 2009. NMDA and benzodiazepine receptors have synergistic and antagonistic effects on precursor cells in adult hippocampal neurogenesis. *Eur J Neurosci* 29:244–252.

Quinlan, E.M., B.D. Philpot, R.L. Huganir, and M.F. Bear. 1999. Rapid, experience-dependent expression of synaptic NMDA receptors in visual cortex in vivo. *Nat Neurosci* 2:352–357.

Radnay, P.A., I. Hollinger, A. Santi, and H. Nagashima. 1976. Ketamine for pediatric cardiac anesthesia. *Anaesthesist* 25:259–265.

Raju, V.K. 1980. Ketamine anesthesia in pediatric ophthalmic surgery. *J Pediatr Ophthalmol Strabismus* 17:292–296.

Ramirez, M., and M. Lamas. 2009. NMDA receptor mediates proliferation and CREB phosphorylation in postnatal Muller glia-derived retinal progenitors. *Mol Vis* 15:713–721.

Ricci, V., G. Martinotti, F. Gelfo, F. Tonioni, C. Caltagirone, P. Bria, and F. Angelucci. 2010. Chronic ketamine use increases serum levels of brain-derived neurotrophic factor. *Psychopharmacology (Berl)* 215(1):143–148.

Roberts, E.B., and A.S. Ramoa. 1999. Enhanced NR2A subunit expression and decreased NMDA receptor decay time at the onset of ocular dominance plasticity in the ferret. *J Neurophysiol* 81:2587–2591.

Romanelli, F., K.M. Smith, and C. Pomeroy. 2003. Use of club drugs by HIV-seropositive and HIV-seronegative gay and bisexual men. *Top HIV Med* 11:25–32.

Rudin, M., R. Ben-Abraham, V. Gazit, Y. Tendler, V. Tashlykov, and Y. Katz. 2005. Single-dose ketamine administration induces apoptosis in neonatal mouse brain. *J Basic Clin Physiol Pharmacol* 16:231–243.

Scallet, A.C., L.C. Schmued, W. Slikker, Jr., N. Grunberg, P.J. Faustino, H. Davis, D. Lester, P.S. Pine, F. Sistare, and J.P. Hanig. 2004. Developmental neurotoxicity of ketamine: Morphometric confirmation, exposure parameters, and multiple fluorescent labeling of apoptotic neurons. *Toxicol Sci* 81:364–370.

Schipke, C.G., C. Ohlemeyer, M. Matyash, C. Nolte, H. Kettenmann, and F. Kirchhoff. 2001. Astrocytes of the mouse neocortex express functional N-methyl-D-aspartate receptors. *FASEB J* 15:1270–1272.

Seeman, P., and H.C. Guan. 2008. Phencyclidine and glutamate agonist LY379268 stimulate dopamine D2High receptors: D2 basis for schizophrenia. *Synapse* 62:819–828.

Seeman, P., and H.C. Guan. 2009. Glutamate agonist LY404,039 for treating schizophrenia has affinity for the dopamine D2(High) receptor. *Synapse* 63:935–939.

Seeman, P., H.C. Guan, and H. Hirbec. 2009. Dopamine D2High receptors stimulated by phencyclidines, lysergic acid diethylamide, salvinorin A, and modafinil. *Synapse* 63:698–704.

Seeman, P., F. Ko, and T. Tallerico. 2005. Dopamine receptor contribution to the action of PCP, LSD and ketamine psychotomimetics. *Mol Psychiatry* 10:877–883.

Seeman, P., and M. Lasaga. 2005. Dopamine agonist action of phencyclidine. *Synapse* 58:275–277.

Semba, J., M. Wakuta, and T. Suhara. 2006. Different effects of chronic phencyclidine on brain-derived neurotrophic factor in neonatal and adult rat brains. *Addict Biol* 11:126–130.

Semple, S.J., S.A. Strathdee, J. Zians, and T.L. Patterson. 2009. Sexual risk behavior associated with co-administration of methamphetamine and other drugs in a sample of HIV-positive men who have sex with men. *Am J Addict* 18:65–72.

Sener, S., C. Eken, C.H. Schultz, M. Serinken, and M. Ozsarac. 2011. Ketamine with and without midazolam for emergency department sedation in adults: A randomized controlled trial. *Ann Emerg Med* 57:109–114.e102.

Shang, Y., Y. Wu, S. Yao, X. Wang, D. Feng, and W. Yang. 2007. Protective effect of erythropoietin against ketamine-induced apoptosis in cultured rat cortical neurons: Involvement of PI3K/Akt and GSK-3 beta pathway. *Apoptosis* 12:2187–2195.

Sharp, F.R., A. Lu, Y. Tang, and D.E. Millhorn. 2000. Multiple molecular penumbras after focal cerebral ischemia. *J Cereb Blood Flow Metab* 20:1011–1032.

Shemesh, E., L. Bat, J. Yahav, Y. Niv, A. Jonas, and P. Rozen. 1982. Ketamine anesthesia for pediatric gastrointestinal procedures. *Isr J Med Sci* 18:760–762.

Shen, L., and J. Zhang. 2007. NMDA receptor and iNOS are involved in the effects of ginsenoside Rg1 on hippocampal neurogenesis in ischemic gerbils. *Neurol Res* 29:270–273.

Sheng, M., J. Cummings, L.A. Roldan, Y.N. Jan, and L.Y. Jan. 1994. Changing subunit composition of heteromeric NMDA receptors during development of rat cortex. *Nature* 368:144–147.

Shi, Q., L. Guo, T.A. Patterson, S. Dial, Q. Li, N. Sadovova, X. Zhang, J.P. Hanig, M.G. Paule, W. Slikker, Jr., and C. Wang. 2010. Gene expression profiling in the developing rat brain exposed to ketamine. *Neuroscience* 166:852–863.

Singh, A., S. Girotra, Y. Mehta, S. Radhakrishnan, and S. Shrivastava. 2000. Total intravenous anesthesia with ketamine for pediatric interventional cardiac procedures. *J Cardiothorac Vasc Anesth* 14:36–39.

Sinner, B., and B.M. Graf. 2008. Ketamine. *Handb Exp Pharmacol* 182:313–333.

Sinor, A.D., and L. Lillien. 2004. Akt-1 expression level regulates CNS precursors. *J Neurosci* 24:8531–8541.

Sircar, R., P. Follesa, and M.K. Ticku. 1996. Postnatal phencyclidine treatment differentially regulates N-methyl-D-aspartate receptor subunit mRNA expression in developing rat cerebral cortex. *Brain Res Mol Brain Res* 40:214–220.

Slikker, W., Jr., X. Zou, C.E. Hotchkiss, R.L. Divine, N. Sadovova, N.C. Twaddle, D.R. Doerge, A.C. Scallet, T.A. Patterson, J.P. Hanig, M.G. Paule, and C. Wang. 2007. Ketamine-induced neuronal cell death in the perinatal rhesus monkey. *Toxicol Sci* 98:145–158.

Smith, D.J., R.L. Bouchal, C.A. deSanctis, P.J. Monroe, J.B. Amedro, J.M. Perrotti, and T. Crisp. 1987. Properties of the interaction between ketamine and opiate binding sites in vivo and in vitro. *Neuropharmacology* 26(9):1253–1260.

Soriano, S.G., Q. Liu, J. Li, J.R. Liu, X.H. Han, J.L. Kanter, D. Bajic, and J.C. Ibla. 2010. Ketamine activates cell cycle signaling and apoptosis in the neonatal rat brain. *Anesthesiology* 112:1155–1163.

Sutton, G., and L.J. Chandler. 2002. Activity-dependent NMDA receptor-mediated activation of protein kinase B/Akt in cortical neuronal cultures. *J Neurochem* 82:1097–1105.

Suzuki, M., A.D. Nelson, J.B. Eickstaedt, K. Wallace, L.S. Wright, and C.N. Svendsen. 2006. Glutamate enhances proliferation and neurogenesis in human neural progenitor cell cultures derived from the fetal cortex. *Eur J Neurosci* 24:645–653.

Sweatt, J.D. 2001. Memory mechanisms: The yin and yang of protein phosphorylation. *Curr Biol* 11:R391–R394.

Szappanyos, G.G., P. Bopp, and P.C. Fournet. 1969. The use and advantage of "Ketalar" (CI-581) as anaesthetic agent in pediatric cardiac catheterisation and angiocardiography. *Anaesthesist* 18:365–367.

Takadera, T., A. Ishida, and T. Ohyashiki. 2006. Ketamine-induced apoptosis in cultured rat cortical neurons. *Toxicol Appl Pharmacol* 210:100–107.

Takai, H., K. Katayama, K. Uetsuka, H. Nakayama, and K. Doi. 2003. Distribution of N-methyl-D-aspartate receptors (NMDARs) in the developing rat brain. *Exp Mol Pathol* 75:89–94.

Tung, A., S. Herrera, C.A. Fornal, and B.L. Jacobs. 2008. The effect of prolonged anesthesia with isoflurane, propofol, dexmedetomidine, or ketamine on neural cell proliferation in the adult rat. *Anesth Analg* 106:1772–1777.

Ulbrich, M.H., and E.Y. Isacoff. 2007. Subunit counting in membrane-bound proteins. *Nat Methods* 4:319–321.

Ulloa-Montoya, F., C.M. Verfaillie, and W.S. Hu. 2005. Culture systems for pluripotent stem cells. *J Biosci Bioeng* 100:12–27.

Urbach, A., C. Redecker, and O.W. Witte. 2008. Induction of neurogenesis in the adult dentate gyrus by cortical spreading depression. *Stroke* 39:3064–3072.

Varju, P., K. Schlett, U. Eisel, and E. Madarasz. 2001. Schedule of NMDA receptor subunit expression and functional channel formation in the course of in vitro-induced neurogenesis. *J Neurochem* 77:1444–1456.

Vescovi, A.L., B.A. Reynolds, D.D. Fraser, and S. Weiss. 1993. bFGF regulates the proliferative fate of unipotent (neuronal) and bipotent (neuronal/astroglial) EGF-generated CNS progenitor cells. *Neuron* 11:951–966.

Viberg, H., E. Ponten, P. Eriksson, T. Gordh, and A. Fredriksson. 2008. Neonatal ketamine exposure results in changes in biochemical substrates of neuronal growth and synaptogenesis, and alters adult behavior irreversibly. *Toxicology* 249:153–159.

Visser, E., and S.A. Schug. 2006. The role of ketamine in pain management. *Biomed Pharmacother* 60:341–348.

Vutskits, L., E. Gascon, G. Potter, E. Tassonyi, and J.Z. Kiss. 2007. Low concentrations of ketamine initiate dendritic atrophy of differentiated GABAergic neurons in culture. *Toxicology* 234:216–226.

Vutskits, L., E. Gascon, E. Tassonyi, and J.Z. Kiss. 2006. Effect of ketamine on dendritic arbor development and survival of immature GABAergic neurons in vitro. *Toxicol Sci* 91:540–549.

Walker, S.M., B.D. Westin, R. Deumens, M. Grafe, and T.L. Yaksh. 2010. Effects of intrathecal ketamine in the neonatal rat: Evaluation of apoptosis and long-term functional outcome. *Anesthesiology* 113:147–159.

Wang, C., W.F. Pralong, M.F. Schulz, G. Rougon, J.M. Aubry, S. Pagliusi, A. Robert, and J.Z. Kiss. 1996. Functional N-methyl-D-aspartate receptors in O-2A glial precursor cells: A critical role in regulating polysialic acid-neural cell adhesion molecule expression and cell migration. *J Cell Biol* 135:1565–1581.

Wang, C., N. Sadovova, X. Fu, L. Schmued, A. Scallet, J. Hanig, and W. Slikker. 2005. The role of the N-methyl-D-aspartate receptor in ketamine-induced apoptosis in rat forebrain culture. *Neuroscience* 132:967–977.

Wang, C., N. Sadovova, C. Hotchkiss, X. Fu, A.C. Scallet, T.A. Patterson, J. Hanig, M.G. Paule, and W. Slikker, Jr. 2006. Blockade of N-methyl-D-aspartate receptors by ketamine produces loss of postnatal day 3 monkey frontal cortical neurons in culture. *Toxicol Sci* 91:192–201.

Wang, C., N. Sadovova, T.A. Patterson, X. Zou, X. Fu, J.P. Hanig, M.G. Paule, S.F. Ali, X. Zhang, and W. Slikker, Jr. 2008a. Protective effects of 7-nitroindazole on ketamine-induced neurotoxicity in rat forebrain culture. *Neurotoxicology* 29:613–620.

Wang, C.Z., S.F. Yang, Y. Xia, and K.M. Johnson. 2008b. Postnatal phencyclidine administration selectively reduces adult cortical parvalbumin-containing interneurons. *Neuropsychopharmacology* 33:2442–2455.

Watanabe, M., M. Mishina, and Y. Inoue. 1994a. Distinct distributions of five NMDA receptor channel subunit mRNAs in the brainstem. *J Comp Neurol* 343:520–531.

Watanabe, M., M. Mishina, and Y. Inoue. 1994b. Distinct spatiotemporal expressions of five NMDA receptor channel subunit mRNAs in the cerebellum. *J Comp Neurol* 343:513–519.

Watanabe, S., H. Umehara, K. Murayama, M. Okabe, T. Kimura, and T. Nakano. 2006. Activation of Akt signaling is sufficient to maintain pluripotency in mouse and primate embryonic stem cells. *Oncogene* 25:2697–2707.

Weatherall, A., and R. Venclovas. 2010. Experience with a propofol-ketamine mixture for sedation during pediatric orthopedic surgery. *Paediatr Anaesth* 20:1009–1016.

Wegner, F., R. Kraft, K. Busse, G. Schaarschmidt, W. Hartig, S.C. Schwarz, and J. Schwarz. 2009. Glutamate receptor properties of human mesencephalic neural progenitor cells: NMDA enhances dopaminergic neurogenesis in vitro. *J Neurochem* 111:204–216.

Whitney, N.P., H. Peng, N.B. Erdmann, C. Tian, D.T. Monaghan, and J.C. Zheng. 2008. Calcium-permeable AMPA receptors containing Q/R-unedited GluR2 direct human neural progenitor cell differentiation to neurons. *FASEB J* 22:2888–2900.

Wilson, M.A., S.L. Kinsman, and M.V. Johnston. 1998. Expression of NMDA receptor subunit mRNA after MK-801 treatment in neonatal rats. *Brain Res Dev Brain Res* 109:211–220.

Wilson, R.D., D.L. Traber, and B.L. Evans. 1969. Correlation of psychologic and physiologic observations from children undergoing repeated ketamine anesthesia. *Anesth Analg* 48:995–1001.

Winkelheide, U., I. Lasarzik, B. Kaeppel, J. Winkler, C. Werner, E. Kochs, and K. Engelhard. 2009. Dose-dependent effect of S(+) ketamine on post-ischemic endogenous neurogenesis in rats. *Acta Anaesthesiol Scand* 53:528–533.

Xia, P., H.S. Chen, D. Zhang, and S.A. Lipton. 2010a. Memantine preferentially blocks extrasynaptic over synaptic NMDA receptor currents in hippocampal autapses. *J Neurosci* 30:11246–11250.

Xia, Y., C.Z. Wang, J. Liu, N.C. Anastasio, and K.M. Johnson. 2010b. Brain-derived neurotrophic factor prevents phencyclidine-induced apoptosis in developing brain by parallel activation of both the ERK and PI-3K/Akt pathways. *Neuropharmacology* 58:330–336.

Yeung, L.Y., M.S. Wai, M. Fan, Y.T. Mak, W.P. Lam, Z. Li, G. Lu, and D.T. Yew. 2010. Hyperphosphorylated tau in the brains of mice and monkeys with long-term administration of ketamine. *Toxicol Lett* 193:189–193.

Yoneyama, M., N. Nakamichi, M. Fukui, T. Kitayama, D.D. Georgiev, J.O. Makanga, N. Nakamura, H. Taniura, and Y. Yoneda. 2008. Promotion of neuronal differentiation through activation of N-methyl-D-aspartate receptors transiently expressed by undifferentiated neural progenitor cells in fetal rat neocortex. *J Neurosci Res* 86:2392–2402.

Young, C., V. Jevtovic-Todorovic, Y.Q. Qin, T. Tenkova, H. Wang, J. Labruyere, and J.W. Olney. 2005. Potential of ketamine and midazolam, individually or in combination, to induce apoptotic neurodegeneration in the infant mouse brain. *Br J Pharmacol* 146:189–197.

Zhang, H., A. Lu, H. Zhao, K. Li, S. Song, J. Yan, W. Zhang, S. Wang, and L. Li. 2004. Elevation of NMDAR after transplantation of neural stem cells. *Neuroreport* 15:1739–1743.

Zheng, W.H., and R. Quirion. 2009. Glutamate acting on N-methyl-D-aspartate receptors attenuates insulin-like growth factor-1 receptor tyrosine phosphorylation and its survival signaling properties in rat hippocampal neurons. *J Biol Chem* 284:855–861.

Zhong, Z., Y. Wang, H. Guo, A. Sagare, J.A. Fernandez, R.D. Bell, T.M. Barrett, J.H. Griffin, R.S. Freeman, and B.V. Zlokovic. 2010. Protein S protects neurons from excitotoxic injury by activating the TAM receptor Tyro3-phosphatidylinositol 3-kinase-Akt pathway through its sex hormone-binding globulin-like region. *J Neurosci* 30:15521–15534.

Zigova, T., V. Pencea, S.J. Wiegand, and M.B. Luskin. 1998. Intraventricular administration of BDNF increases the number of newly generated neurons in the adult olfactory bulb. *Mol Cell Neurosci* 11:234–245.

Zook, E.G., R.P. Roesch, L.W. Thompson, and J.E. Bennett. 1971. Ketamine anesthesia in pediatric plastic surgery. *Plast Reconstr Surg* 48:241–245.

Zou, X., T.A. Patterson, R.L. Divine, N. Sadovova, X. Zhang, J.P. Hanig, M.G. Paule, W. Slikker, Jr., and C. Wang. 2009a. Prolonged exposure to ketamine increases neurodegeneration in the developing monkey brain. *Int J Dev Neurosci* 27:727–731.

Zou, X., T.A. Patterson, N. Sadovova, N.C. Twaddle, D.R. Doerge, X. Zhang, X. Fu, J.P. Hanig, M.G. Paule, W. Slikker, and C. Wang. 2009b. Potential neurotoxicity of ketamine in the developing rat brain. *Toxicol Sci* 108:149–158.

chapter five

Ketamine—Epidemiology of misuse and patterns of acute and chronic toxicity

Shwetha S. Rao, David M. Wood, and Paul I. Dargan

Contents

5.1 Introduction

Ketamine was developed in 1962 by the Parke Davis Company in the United States (Copeland and Dillon 2005). It was extensively used as a field anesthetic during the Vietnam War and then introduced for civilian use in 1970 (Copeland and Dillon 2005; Curran and Morgan 2011). Ketamine now has an established role as an anesthetic in prehospital and military settings (Jennings et al. 2011; Jensen et al. 2010; Mellor 2005). As hospital medicine, ketamine is particularly used for pediatric sedation; it is less commonly used in adults for dissociative sedation (CEM 2009; RCoA/CEM Report and Recommendations 2012). As an anesthetic, ketamine provides clinicians with an anesthetic that is potent and rapidly acting, with a short duration of action providing good analgesia without compromising cardiovascular and respiratory stability (Jensen et al. 2010; Mellor 2005). However, these excellent properties come at the price of unpleasant hallucinations, delirium, and vivid dreams during awakening (these are often referred to as "emergence phenomena") (Corssen and Domino 1966; Morgan et al. 2012). Ketamine is widely used in veterinary medicine as an anesthetic and its use as an analgesic has also been explored; however, recent Cochrane Reviews have suggested that there is as yet insufficient evidence for ketamine to be recommended for the routine management of chronic pain (Bell et al. 2012).

The first reports of nonmedical ketamine misuse were in 1967 (Jansen 2001a,b) with increasing misuse in the 1970s through the 1990s when it emerged as a recreational drug, particularly in the rave party or nightclub scene in Europe and the United States (Dalgarno and Shewan 1996; Jansen 1993); more recently, it has become a common drug of misuse in Hong Kong and China (Central Registry of Drug Abuse Sixty-Second Report 2013; Patterns and Trends of Amphetamine-Type Stimulants and Other Drugs: Asia and the Pacific 2013). Unlike many other recreational drugs, in addition to evidence of acute toxicity, ketamine is now also recognized to be associated with chronic toxicity in regular long-term frequent users—particularly involving the lower urinary tract, bladder, kidneys, and central nervous system. In this chapter, we will discuss the epidemiology of ketamine misuse and the literature describing the patterns of acute and chronic toxicity associated with ketamine misuse.

5.2   Epidemiology of ketamine misuse

The first reports of the nonmedical misuse of ketamine were in the late 1960s, but it was not until the late 1970s and 1980s that more widespread misuse was reported in North America, followed by Europe in 1990, and on to Asia in the latter part of the 1990s and in the 2000s (Dalgarno and Shewan 1996; Jansen 1993, 2001a,b; Li et al. 2011). Although nonmedical misuse (recreational use) of ketamine is now documented worldwide, the prevalence of its misuse varies around the world. Data on the prevalence of ketamine misuse are available from a variety of sources and these will be summarized in this section. We will include data from the United Nations Office on Drugs and Crime (UNODC) World Drug Report, UK Crime Survey for England and Wales (CSEW), and other national data sets including Hong Kong and Australia; this will be followed by data on ketamine use in subpopulations including those attending nightclubs, school-age and college-age students, and internet drug discussion forums.

5.2.1   Prevalence of recreational ketamine use

5.2.1.1   National level data

Ketamine is legal in many countries and so it is not included in national population level surveys on recreational drug use in all countries; however, data from the United Kingdom, Spain, the United States, Canada, Argentina, Brazil, Australia, New Zealand, China, and Hong Kong are available, as well as some limited data from the Middle East.

5.2.1.1.1   Europe

5.2.1.1.1.1   *United Kingdom—England and Wales.*   Data regarding population level prevalence rates for ketamine misuse in the United Kingdom are available in the CSEW; this was formerly the British Crime Survey (Home Office National Statistics on Drug Misuse 2012).

CSEW statistics for 2012/2013 reported that lifetime prevalence of ketamine use in 16- to 59-year-olds was 2.3% and that for 16- to 24-year-olds was 3.3%. Last year prevalence of ketamine use was 0.4% in 16- to 59-year-olds and 0.8% in 16- to 24-year-olds. Last month prevalence was 0.9% in 16- to 59-year-olds and 0.5% in 16- to 24-year-olds. Ketamine users were more likely to be male, single, in the age group of 20–24 years, unemployed or a student, and from Chinese or mixed race (Home Office Statistical Bulletin 2010, 2012).

Data from the CSEW suggests an association between higher prevalence of ketamine use and frequency of attendance at nightclubs and pubs. The last year use of ketamine was around 20 times higher among those who had visited a nightclub four or more times in the last month (6.6%) as compared to those who did not (0.3%) (CSEW Drug Misuse Data 2012).

5.2.1.1.1.2 United Kingdom—Scotland. Data collected as part of the Scottish Crime and Justice Survey for 2012 on the prevalence of ketamine use in those over 16 years of age in Scotland found that lifetime prevalence of ketamine use was 1.2%, last year prevalence of use was 0.3%, and last month prevalence of use was 0.1% (Scottish Crime and Justice Survey: Drug Use 2012).

5.2.1.1.1.3 Spain. Spain's national household survey in 2012 showed a lifetime prevalence of ketamine use of 1% and last year prevalence of ketamine use of 0.2% among those aged 15 to 64 years. Among those aged 14–18 years, the lifetime prevalence of ketamine use was 1.1% more than the prevalence rates for methamphetamine or heroin. Another study in Spain among 14- and 15-year-old students in 2010 suggested that ketamine had a lifetime prevalence of use of 1.1% in this age group and was almost three times as commonly used as piperazines or mephedrone (0.4%) (World Drug Report 2013).

5.2.1.1.2 North America and South America

5.2.1.1.2.1 United States. In the United States, Michigan University conducts an annual survey of high school students through the Monitoring the Future project, which is funded by the National Institute of Drug Abuse. Their results from 2012 reveal that the annual prevalence rate for ketamine in 12th graders (aged 17–18 years) was 1.5%; the data were not collected for the 8th and 10th graders because of low prevalence rates. In comparison, the annual prevalence of cocaine and 3,4-methylenedioxy-N-methylamphetamine (MDMA) use among the 12th graders was 2.7% and 3.8%, respectively (Monitoring the Future National Survey Results 2012). The annual prevalence rate of use of ketamine in 2012 was found to be 12% less than that for the year before (World Drug Report 2013).

5.2.1.1.2.2 Canada. Lifetime prevalence of use of ketamine in 15- to 16-year-olds was 1.6% in 2012 (Canadian Alcohol and Drug Use Monitoring Survey 2012; World Drug Report 2013).

5.2.1.1.2.3   Argentina.   In Argentina, among the population aged 12 to 65 years, lifetime prevalence of ketamine use was 0.3% in a national household survey conducted in 2010 (World Drug Report 2013).

5.2.1.1.2.4   Brazil.   The latest data available for Brazil are from a household survey conducted in 2005, and this revealed a lifetime prevalence of ketamine use of 0.2% among those aged 12 to 65 years (World Drug Report 2013).

5.2.1.1.3   Australasia

5.2.1.1.3.1   Australia.   The Australian Institute of Health and Welfare is Australia's national health and welfare statistics and information agency. They reported that there was a decrease in the use of ketamine from 2004 to 2010. In their summary in 2010, they reported a lifetime prevalence of ketamine use of 1.4% for those 14 years or older and prevalence rates for last year use of 0.2% (National Drug Strategy Household Survey Report 2010).

5.2.1.1.3.2   New Zealand.   The most recent data from New Zealand come from the 2007/2008 New Zealand Alcohol and Drug Use Survey (2010). Lifetime prevalence of ketamine use in adults aged 16–64 years was 1.2%. Men (2.2%) were more likely to use ketamine for recreational reasons than women (0.9%).

5.2.1.1.4   Asia.   Although there are limited data on ketamine use in many Asian countries, it appears that ketamine use is more prevalent in Asian and Southeast Asian countries compared to Europe and North America. The UNODC reports an emerging trend of increased ketamine use especially in Southeast and East Asian countries including Hong Kong, China, Indonesia, Brunei Darussalam, the Philippines, Thailand, and Malaysia. More detailed data on the prevalence of ketamine use are available from China and Hong Kong.

5.2.1.1.4.1   China.   Ketamine was the third most common drug of abuse in China in 2011; in 2012, more than 7.6% of registered drug users were using ketamine (Asia & Pacific Amphetamine-Type Stimulants Information Centre; World Drug Report 2013).

5.2.1.1.4.2   Hong Kong.   In 2012, the drug abuse registry for Hong Kong found that ketamine was the most popular psychotropic substance used, with 29% of users reporting its use; it was particularly popular among younger adults with 61% of users less than 21 years old reporting use of ketamine (Central Registry of Drug Abuse Sixty-Second Report 2013).

5.2.1.1.4.3   Middle East.   Use of ketamine has been reported in the Middle East; however, there are no detailed data available. In 2010, ketamine was recognized as the seventh most widely used drug in Saudi Arabia. Israel ranked it as the ninth most common drug of abuse in 2011 (World Drug Report 2013).

5.2.1.1.4.4   Data on ketamine seizures in Asia.   Patterns of seizures of ketamine can be used as a surrogate marker of availability and therefore potentially of use; this is particularly important in Asia where there are less data available from population surveys on the prevalence of use. As mentioned previously, Asia accounts for 95% of ketamine seizures globally (with 86% in East and Southeast Asia and 9% in South Asia) (World Drug Report 2013). In the East Asian region alone, one of the largest seizures of ketamine of approximately 6 tons was reported from China in 2011. In 2012, China accounted for 82% (4716.6 kg) of the seizures followed by Hong Kong at 12.6% (723.9 kg) and Malaysia at 4% (238.9 kg). There was an increase in seizures of ketamine of 162% in Hong Kong and China and 18% in Malaysia in 2012 (Patterns and Trends of Amphetamine-Type Stimulants and Other Drugs: Asia and the Pacific 2013). Other countries reporting ketamine seizures have been Indonesia, Thailand, Brunei Darussalam, Burma, and Singapore. India reported 883 kg of ketamine seizure in 2012 with the maximum amount seized from the country being 1.4 tons in 2011. China and India are thought to be the source countries for ketamine for illicit ketamine seized from China, Hong Kong, Malaysia, and the Philippines (Asia & Pacific Amphetamine-Type Stimulants Information Centre). In many Asian countries, ketamine has been found in pills that have been sold as ecstasy in addition to being found in powder form (Asia & Pacific Amphetamine-Type Stimulants Information Centre).

5.2.1.2   Subpopulation level data

Most of the subpopulation data available on ketamine misuse come from the United Kingdom; this offers insight into the prevalence of use in higher using groups (e.g., those who attend nightclubs) and in school-age and college-age children.

5.3   Ketamine use by clubbers

5.3.1   MixMag/Global Drug Survey

The former MixMag, now known as the Global Drug Survey, is one of the largest subpopulation drug surveys. It currently not only includes those individuals who read the MixMag magazine in paper form and online (a publication aimed at those who frequent nightclubs in the United Kingdom) but also has respondents from a range of social networking sites

Table 5.1 Country and Last Year Prevalence of
Ketamine Use in the 2014 Global Drug Survey

Country	Ketamine use in the last year (% total responses)
Australia	4.1
Belgium	6.5
Denmark	7.3
France	5.3
Germany	4.7
Hungary	1.7
Netherlands	12.0
Republic of Ireland	10.4
Scotland	14.7
Spain	5.5
United Kingdom	19.8

and is advertised in the *UK Guardian* newspaper. In the 2013 survey, 50.6% of UK respondents (n = 7700) reported lifetime use of ketamine and 31.5% reported ketamine use in the previous month (MixMag 2014). Ninety-six percent reported that they usually used the drug by nasal insufflation, and 42% reported that they usually drank alcohol during the same use episode. In 2014, the Global Drug Survey received almost 80,000 respondents in 18 different countries in Europe and North America. Ketamine was among the top 20 drugs used, with 5.7% of the total sample reporting its use in the last 12 months. The last 12-month prevalence of ketamine use in each country is detailed in Table 5.1 (Global Drug Survey 2014).

5.3.2   Other UK surveys

Ongoing annual surveys from our group are undertaken in South London, UK gay-friendly dance clubs in the summer of each year. Surveys conducted in 2010, 2011, 2012, and 2013 had 308, 315, 330, and 397 respondents, respectively (Measham et al. 2011; Wood et al. 2012a,b). Data are shown in Table 5.2; it shows that ketamine use is much higher than that

Table 5.2 Data from London Gay Nightclub Surveys

	2013 (n = 397)	2012 (n = 330)	2011 (n = 315)	2010 (n = 308)
Ketamine lifetime use	70%	67%	60%	57%
Ketamine last year use	51%	44%	49%	46%
Ketamine last month use	35%	24%	35%	30%

Source: Wood, D.M. et al., *J Subst Use* 17:91–7, 2012b.

in the general population (CSEW Drug Misuse Data 2012). Lifetime ketamine use was as high as 70% in 2013, with last month use of 35% (compared to comparative population level data of 2.2%–3.3% and 0.5%–0.9%,
respectively).

In a similar study done in 2012, mainstream nightclubs in Lancashire,
UK, town, and city centers were surveyed. Lifetime ketamine use was 18%,
last year use was 11%, and last month use was 3% (Measham et al. 2012).

5.3.3 Ketamine use in school- and college-age students

5.3.3.1 England

The "Smoking, Drinking and Drug Use Among Young People Survey,"
conducted yearly by "NatCen" (National Centre for Social Research), an
independent social research agency, includes schoolchildren aged 11–15
years in England (Smoking, Drinking and Drug Use Survey 2012). Their
latest report from 2012 included 7589 pupils in 254 schools. Thirty-five
percent of pupils responding to the survey had heard about ketamine and
0.5% had used it in the last year. It was found that use increased with age
and 0.4% of 13-year-olds and 0.9% of 15-year-olds reported lifetime use of
ketamine.

5.3.3.2 Scotland

A total of 37,307 thirteen- and fifteen-year-old pupils completed the 2010
Scottish Schools Adolescent Lifestyle and Substance Use Survey (SALSUS).
Two percent of 15-year-old boys and 1% of 15-year-old girls reported ever
taking ketamine and 1% of 13-year-old boys and less than 1% of 13-year-
old girls had used it previously. Three percent of 15-year-old boys and 1%
of 15-year-old girls reported that they had previously been offered ketamine, and among 13-year-olds, 1% of both boys and girls reported being
offered ketamine (SALSUS National Report 2011).

5.4 Patterns of ketamine misuse

Data from a number of studies provide information on the patterns of
ketamine misuse including doses used, frequency, and routes of use. The
2009 MixMag survey reported on 1285 individuals who had used ketamine in the previous year (Winstock et al. 2012). They were asked about
doses used in a typical use session. Thirty-one percent reported using less
than 0.125 g, 35% reported using between 0.25 and 0.5 g, and 34% reported
using more than 1 g in a single session, with 5% regularly using more than
3 g per typical session. The mean number of days of consecutive use of
ketamine was 3.5 days, with 11% reporting to have ever used it consecu-
tively for 7 days or more. Last-month users reported that they had used
ketamine for a mean of 4.2 days in the last month. Of these individuals,

13.4% were found to be high-frequency users who used ketamine on nine or more days per month (Winstock et al. 2012).

5.4.1   Ketamine and polysubstance use

Ketamine is commonly used together with other drugs, and the small amount of data available from the United Kingdom suggests that ketamine use is among the drugs that are most likely to be associated with polysubstance use. The UK CSEW 2012 (Home Office National Statistics on Drug Misuse 2012) data suggested that ketamine was among the top three drugs to be used simultaneously with other drugs (ketamine, 48%; MDMA, 49%; and methadone, 58%). A cross-sectional survey of 100 ketamine users in Australia (Dillon et al. 2003) revealed the following results: 74% preferred to use MDMA with ketamine rather than ketamine alone; MDA (3,4-methylenedioxyamphetamine) (37%), cannabis (19%), cocaine (19%), amphetamine (16%), and LSD (6%) were the other drugs commonly used along with ketamine in this cohort. Ng et al. (2010) also reported coingestion of other drugs in 35% (81) of their sample of 233 patients presenting to their Hong Kong Emergency Department with acute ketamine toxicity; coingestion substances were alcohol, cocaine, MDMA, methamphetamine, and zopiclone (Ng et al. 2010).

5.4.2   Ketamine routes of use

Recreational ketamine users appear to predominantly use it in powder form by nasal insufflation; less commonly users dissolve the powder in water and use it orally and there are rarer reports of injection use of ketamine (Erowid Ketamine Vault). There are also reports of the availability of ketamine in tablet form (often sold to users as ecstasy) (Asia & Pacific Amphetamine-Type Stimulants Information Centre).

5.5   Ketamine toxicity

5.5.1   Acute toxicity

Ketamine predominantly causes neuropsychiatric and cardiovascular features of acute toxicity. Some of the neuropsychiatric and behavioral symptoms can be unpleasant and frightening, particularly for the novice user (Kalsi et al. 2011), but the acute toxic effects of ketamine are generally not clinically severe and rarely result in severe life-threatening features. There is the potential for individuals to put themselves at risk of harm as a result of the dissociative and hallucinogenic effects of ketamine, which can lead to an increased risk of physical trauma, which may be associated with significant morbidity and potentially mortality.

The main sources of data on the patterns of acute toxicity associated with ketamine use include the following: (i) user reports on the Internet and in surveys, (ii) poison center case series, and (iii) case series from pre-hospital and emergency department presentations with acute ketamine toxicity. We shall discuss information from each of these data sources in detail. In addition, we will draw where applicable on data from animal studies and the use of ketamine as an anesthetic/analgesic.

5.5.1.1　Acute physical effects

5.5.1.1.1　Cardiopulmonary effects. Ketamine is a unique anesthetic agent in that it does not cause cardiovascular depression. It stimulates the cardiovascular system leading to tachycardia, hypertension, and an increase in cardiac output (Haas and Harper 1992). A study assessing hemodynamic response after a single dose of ketamine in 12 patients who were undergoing routine cardiac catheterization and angiography showed that after a 2-mg/kg dose of intravenous ketamine, there was a 33% increase in heart rate (from 73.5 ± 14.2 to 97.7 ± 28.9 beats per minute [bpm]), a 29% increase in cardiac index (from 2.97 ± 0.66 to 3.83 ± 0.63 L/min/m^2), a 28% increase in mean aortic pressure (from 93.3 ± 14.0 to 119.3 ± 10.1 torr), and a 49% increase in pulmonary arterial pressure (from 17.0 ± 4.6 to 24.5 ± 5.8 torr) (Tweed et al. 1972). The cardiovascular effects of ketamine are thought to be attributed to stimulation of the central sympathomimetic pathway and inhibition of reuptake of catecholamines (Reich and Silvay 1989).

Animal studies suggest that ketamine does not cause cardiovascular depression at therapeutic doses used in anesthesia (Bettschart-Wolfensberger et al. 2003; Jacobson and Hartsfield 1993). In a horse study, six ponies were anesthetized using ketamine and medetomidine. Sedation consisted of 7 µg/kg body weight intravenous medetomidine followed after 10 min by a bolus of 2 mg/kg body weight intravenous ketamine and maintenance of anesthesia was with an infusion of propofol at rates of 0.89–1.0 mg/kg/min. It was noted that ketamine induction was associated with stable cardiovascular and pulmonary function compared to propofol (Bettschart-Wolfensberger et al. 2003). In another study, 12 healthy dogs were given 10 mg/kg of ketamine and 0.5 mg/kg midazolam intravenously for induction of anesthesia; in six dogs, this was given as a bolus, and in the other six, it was given by an infusion over 15 min. Both the bolus and infusion ketamine–midazolam combination were associated with a significant ($P < 0.05$) increase in mean heart rate and rate–blood pressure product. There were no specific details in the paper in terms of the absolute physiological parameters. Mild and transient respiratory depression was observed in dogs of both groups immediately after administration but was more pronounced in the bolus group;

again, no specific information on this was included in the paper (Jacobson and Hartsfield 1993).

   5.5.1.1.2 Human studies.   A review conducted by Miller et al. (2011) included a total of 20 studies looking at the hemodynamic and pulmonary effects of ketamine on mechanically ventilated patients treated with a continuous intravenous infusion of ketamine (for a minimum of 2 h or more) at doses ranging from 0.01 to 5 mg/kg/h. Of the 281 patients included in the review, 223 had respiratory variables assessed; ketamine was found to reduce airway resistance, improve chest compliance, increase partial pressure of oxygen, and improve oxygen saturation. In patients with refractory bronchospasm, ketamine was found to decrease wheezing and reduce requirements for bronchodilator therapy (Miller et al. 2011). Other studies have also shown that ketamine causes bronchodilation and improves respiratory function in asthmatic patients and, hence, is also recommended in the emergency intubation of asthmatic patients (Hemmingsen et al. 1994; Petrillo et al. 2001). Ketamine is generally found to preserve airway reflexes such as the pharyngeal and laryngeal reflexes; laryngospasm and aspiration are very rare even at anesthetic doses (Green et al. 1998; Hemmingsen et al. 1994; Strayer and Nelson 2008). In one retrospective study of 1022 pediatric cases over a 9-year period, children given intramuscular ketamine for sedation were included. Airway complications were reported in only 1.4% of cases, laryngospasm in 0.4%, apnea in 0.2%, and respiratory depression in 0.1%. All the adverse respiratory effects were transient and managed conservatively without needing intubation (Green et al. 1998). A literature review in 2008 that included 87 papers where ketamine was used for sedation concluded that events such as laryngospasm, apnea, and respiratory depression were rare with good preservation of pharyngeal reflexes particularly when ketamine was used as a single agent (Strayer and Nelson 2008).

   5.5.1.1.3 Ketamine and pulmonary edema.   There are no reports of pulmonary edema related to the recreational use of ketamine; however, there have been at least five case reports of pulmonary edema caused by the use of ketamine in anesthetic practice (Baduni et al. 2010; Pandey et al. 2000; Parthasarathy et al. 2009; Tarnow and Hess 1978). The first reported case was in a patient with coronary artery disease receiving ketamine as an induction agent at 1.5 mg/kg intravenously. Immediately after induction, there was a rise in mean pulmonary arterial pressure from 27 to 65 mm Hg; this was associated with hypoxia due to pulmonary edema (Tarnow and Hess 1978). Another case reported was of an 8-year-old girl who was given 6.25 mg/kg ketamine intramuscularly for the purpose of sedation; she had no previous medical problems but developed pulmonary edema needing intubation 10 min after ketamine administration

(Pandey et al. 2000). The treating physicians noted pink frothy sputum through the endotracheal tube. She was treated with intravenous furosemide, and positive pressure ventilation was performed. Her condition improved after 3 h with no further complications. Parthasarathy et al. reported two cases, a 27-year-old and a 2-year-old, who were given ketamine at doses of 2 mg/kg; they developed hypoxia and frothy secretions (noted on intubation) soon after ketamine administration. In both instances, the clinical situation improved within an hour with the help of diuretic and positive pressure ventilation (Parthasarathy et al. 2009). The most recent report from 2010 was in a 7-year-old child who developed recurrent laryngospasm after ketamine induction (1 mg/kg intravenous). He later developed pulmonary edema during the postoperative period; the authors suggested that the laryngospasm might have contributed to negative pressure pulmonary edema in this case (Baduni et al. 2010).

It has been thought that the pathophysiology of ketamine-related pulmonary edema could be due to a direct stimulation of the sympathetic nervous system leading to elevated catecholamine levels and causing increased vascular pressure leading to pulmonary edema. Another potential hypothesis is that this is neurogenic edema; in vitro studies on human lungs have shown that ketamine inhibits lung epithelial sodium channel function; this would lead to a reduction in alveolar fluid clearance and resultant pulmonary edema (Cui et al. 2011).

5.5.1.2 *Acute effects of ketamine on the urogenital system*

There are little data on the acute effects of ketamine on the renal tract. Ng et al. in their case series of 233 patients presenting to the emergency department with acute ketamine toxicity reported that 32 (13%) reported any urological symptoms, which included dysuria, urgency, or frequency. There is the potential that these symptoms relate to chronic rather than acute effects of ketamine; however, there is no information on the chronicity of ketamine use or of these symptoms to be able to determine the likelihood of this. The same study found deranged renal function (abnormal creatinine in 5 [3%]; range, 103–200 µmol/L [1854–3600 mg/dL]). In a retrospective review of patients registered on the Hong Kong poison center database from 2008 to 2011, 188 cases were identified as acute cases who presented with ketamine-related problems to the emergency department. Sixty (32%) of the acute cases had urinary symptoms in keeping with symptoms suggestive of ketamine cystitis; it is unclear whether these individuals were frequent or infrequent users of ketamine; we feel that these reported symptoms were more likely to be a reflection of the chronic ketamine effects being reported in the acute setting. There are no other data to be able to determine whether single-use ketamine is associated with acute urological/renal effects, but as discussed in more detail in Chapter 10, animal studies suggest that chronic ketamine use for a period of a minimum of a few months is required for these effects to occur (Yiu-Cheung 2012).

5.5.1.3   User self-reports of acute physical effects of ketamine

5.5.1.3.1   User reports on internet discussion fora.   Bluelight and Erowid are Internet fora used internationally by drugs users to share experiences and information on drug use. The dose of ketamine reported by recreational users are generally small (e.g., 75–100 mg intranasal), although users report use of multiple doses per use session, with doses of up to 1–3 g per session. Common symptoms reported by users include "racing of the heart," a sense of euphoria, dizziness, agitation, irritability, hallucinations, unusual thought content, muscle cramp, chest pain, abdominal pain, nausea, vomiting, and dissociation.

5.5.1.3.2   User surveys.   The largest published user survey on the effects of ketamine was from Sydney, Australia, in which 100 ketamine users were recruited between January 1998 and October 1999 (Dillon et al. 2003). Of the 100 participants, 58% had used ketamine on more than 10 occasions and 26% were regular users. Participants were interviewed using a structured predetermined questionnaire, which collected data on experience of the physical and psychological effects of ketamine and their attitudes toward these effects as to whether they were positive or negative. The positive effects reported included blurred vision (21%), increased heart rate (17%), temporary paralysis (16%), lack of coordination (14%), and feeling of no pain (7%). The negative effects reported were inability to speak (30%), "increased breathing" (11%), difficulty breathing (2%), nausea (9%), vomiting (3%), and convulsions (<1%). Fifty-six percent reported having experienced the K-hole; it is interesting to note that symptoms such as inability to speak, incoordination, and blurred vision seen as negative symptoms by some were seen as positive effects by other users (Dillon et al. 2003).

5.5.1.4   Poisons information service data

Poison center data are a rich source of information on the physical effects of ketamine.

The UK National Poisons Information Service is a service providing information and advice to health professionals regarding management of patients with suspected poisoning, which can be accessed through their Internet site TOXBASE or their telephone enquiry service. In 2012/2013, there were approximately 550,000 TOXBASE accesses and 54,000 telephone consultations; of these, 6% and 4% were ketamine related, respectively (National Poisons Information Service Annual Report 2012/2013).

In the United States, poison centers also provide information to members of the public in addition to health care professionals. The American Association of Poison Control Center's National Poisons Data System in their annual report for 2012 reported 308 ketamine-related human exposures with about a half of them presenting to a health care facility. They also reported one death attributable to intentional use of ketamine during the same period (details about the circumstances of death were not available on the report) (Mowry et al. 2013).

The Hong Kong Poisons Information Centre (HKPIC) not only provides a 24-h information service but also collects poisons data from emergency departments. In their annual report for 2010, ketamine was ranked as the fourth most common poison exposure in the country with just more than 300 cases reported (Chan et al. 2012).

The Connecticut Poison Control Center performed an observational study of 20 ketamine-related calls to their service in 1997 (Weiner et al. 2000); the common symptoms reported by patients included anxiety (40%); palpitations (15%); chest pain (10%); and confusion, vomiting, and memory loss (all 5%). Fifty percent of patients were asymptomatic at the time of their initial evaluation in the emergency department. On physical examination, patients were found to have altered consciousness (30%), tachycardia (defined as heart rate >100 bpm, 60%), slurred speech (15%), hallucinations (15%), nystagmus (15%), mydriasis (15%), and hypertension (undefined; 10%). Five patients were treated with benzodiazepines for their agitation, two of whom developed mild rhabdomyolysis (creatinine kinase, 200–600 IU/L); they were treated with sedation (further doses of benzodiazepine) and intravenous fluids before discharge home. These effects were found to be short lived and 18 of the 20 patients were discharged within 5 h of presentation to the emergency department. Two patients had a prolonged period of observation in the emergency department. One of them developed chest pain with hypertension of up to 180/100 mm Hg and, hence, required intravenous metoprolol and further cardiac investigations (which showed negative results). Another had prolonged paranoid behavior but was later found to have a previous history of schizophrenia.

5.5.1.5 *Prehospital case series*

Prehospital emergency services are frequently used in managing patients with recreational drug toxicity and include ambulance services, first aid medical facilities in nightclubs, and telephone advice services provided for members of the public.

As part of the 2010/2011 online MixMag survey, individuals were asked about their utilization of prehospital and hospital emergency services after the use of recreational drugs. This study included 2474 respondents from the global survey who were United Kingdom based; 462 reported accessing an emergency medical service after the use of recreational drugs. A total of 259 (57%) individuals admitted to using the hospital emergency department, 196 (26%) had used the ambulance services,

65 (11%) had accessed NHS Direct (the UK telephone medical assistance for members of the public), and 120 (26%) had accessed other first aid facilities at the club/festival or other venue. Some individuals had previously used more than one type of emergency service. Of those who had accessed emergency medical services previously, 115 provided details of their most recent access episode, including the drug(s) used before becoming unwell; ketamine was the second most common self-reported drug used after ethanol (ethanol, 89 [77%]; ketamine, 26 [23%]) (Archer et al. 2013).

In another study recruiting participants over a 5-month period from September 2007 to January 2008, 173 participants were found to be utilizing the medical facility at a nightclub (Wood et al. 2009). The majority (131 [75.7%]) of the presentations were due to effects of recreational drugs, while 30 (17.3%) were due to recreational drugs with ethanol; the remainder were not due to alcohol or drugs. Ketamine was found to be the second most common drug (62 [37.9%] after gamma-hydroxybutyrate [GHB], 107 [66.5%]) that was associated with an individual becoming unwell in the nightclub environment.

5.5.1.6   Emergency department data

Ng et al. (2010) published a retrospective case series of 233 ketamine-related visits to 15 emergency departments in Hong Kong from July 1, 2005, to June 30, 2008. Individuals were included if they either self-reported the use of ketamine or had a urine test positive for ketamine. The majority of patients (152 [65%]) had used ketamine as a single drug while the remaining 35% had coingested ketamine with other drugs (data represented in Section 5.2). The symptoms reported by patients included abdominal pain in 49 (21%), nausea and vomiting in 23 (12%), dysuria or urinary frequency in 20 (8%), and cardiovascular symptoms of either palpitations or chest pain in 25 (5%). Clinical features seen on examination in the emergency department were low GCS (GCS less than 8) in 7%, tachycardia (defined as heart rate >100 bpm) in 39%, hypertension (not defined) in 40%, abdominal tenderness in 14%, and hyperthermia (not defined) in 14% (Ng et al. 2010).

In a further case series from the HKPIC database from June 2008 to July 2011, 284 cases of acute and chronic toxicity were identified and described (Yiu-Cheung 2012). Patients presenting acutely to the emergency department were classified as acute toxicity (186 cases), and those presenting to the outpatient clinics were classified as chronic toxicity (96 cases). While the above study from Ng et al. study included 233 acute cases, this study from Yiu-Cheung included both acute and chronic presentations. Among the acute presentations, 90 (48%) presented with neurological features such as drowsiness, confusion, and transient loss of consciousness; 60 (32%) presented with lower urinary tract symptoms such as frequency, dysuria, urgency; 50 (27%) presented with cardiovascular features such as hypertension (systolic >160 mm Hg and diastolic >90 mm Hg) and tachycardia (heart rate >100 bpm); 49 (26%) presented with abdominal pain; and 19 (10%) presented with features of acute psychosis (Yiu-Cheung 2012). The chronic presentations are discussed in Section 5.5.4.

Another case series reported on 116 presentations to a South London emergency department with acute toxicity related to self-reported ketamine use in 2006 ($n = 58$) and 2007 ($n = 58$) (Wood et al. 2008). Of these, 45 (38.2%) had hypertension (systolic blood pressure >140 mm Hg), 34 (29.3%) had tachycardia (heart rate ≥100 bpm), and 29 (25%) had agitation/aggression. Only 13 (11.2%) were self-reported lone use of ketamine, and therefore some of the effects may have been due to coused drugs in addition to ketamine. The coused drugs included ethanol (38.8%), GHB/GBL (47.3%), cocaine (19.0%), and MDMA (52.6%). No patients with lone ketamine use were admitted to critical care (Wood et al. 2008).

5.5.1.7   Acute psychological effects
Although ketamine is still widely used as an anesthetic, it fell out of favor as a mainstream anesthetic agent in the 1970s because of its neuropsychiatric side effects such as emergence phenomena, hallucinations, and delusions (Jansen 2001a,b). It is these same effects that are favored by recreational users and led to the increase in ketamine misuse. Even at the doses that are used when ketamine is used as an analgesic (0.1–0.5 mg/kg), patients have reported adverse effects such as hallucinations, dysphoria, and nightmares (Mercadante et al. 2000; Visser and Schug 2006). A Cochrane Review published in 2005, and updated in 2012, evaluated the literature on the adverse effects associated with the use of ketamine as an analgesic for acute postoperative pain (Bell et al. 2012; Elia and Tramer 2005). This review included 37 trials involving a total of 2240 patients; neuropsychological adverse effects reported were pleasant dreams in 18.3%, hallucinations in 7.4%, and nightmares in 4%.

5.5.1.8   Human volunteer studies
Pomarol-Clotet et al. (2006) performed a study involving 15 healthy volunteers who were administered ketamine infusion intravenously to achieve a plasma ketamine concentration of 200 ng/mL, using a computerized infusion pump, and then performed a Present State Examination (PSE). They concluded that ketamine appeared to have four main effects:

1. A central nervous system depressant or intoxicating effect
    Ten volunteers receiving the ketamine infusion reported a sensation of tiredness with poor concentration and lack of clarity in thought; some reported a sensation of "feeling drunk," which seemed to be similar to the effect of any anesthetic/sedative drug.

2. Perceptual alterations

Symptoms reported by volunteers based on the PSE ratings were poor concentration (14 subjects); inefficient thinking (14 subjects); distorted perception in the visual, auditory, and somatosensory areas (13 subjects); change in perception of time (11 subjects); depersonalization (4 subjects); and derealization (4 subjects). No participants in this study reported hallucinations.

3. Referential ideas or delusions, plus other subjective changes in thinking

Ketamine-induced delusions occurred in 6 of the 15 volunteers, and these mainly were delusions of reference and misinterpretation. Some subjects also reported thought insertion (3 subjects) and poverty of speech (11 subjects).

4. Negative-type symptoms

In this study, the main negative symptoms noted were poverty of speech (11 subjects) and an increased latency of response (13 subjects).

Similar results were seen in another placebo-controlled, double-blind, randomized controlled trial by a group of psychiatrists studying the effects of subanesthetic doses of ketamine on 19 healthy volunteers (Krystal et al. 1994). Placebo or ketamine at 0.1 and 0.5 mg/kg was injected intravenously over a 40-min period for three test days and the participants were subjected to a series of behavioral and cognitive testing. The participants acted as their own controls over the 3-day period. The study found that ketamine produced

1. Symptoms similar to schizophrenia
2. Alterations to perception
3. Reduced verbal fluency and vigilance
4. Symptoms similar to dissociative states
5. Disruption in delayed word recall but spared immediate recall and postdistraction recall (Krystal et al. 1994)

5.5.1.8.1 Positron emission tomography.   Positron emission tomography (PET) studies on human subjects have shown a reduction in the dopamine receptor availability and increased prefrontal cortex metabolic activity in healthy volunteers given subanesthetic doses of ketamine. Breier administered a 0.12-mg/kg IV bolus followed by 0.65 mg/kg dose maintenance over an hour to 17 healthy volunteers. PET scans done subsequently showed increased metabolic activity in the prefrontal cortex; subjects also had an interview using the Brief Psychiatry Rating Scale before and after ketamine infusion, and it was found that ketamine induced acute psychosis (in the form of conceptual disorganization, unusual thought content, hallucinatory behavior, and suspiciousness). This difference between placebo and ketamine effect was found to be statistically significant ($P = 0.0001$) (Breier et al. 1997).

Vollenweider studied the effects of ketamine on 15 healthy volunteers who were initially given a 20-mg IV loading dose of ketamine followed by a 0.02- to 0.03-mg/kg/min IV infusion for 60 min. PET scans done later showed increased metabolic activity in the anterior cingulate and medial prefrontal cortex of the brain (Vollenweider et al. 1997). These results provide in vivo evidence of striatal increase in dopamine concentrations and confirm the acute psychosis produced by ketamine via the N-methyl-D-aspartate (NMDA) receptor.

5.5.1.9 User report surveys

5.5.1.9.1 Internet user discussion fora. Erowid and Bluelight are Internet drug fora used by drug users globally. In these fora, ketamine users describe the effects of ketamine as being on a "trip"; they describe an initial sense of defragmentation and then chaos (Erowid Ketamine Vault; The Ketamine Discussion Thread). Other effects commonly reported include dizziness, euphoria, and a "floaty dream-like state." Other effects reported include a "scary sensation of feeling numb" and "a loss of self-control," abnormal perception of music and objects, loss of fear of death, and dissociative effects such as a feeling of being lifted away from the body that some also describe as an "out of body experience." Some users mention having "experienced death" or a "near-death experience." These effects are unpleasant to some users, but others describe them as exciting experiences—they are often referred to by users as traversing through the K-hole. Some users describe this as reexperiencing old memories, seeing visions, and entering a spiritual state.

5.5.1.9.2 Data from user report surveys. Dillon et al. (2003) reported a range of psychological and behavioral effects experienced by 100 ketamine users. The effects reported in this user report survey from Australia were altered thought content (60%), a sensation of separation from the body (57%) and environment (53%), absence of time (74%), visual (49%) and auditory hallucinations (46%), euphoria (55%), confusion (45%), and excitement (46%). Fifty-six percent of participants reported one or more previous experiences of "falling into a K-hole." The "K-hole" is a term that is often used by users to describe the unpleasant hallucinations and in particular the dissociative features that are seen with high-dose ketamine use (Dalgarno and Shewan 1996; Li et al. 2011). Dillon et al. (2003) found that the risk of developing a K-hole was reported to be greater with increased use of ketamine, particularly in those who had used ketamine more than 20 times. The K-hole is a desired effect for some users, but it can be an unpleasant and even frightening effect for others and there is the potential that it can place users in a vulnerable position where they are more at risk of rape or assault. In the Dillon et al. study, 56% reported experiencing the K-hole while 45% of ketamine users feared the experience.

5.5.1.10   Emergency department data

The Ng et al. series included 233 patients attending Hong Kong emergency departments with acute ketamine toxicity (Ng et al. 2010). On evaluation, patients were found to have the following neuropsychological features: impaired conscious level in 106 (45%), dizziness in 28 (12%), agitation/irritability in 10 (4%), hallucinations/delusions in 10 (4%), seizures in 2 (8%), and other neurological symptoms in 7 (3%). In their follow-up study, they found that 19 (10%) of the 188 acute ketamine-related presentations had acute psychosis (Yiu-Cheung 2012).

5.5.1.11   Treatment of acute ketamine toxicity

Generally, individuals presenting with acute ketamine toxicity do not need pharmacological therapy. Often, all that is required, particularly for those with hallucinations and other neuropsychological effects, is observation in a quiet area with minimal auditory and visual stimulation and reassurance (Kalsi et al. 2011). Agitated and aggressive patients can potentially injure themselves and may require treatment with a benzodiazepine such as diazepam or lorazepam (Weiner et al. 2000). Emergence phenomena especially seen in the setting of ketamine used for procedural sedation can be managed with small doses of benzodiazepines and an intravenous short-acting agent such as midazolam is a good choice in this context (CEM 2009; Kalsi et al. 2011). If hallucinations, tachycardia, and hypertension do not settle within 2 to 3 h with the above treatment, then coingestion of other drugs or a medical cause should be suspected and treated accordingly (Kalsi et al. 2011; Ng et al. 2010; Weiner et al. 2000).

5.5.2   Risk of physical harm

One of the acute risks associated with ketamine use is the potential increased risk of physical injury owing to decreased awareness of surroundings. This is a direct result of the neuropsychiatric effects of ketamine, in particular its dissociative effects. Ketamine causes reduced awareness, depersonalization, and reduced perception of pain. Users report lack of coordination, temporary paralysis, and near-death experiences (Dillon et al. 2003). In addition, there is the potential that the effects of ketamine may put an individual at an increased risk of physical assault or sexual assault (Kalsi et al. 2011; Li et al. 2011). However, reports of ketamine detected in the scenario of drug-facilitated sexual assault are rare. In a UK forensic study looking at 1014 cases of drug-facilitated sexual assault, only 3 (0.5%) had ketamine detected on analysis of blood and urine specimens (Scott-Ham and Burton 2005). A similar study in Canada found only 2 (1.1%) of the 178 toxicologically tested samples positive for ketamine (Du Mont et al. 2010). This low detection rate may reflect that individuals may present early enough for ketamine to be still detectable in biological specimens.

In a study of 90 ketamine users, 13% reported being involved in an accident as a direct result of taking ketamine and 83% knew someone who had (Muetzelfeldt et al. 2008). In a longitudinal study by the same authors including 150 volunteers undertaken to study the long-term neuropsychiatric effects of ketamine, 2 deaths were reported on a 12-month follow-up among frequent ketamine users as a result of their acute ketamine use (Morgan et al. 2010). A Hong Kong study designed to evaluate the correlation between fatal vehicle crashes and consumption of alcohol or drugs among drivers between 1996 and 2000 reported that ketamine was detected in 18 (9%) of 197 fatal drug- and alcohol-related single motor collisions (Cheng et al. 2005).

5.5.3   Risk of death

Ketamine is rarely associated with fatalities when used alone, and where deaths occur, these are often caused by accidents/trauma rather than directly related to acute ketamine toxicity. Ketamine has a wide therapeutic range and the median lethal dose in animals is 100 times the average therapeutic intravenous dose (Sinner and Graf 2008). Ketamine is therefore generally considered a safe drug; however, fatalities are related to the increased risk of physical harm related to its effects.

Gill and Stajíc reviewed ketamine-positive deaths in New York City over 2 years from 1997 to 1999 and found 15 deaths owing to nonmedical use of ketamine; 12 of these were attributed to "multidrug" intoxications—opiates were found in 10 cases, amphetamines in 7, and cocaine in 6 of these as cointoxicants. Two of the 15 died because of physical injuries (one because of burns and another because of blunt injury) and another died because of sarcoidosis (Gill and Stajíc 2000).

Another report from Portugal reported a 29-year-old man who died after hanging himself while intoxicated with ketamine and ethanol. The blood concentration of ketamine and ethanol at postmortem toxicology analysis was found to be 1.3 and 0.66 mg/L, respectively (Dinis-Oliveira et al. 2010).

Karl Jansen reported in his book a case of a 34-year-old male from California who was found dead in his bathtub in 1997; his death was attributed to the use of ketamine (Erowid Ketamine Vault; Jansen 2001a,b).

5.5.3.1   More recent statistics on fatalities related to ketamine

The UK National Programme on Substance Abuse Deaths (np-SAD) system collects information from coroners (medical examiners) on drug-related deaths. In a recent study looking at data in the np-SAD database, 93 deaths were identified as involving ketamine for the period 1993 to 2013 (Ghodse et al. 2013). Ketamine-related deaths from medical use were excluded. Of these 93 deaths, 86% were male and the mean age at death was 30.9 (range, 15.8–60.6) years. The majority were employed (60%), had a history of previous drug misuse (75%), and lived with others (54%). Drowning occurred in nine instances and there were three road traffic accidents where use of ketamine may have impaired judgement. In 23 of the 93 cases, ketamine was the sole substance found during postmortem toxicological analysis; however, on the basis of the data included in the paper, it is not possible to be certain whether ketamine was the cause of death. In the remaining 70 cases, ketamine was detected with alcohol or other drugs and the cause of death was reported as "accidental poisoning not solely relating to ketamine."

The UK Office for National Statistics Report for 2012 found 3 deaths with only ketamine mentioned on the death certificate and 12 deaths with any mention of ketamine on the death certificate. It is not clear from these data whether ketamine was responsible for the deaths. Approximately half of these deaths involved people under the age of 30 (UK Office for National Statistics 2012).

It is clear from the above data that ketamine is linked with a low mortality risk among users and that the cause of death is mostly attributed to trauma and risky behavior rather than direct acute ketamine toxicity.

5.5.4   Chronic effects of ketamine

Numerous animal studies (see Chapters 8, 10, and 14) have demonstrated that chronic ketamine administration associated with chronic toxicity in a range of organs. Chronic ketamine use predominantly results in chronic toxicity that affects the renal and lower urinary tract, gastrointestinal and hepatobiliary systems, and neurocognitive function.

5.5.4.1   Effects on kidneys, bladder, and urinary tract

Over the last few years, there has been increasing awareness of the renal, bladder, and urinary tract problems caused by chronic ketamine use. The animal models that have investigated and confirmed that ketamine causes these effects are summarized in Chapter 10. Several case series and reports linking bladder symptoms to chronic ketamine use have been published (Chu et al. 2008; Cottrell et al. 2008; Shahani et al. 2007). The effects involve the bladder, the whole urinary tract, and kidneys (Chu et al. 2008; Cottrell et al. 2008; Shahani et al. 2007). There have been no studies that have investigated the incidence of ketamine-related urinary tract problems. However, recent UK studies of ketamine users have provided some information not only on the patterns of symptoms experienced by users but also on the time course and likely dose-dependent nature of these effects.

5.5.4.1.1 Data on ketamine-related bladder effects from ketamine user surveys. The first of the studies using survey data to investigate urinary tract problems in ketamine users was a follow-up to the 2009/2010 MixMag study in the United Kingdom (Winstock et al. 2012). This study asked about the presence of urinary symptoms in 1285 ketamine users from the MixMag survey. Of the 1285 ketamine users, 340 (26.5%) reported experiencing urinary symptoms, which included pain in the lower abdomen (145 [11.3%]), burning or stinging when passing urine (104 [8.1%]), urinary frequency (224 [17.4%]), urinary incontinence (43 [3.3%]), or blood in the urine (19 [1.5%]). Another UK study that included 90 participants (30 frequent [more than four times per week] ketamine users, 30 infrequent [two times per month to four times per week] ketamine users, and 30 ex-ketamine users) also reported on the prevalence of urinary tract symptoms (Muetzelfeldt et al. 2008). Twenty percent of the frequent ketamine users reported "cystitis" or bladder problems as compared to 6.7% of infrequent users and 13.3% of ex-ketamine users (Muetzelfeldt et al. 2008).

The most recent study, also from the United Kingdom, was a questionnaire survey of 48 ketamine users in Bristol (Weinstock et al. 2013); 36 (75%) of these individuals reported urological symptoms.

Data from these studies have also demonstrated that ketamine-related urinary tract problems appear to be related to the amount (dose), frequency, and duration of ketamine use (Chu et al. 2008; Cottrell and Gillat 2008; Cottrell et al. 2008; Shahani et al. 2007). In the 2010 MixMag follow-up survey (Winstock et al. 2012), those that were using doses of ketamine of more than 1 g during a typical session were more likely to develop urological symptoms. In the Bristol study, all eight ketamine users who were taking more than 5 g ketamine per day had urological symptoms; however, only 10 (50%) of those reporting that they were using less than 2 g of ketamine per day had urological symptoms (Weinstock et al. 2013).

A number of reports suggest that ketamine cessation is associated with a decrease and potential reversal of symptoms in some users. In one study of 251 users who reported urinary symptoms after ketamine use, 128 (51%) reported that abstinence form ketamine had improved their symptoms, 8 (3.8%) reported ongoing deterioration, and 108 (43%) showed no improvement while they had continued to use ketamine (Weinstock et al. 2013). Cottrell et al. (2008) reported that only a third of patients with ketamine-related vesiculopathy improved after cessation of ketamine use (Cottrell et al. 2008).

5.5.4.1.2   Urinary tract effects associated with analgesic use of ketamine.
There are five cases of significant bladder effects from therapeutic use of
ketamine as analgesic reported in the literature. Storr and Quibell reported
three patients in palliative care practice who were treated with ketamine
for its analgesic properties; the doses used in these patients were in the
range of 200–600 mg in a 24-h period. Two cases developed debilitating
urinary tract symptoms including suprapubic pain, frequency, hematuria,
and nocturia 5 months after initiation of ketamine; the time of onset is
unclear in the third. The symptoms resolved in 3 weeks' time in one case
and persisted but decreased in the other two patients after cessation of
ketamine therapy (Storr and Quibell 2009). Gregoire et al. reported a case
of a 16-year-old with complex regional pain syndrome who developed
clinical symptoms suggestive of ketamine cystitis (dysuria, frequency,
urgency, and incontinence) 9 days after treatment with oral ketamine was
initiated at a dose of 8 mg/kg; the symptoms subsided when dose was
subsequently reduced to 5 mg/kg and then 2 mg/kg (Gregoire et al. 2008).
A recent report described a 52-year-old man who was prescribed 240 mg/
day of ketamine for chronic back pain and who developed severe bladder
damage. These effects did not improve with ketamine cessation and the
patient had to have a cystectomy (Shahzad and Svec 2012).
    It is worth noting that doses used in pain and palliative care practice
are generally lower than the doses used in recreational ketamine users,
and despite this, bladder effects and debilitating urinary symptoms have
been reported.

*5.5.4.1.3   Ketamine-related urinary tract damage in recreational ketamine
users.*   Ketamine-associated ulcerative cystitis in recreational/dependent
ketamine users was first described by Shahani et al. in 2007 (Shahani et
al. 2007). Their case series included nine patients who were daily users of
ketamine; their index case was a 28-year-old male who presented with a
history of painful hematuria, post-micturition pain, dysuria, and urgency.
His urine cytology showed negative results and bladder biopsy confirmed
features of chronic cystitis. The subsequent eight patients presented
with similar clinical features and patterns of ketamine use; five of these
patients had bladder biopsies, which showed features of cystitis. All but
one patient improved after cessation of ketamine.
    Ketamine-induced cystitis and lower urinary tract symptoms have
been reported by several groups since this initial report (Chang et al. 2012;
Chen et al. 2012; Chiew and Yang 2009; Chu et al. 2007, 2008; Colebunders
and Van Erps 2008; Dhillon et al. 2008; García-Larrosa et al. 2012; Ho et
al. 2010; Lai et al. 2012; Lieb et al. 2012; Noorzurani et al. 2010; Tsai et al.
2008; Wu et al. 2012). Many of these reports are from the Southeast and
East Asia, reflecting the high prevalence of use of ketamine in the region,
particularly Hong Kong. Typically, these patients present with symptoms of

severe dysuria, painful hematuria, urinary urgency, urge incontinence, frequency, and nocturia. Urine cultures and microbiology are typically sterile, although individuals may subsequently develop concomitant urinary tract bacterial infections (Cottrell et al. 2008; Lai et al. 2012; Shahani et al. 2007).

The largest series of patients with ketamine-associated urinary symptoms was published in 2008 by Chu et al. who described 59 patients who presented to two urology units in Hong Kong (Chu et al. 2008). This cohort had a mean duration of ketamine use of 3.5 years. Severe lower urinary tract symptoms reported by these patients included pelvic pain, urgency, frequency, dysuria, urge incontinence, and hematuria. Frequency and volume charts showed that these patients had a functional voiding capacity of only 20–200 mL (normal volumes >600 mL) and needed to void urine every 15–90 min. This frequency of urination can be extremely disabling, particularly when combined with dysuria and pelvic pain. There are reports of ketamine users increasing their ketamine use in an effort to manage their pelvic pain—this results in a vicious cycle of worsening bladder and urinary tract damage and associated pain resulting in further increases in ketamine use.

5.5.4.1.4 Pathogenesis of ketamine bladder effects.   The changes that occur in the bladder are a direct result of ketamine and its metabolites being excreted via the kidneys into the urine (Moore et al. 2001) and the prolonged contact of these metabolites with the lining of the bladder (Chu et al. 2008; Cottrell et al. 2008). Various animal studies have shown effects of ketamine on the bladder mucosa and the muscular wall causing destruction of the lining leading to lower urinary tract symptoms (Tan et al. 2011; Yeung et al. 2009); these are discussed in more detail in Chapter 10.

The pathogenesis of "ketamine bladder" starts with the inflammation and ulceration of the bladder lining resulting in hematuria and denudation of bladder epithelium (Cottrell et al. 2008; Tsai et al. 2009). Prolonged exposure results in further inflammatory changes and scarring of the bladder, leading to collagen deposition and a reduction in the bladder volume. This causes urinary frequency, incontinence, pelvic pain, and pain on passing urine.

As demonstrated in the animal studies investigating ketamine-related urinary tract damage, this damage seen in the bladder can extend to the upper urinary tracts and the kidneys (Chu et al. 2008; Wei et al. 2013). The well-documented direct effects of ketamine on the kidney and urinary tract are seen as hemorrhagic inflammation and subsequently scarring, fibrosis, or narrowing of the urinary tract; hydronephrosis; and papillary necrosis (Chu et al. 2008; Mason et al. 2010). In the Chu et al. case series, 51% of the 59 patients had hydronephrosis (44% bilateral and 7% unilateral) and papillary necrosis was noted in 13% of cases. This was attributed to reduced bladder compliance leading to vesicoureteric reflux causing secondary hydronephrosis. This picture of obstructive uropathy causes derangement in renal function, ultimately leading to renal failure. In their series, 14% of patients had renal dysfunction with a raised creatinine (126–554 μmol/L). It is therefore essential that clinicians recognize symptoms of bladder dysfunction early and arrange for appropriate imaging and renal function testing.

5.5.4.1.5 Investigation and management of ketamine-related urinary tract pathology. In the next section, we discuss the investigational modalities used in the identification and the treatment options for ketamine-related urinary tract symptoms.

5.5.4.1.5.1 Urine culture. Urine culture commonly shows sterile pyuria with negative cultures (including cultures for acid-fast bacilli) (Chu et al. 2008; Shahani et al. 2007; Tsai et al. 2009). In the Chu et al. series, 2 of the 59 patients (3%) had subsequent positive urine cultures suggesting secondary bacterial infection owing to urinary stasis (Chu et al. 2008).

5.5.4.1.5.2 Imaging studies

Ultrasound. Ultrasound of the renal tract is one of the common investigations reported in the published case series of ketamine-related urinary tract damage. It typically shows evidence of small bladder volume and bladder wall thickening; in more severe cases, changes in the kidney such as hydronephrosis can also be seen (Chu et al. 2008; Mason et al. 2010). In the Chu et al. case series, all 59 patients had a renal ultrasound, and 30 (51%) were found to have hydronephrosis; this was unilateral in 4 (6%) and bilateral in 26 (44%). In Mason et al.'s series of 23 patients, 12 had ultrasonography and this was abnormal in 8 (67%) of these cases; abnormalities seen included thickening of the bladder wall and small bladder volume. It is important to note that four of the patients in this series developed extreme discomfort as they were unable to fill their bladder for the examination and the ultrasound had to be halted for patients to empty their bladder (Mason et al. 2010); this may limit the ability to undertake an adequate ultrasound in these patients.

Intravenous urography. This typically demonstrates a small-capacity bladder, irregular bladder outline, narrowing of the upper urinary tract, and hydronephrosis (Chiew and Yang 2009; Mason et al. 2010; Wei et al. 2013).

Computerized tomography (CT) of the kidney ureter and bladder. CT typically shows thickening of the bladder wall with perivascular inflammation and pseudo-diverticulum formation (Mason et al. 2010; Shahani et al. 2007; Wei et al. 2013). In the Mason et al. series, 12 (52%) of the 23 patients with ketamine-related urological symptoms underwent CT; abnormal findings were reported in 9 (75%) of these patients. Two also had involvement of the upper urinary tract with upper ureteric wall thickening, dilatation, and enhancement. In a recent case series reported from Taiwan, CT findings in 27 chronic ketamine users with urological symptoms who underwent CT urography between January 2006 and December 2011 were described. In this series, the most common findings were diffuse bladder wall thickening in 24 (88.9%) and small bladder volume in 18 (66.7%). The other abnormalities detected were perivesical inflammation in 12 (44.4%), hydronephrosis in 12 (44.4%), and ureteral abnormalities such as wall thickening in 9 (33.3%) (Huang et al. 2014). Hopcroft et al. (2011) reported upper urinary tract damage seen on CT in users with ketamine-related urinary symptoms, including ureteric thickening, periureteric fat stranding, ureteric strictures, and hydronephrosis.

Urodynamic and detrusor studies. Urodynamic studies can be useful, particularly in those with significant symptoms. Common findings include detrusor overactivity and decreased bladder compliance with or without vesicoureteric reflux; significant reduction in functional bladder capacity is frequently reported with reductions to 30–200 mL (normal >600 mL) (Chu et al. 2008; Middela and Pearce 2011).

In the Chu et al. case series, 47 (80%) of the 59 cases underwent video-cystometrogram, all of whom had abnormal findings such as reduced bladder volume and detrusor overactivity. The mean bladder capacity was 154.5 mL, with 51% (24) having capacities less than 100 mL. Six (13%) patients were noted to have vesicoureteric reflux secondary to the contracted bladder (Chu et al. 2008). In a similar series reported by Chang et al. (2012), 10 of the 20 patients underwent video-cystometrography and the mean bladder capacity was 70.8 ± 53.4 mL. It is worth noting that urodynamic studies can be poorly tolerated as patients with ketamine-related urological damage often report that urethral catheterization and bladder distension are uncomfortable and this can lead to having to abandon the investigations (Chang et al. 2012; Cottrell and Gillatt 2008; Wei et al. 2013).

Cystoscopy. Cystoscopy findings in individuals with ketamine-related bladder damage include a large range of findings such as a small bladder with epithelial inflammation similar to carcinoma in situ or chronic infection, a contracted bladder that bleeds on distension, an "erythematous bladder," ulceration, and hemorrhage in the lumen (Mason et al. 2010; Shahani et al. 2007; Tsai et al. 2009; Wei et al. 2013).

In Chu's case series, 42 (71%) of the 59 patients underwent cystoscopy, all of whom had structural changes to their bladder with epithelial inflammation resembling interstitial cystitis and neovascularization (Chu et al. 2008). In the Mason et al. series, 17 (74%) of the 23 patients underwent cystoscopy, which showed erythematous bladder with or without ulceration (Cottrell and Gillatt 2008). In a review on ketamine-induced vesiculopathy, 110 cases of ketamine-induced urinary tract symptoms were reviewed and the cystoscopic findings commonly seen were described as ulcerative cystitis with inflammation, which could be severe with denudation of the epithelium, and also features suggestive of eosinophilic cystitis (Middela and Pearce 2011). In the Chang et al. (2012) series, 19 (95%) of 20 patients had erythematous patches, mucosal neovascularization, and glomerulation reported on cystoscopy. They describe that in severe cases, hydrodistension of the bladder during cystoscopy resulted in diffuse mucosal oozing and ecchymosis. There are other reports of similar changes noted in the ureters causing ureteric strictures and in some cases also the presence of vesicoureteric reflux (Chang et al. 2012; Chu et al. 2008; Mason et al. 2010).

5.5.4.1.5.3   Bladder biopsy.   Histological specimens of the bladder in individuals with ketamine-related urological symptoms demonstrate chronic inflammatory changes similar to those seen in rodent studies (Wai et al. 2012; Yeung et al. 2009). The changes that have been reported range from erythema to ulceration, severe denudation of mucosa with interstitial infiltration of granulocytes (predominantly eosinophils) and mast cells, bladder tissue fibrosis without dysplasia or malignancy, loss of nerve fibers among the muscles of the bladder, and calcification (Chu et al. 2008; Shahani et al. 2007; Shahzad and Svec 2012; Wei et al. 2013).

One case series included 17 patients with ketamine-induced lower urinary tract symptoms and reported the histopathological features in greater detail. The morphological features noted were ulceration in 12 (70%), presence of eosinophils in 12 (70%), and urothelial atypia in 12 (70%). Immunohistochemistry markers such as p53, Ki67 reactivity, and cytokeratin 20 have previously been used to demonstrate urothelial atypia and carcinoma in situ; in this series, 10 patients (59%) had immunohistochemistry performed and 9 of these 10 showed high p53 immunoreactivity and 7 of the 10 had moderate to high levels of Ki67 reactivity, but all were negative for cytokeratin 20. The authors hence concluded that ketamine cystitis could mimic carcinoma in situ (Oxley et al. 2009). However, the long-term risks of cancer remain unknown.

Histopathological lesions such as ulceration of the ureteral mucosa with inflammatory infiltrate extending throughout the ureter were reported in a chronic ketamine user who had used 1 g of ketamine per day for 12 years (Hopcroft et al. 2011). Ureteric biopsies performed in later case series have shown chronic inflammatory changes similar to the bladder changes with granulation tissue formation and inflammatory infiltrates progressing to periureteric fibrosis (Chang et al. 2012; Wei et al. 2013).

5.5.4.1.6 Management of ketamine-related urinary tract effects. As described previously, ketamine causes destruction to the lower urinary tract, but it can also involve the upper urinary tracts. Several authors have described the relationship of the frequency and intensity (using >1 g in a typical session) of ketamine use to the initiation and progression of bladder symptoms (Chu et al. 2008; Storr and Quibell 2009; Tsai et al. 2009). Pelvic pain, urgency, and frequency seem to be the debilitating symptoms in ketamine cystitis and the management of this condition requires both symptom control in the form of pain management and improvement in bladder capacity along with psychological support for cessation of ketamine use.

5.5.4.1.6.1 Cessation of ketamine use. Cessation of ketamine use has been shown to result in symptomatic relief and improvement in many patients; it is likely that this will be most successful in those with early effects before the onset of fibrosis (Mason et al. 2010; Shahani et al. 2007; Tsai et al. 2009; Winstock et al. 2012). The bladder symptoms in 6 of the 11 cases in one series improved after cessation of ketamine (Tsai et al. 2009). In another series, 7 of the 59 patients with minimal damage to the bladder had recovery of symptoms after cessation or reduction in frequency of ketamine use (Chu et al. 2008). In one of the largest user surveys of 251 ketamine users with urinary symptoms, 51% stated that their symptoms had improved with cessation of ketamine use and only 8 (3.8%) reported that their symptoms worsened after cessation of use. However, the biggest hindrance to stopping the use of ketamine is the persistence of debilitating bladder pain and dysuria. Some users may resort to taking higher doses to get relief from this pain, causing further damage to the urinary tract. It is therefore important that, in addition to providing the necessary psychological and addiction/dependence treatment, an alternative analgesic regime is considered during the early phases of treatment (Wood et al. 2011). Therefore, close working between urologists, dependence services, and pain services can be helpful in the management of these patients.

In addition to ketamine cessation and adequate analgesia, other treatments for ketamine-related bladder and urinary tract damage are generally supportive/symptomatic treatments rather than curative. Anticholinergic drugs such as amitriptyline may be used to reduce urgency and frequency (Tsai et al. 2009).

Bladder protective agents such as local hyaluronic acid and oral pentosan sulfate have been tried with unclear benefits. There are reports of the use of intravesical hyaluronic acid in those with interstitial cystitis (Kallestrup et al. 2005). Hyaluronic acid is a mucopolysaccaride that acts as a protective lining to the bladder epithelium. Its use has been explored in ketamine cystitis because of its histopathological resemblance to interstitial cystitis. In one series, six patients managed with hyaluronic acid (together with ketamine cessation) had immediate relief in symptoms such as pain, frequency, and dysuria. It is not clear from the paper whether this improvement was seen at cystoscopy as well (Tsai et al. 2009). Case series from China and Taiwan have also reported symptomatic improvement in patients treated with intravesical hyaluronic acid (Chen et al. 2012; Lai et al. 2012). All patients mentioned in these series had stopped ketamine use; therefore, it is not possible to determine the degree to which the improvement relates to hyaluronic acid alone versus ketamine cessation. Oral pentosan sulfate was used in seven of nine patients who did not benefit with cessation of ketamine alone; all improved with the use of pentosan sulfate (Shahani et al. 2007). Pentosan sulfate acts by supplementing the glycosaminoglycan layer of the bladder (Chiang et al. 2000), and benefits may occur by reducing the irritative effects of ketamine on lower urinary tract symptoms.

5.5.4.1.6.2 Oral antibiotic therapy. Many studies have shown that urine cultures in patients with ketamine cystitis are sterile and therefore antibiotic therapy is unlikely to help with the management of symptoms or to have an impact on progression. However, in the Chu et al. series, 2 (3%) patients developed pyuria subsequently (Chu et al. 2008) and prompt recognition and treatment of secondary bacterial infection in those with ketamine-related urinary tract damage is important.

5.5.4.1.6.3 Hydrodistension. Cystoscopic hydrodilatation of the bladder wall under general anesthetic has been tried in 20 patients, with 14 reporting some relief in symptoms; the mechanism of its benefit in ketamine cystitis is still unclear (Chang et al. 2012).

5.5.4.1.6.4 Surgical treatment. Persistent debilitating symptoms such as severe frequency or urgency, bladder pain, and incontinence owing to reduced bladder capacity, despite abstinence and medical therapy, are all indications for consideration of surgical intervention. Surgical management may include nephrostomy (in patients with obstructive uropathy), cystectomy, and augmentation cystoplasty (Chu et al. 2008; Chung et al. 2013; Shahani et al. 2007; Tsai et al. 2009). The latter involves removal of a part of the bladder and replacement with an intestinal segment. A recent pilot study including 14 refractory cases of chronic ketamine-related bladder damage has demonstrated effective reduction in symptoms and improved results (Chung et al. 2013). All patients in this series had a contracted bladder, 9 had hydronephrosis, and 8 had vesicoureteric reflux; after the procedure, there was increased bladder capacity in all 14 patients, hydronephrosis improved in all 9 patients, and vesicoureteric reflux improved in 5 of the 8 patients. Though surgery may be beneficial, it has to be borne in mind that this is radical surgery and cystectomy or insertion of an intestinal segment (particularly in young individuals) has long-term consequences on lifestyle and self-esteem. In addition to the cosmetic concerns, complications include renal problems, continence issues, and the unknown risks of cancer. In those treated with an intestinal segment, the effects of ketamine on the intestinal mucosa are unknown and so ketamine cessation is important.

In summary, the mainstay of the management of ketamine-related urinary tract damage is early recognition and cessation of ketamine use. This requires increased awareness among clinicians and also a multidisciplinary approach to care including experts in dependence management, urologists, social workers, pain specialists, and assistance from drug rehabilitation specialists (Morgan et al. 2012; Shahani et al. 2007; Wei et al. 2013).

5.5.4.2   Ketamine-related chronic effects on the gastrointestinal tract and liver

Abdominal cramps is one of the most common symptoms reported by ketamine users; this is often referred to as "K-cramps." One study that interviewed 90 ketamine users found that "stomach problems" (including K-cramps) were reported by 33.3% of frequent users (ketamine usage four times or more per week), 3.3% of infrequent users (ketamine use less than four times a week but at least once a month), and 6.7% of ex-users (Muetzelfeldt et al. 2008). In a series of 233 patients presenting to the emergency department after ketamine use, 21% complained of abdominal pain (Ng et al. 2010). Another series from Hong Kong following the case series by Ng et al. reported on 188 acute cases and 96 chronic cases. Twenty-six percent (49) of acute cases presented with abdominal pain and 66% (63) of chronic cases presented with chronic abdominal pain (Yiu-Cheung 2012). The cause of this abdominal pain and relationship with other features in these reports are not clear.

In a single series looking at chronic abdominal effects from ketamine, Poon et al. identified 37 cases retrospectively between 2001 and 2008; 28 (75.7%) had upper gastrointestinal symptoms, 23 (62.2%) had epigastric pain, 4 (10.8%) had epigastric pain with vomiting, and 1 (2.7%) presented with anemia. Of these 37 patients, 14 had endoscopy, and of these, 12 (85.7%) had endoscopic evidence of gastritis, 1 (7.1%) had gastroduodenitis, and 1 (7.1%) had a normal endoscopy. The gastritis was found to be *Helicobacter pylori* negative (Poon et al. 2010).

In a case series from Hong Kong including 233 patients presenting to the emergency department with ketamine toxicity, 29 (16%) had abnormal liver function. This was noted in the form of raised alkaline phosphatase (ALP) in 20 patients (100–508 IU/L; reference range, 35–98 IU/L) or raised alanine transaminase (ALT) in 16 patients (52–180 IU/L; reference, <41 IU/L). Ng et al. (2009) followed up two patients who had further investigations in the form of ultrasound and CT; both the patients were found to have dilated common bile ducts. Another report from Hong Kong described three individuals with chronic ketamine use who developed recurrent upper abdominal pain and were found to have abnormal liver function in the form of raised ALP and ALT results. CT showed dilatation of the common bile duct in all three; the findings were described as being similar in appearance to choledochal cysts. This was later confirmed by cholangiopancreatography in two cases (Wong et al. 2009). It has been postulated that ketamine-related upper abdominal pain could be caused by injury to the hepatobiliary system owing to the direct hepatotoxic effect of ketamine (Lee et al. 2009) and the biliary tree dilatation could be related to sphincter of Oddi dysfunction (Wong et al. 2009).

Abnormal liver function has also been reported in relation to therapeutic use of ketamine for chronic pain and as an anesthetic. In one study of patients in whom ketamine was used as an anesthetic for minor procedures, 14 (41%) of 34 patients developed deranged liver enzymes (Dundee et al. 1980). In another study, 100-h ketamine infusions were used as part of the treatment regimen for chronic regional pain syndrome in six patients; three of these patients developed elevated levels of ALP, ALT, aspartate transaminase, and gamma glutamyl transferase, suggestive of liver injury; this settled over a 2-month period once ketamine therapy was withdrawn (Noppers et al. 2011).

Experimental studies on hepatoma G2 cells has demonstrated that exposure to ketamine increases DNA fragmentation and cellular apoptosis (Lee et al. 2009). This suggests that hepatic dysfunction noted with chronic ketamine use could be directly related to its hepatotoxic effects in addition to effects on the biliary tree.

5.5.4.2.1 Management of ketamine-related abdominal effects. All of the reports of ketamine-related derangement of liver function have shown improvement in liver function after cessation of ketamine use (Dundee et al. 1980; Noppers et al. 2011). In the Wong et al. case series, there was improvement in liver function in all three patients with ketamine-related liver dysfunction; one of the three patients restarted ketamine use, which resulted in recurrence of epigastric pain with elevated ALP, indicating a causal relation to ketamine. It appears that ketamine-related biliary duct dilatation may improve with ketamine cessation, but some patients may require biliary stenting or other procedures (Wong et al. 2009).

5.5.4.3   Chronic ketamine-related neuropsychiatric effects

Acute ketamine toxicity predominantly results in effects on the central nervous system and cognitive features and hence it would be plausible to expect neurocognitive changes related to chronic ketamine use.

5.5.4.3.1 Depression. A 1-year longitudinal study of 150 individuals looked at the relationship between ketamine use and depression. Participants were divided into five groups of 30 each: frequent users (ketamine use four times or more per week), infrequent users (use at least four times a month), ex-ketamine users (abstinent from ketamine use for at least a month), polysubstance users, and non–drug users (Morgan et al. 2010). At 1 year follow-up, increased depression was noted in ex-ketamine users and frequent ketamine users as assessed by the Beck Depression Inventory (frequent user: initial score, 7.08 ± 5.27; at follow-up, 10.67 ± 7.99; ex-user: initial score, 4.96 ± 3.52; at follow-up, 9.00 ± 8.48). However, a review by the same authors a year later mentioned that this depression was not clinically significant (Curran and Morgan 2011). No changes were seen in current infrequent ketamine users of the drug.

5.5.4.3.2 Psychosis. It has been established that acute ketamine toxicity is associated with delusions, hallucinations, and other positive and some negative symptoms of schizophrenia (Krystal et al. 1994; Pomarol-Clotet et al. 2006). A parallel design study compared 20 ketamine users to 19 nonketamine polydrug users on day 0 and again on day 3. The mean ketamine dose used on day 0 was 0.14 (standard deviation, 0.16 g), and the participants abstained from ketamine use in the next 3 days (Curran and Morgan 2000). Results showed that ketamine users had higher dissociation scores (mean score of 44.00 in ketamine users vs. 5.47 in controls) and schizotypal symptomatology scores (mean score of 87.4 in ketamine users vs. 33.37 in controls) with poor immediate recall on day 0 as compared with controls. On day 3, despite not using ketamine, the ketamine user group showed persistently higher scores for dissociative (mean score of 17.75 in ketamine users vs. 5.37 in controls) and schizotypal symptomatology (mean score of 62.35 in ketamine users vs. 26.42 in controls) such as depersonalization, derealization, and amnesia along with poor immediate recall as compared to controls. Though this difference was less than when tested on day 0, it remained significant. The authors commented that this could be attributed to the chronic effects of the drug, impairing cognitive and memory functions. The same authors did a follow-up study 3 years later, this time with 18 lone ketamine users and 10 controls (nonketamine polydrug users) from the previous study. The ketamine users had reduced their usage by an average of 88.3%, using the drug less frequently than 3 years previously. There was a decrease in schizotypal scores over the 3-year period among both groups, but the ketamine user group still had higher scores than controls (mean schizotypal symptom score based on a questionnaire; ketamine users: baseline, 51.5; 3 years later, 18.67; controls: 24.8 and 4.0) (Morgan et al. 2004c).

The same group undertook a longitudinal 1-year study in the United Kingdom assessing psychological well-being in ketamine users (Morgan et al. 2010). In this study, dissociative scores and mild delusional ideation were shown to be highest among frequent ketamine users followed by infrequent users and then ex-users, suggesting a dose-related effect. Over the time course of the study, schizotypal symptom scores decreased in infrequent users and polydrug controls but not in frequent users.

These studies suggest that psychotic symptoms can occur in chronic ketamine users especially frequent, daily users; however, these features are subclinical, and with the current level of evidence, chronic ketamine use cannot be linked to the diagnosis of a psychotic disorder.

5.5.4.3.3 Cognitive impairment. The NMDA receptor has a central role in brain functions of learning and memory involving a form of synaptic modification called long-term potentiation (Newcomer et al. 2000); the principal action of ketamine is on the NMDA receptor and there has therefore been an interest in the potential for ketamine to affect memory.

In healthy volunteers, a single dose of ketamine has been shown to result in impairment in cognitive function, which has been described in detail in Section 5.5.1 (Krystal et al. 1994). Similarly, impairments to working and episodic memory have also been shown in a double-blind placebo-controlled study in 54 healthy volunteers. In this study, ketamine at doses of 0.4 or 0.8 mg/kg or vehicle placebo (0.9% sodium chloride saline) was injected intravenously for 80 min. After this, volunteers were subjected to a range of memory tests. The participants showed impaired working and episodic memory with increasing dose and impaired semantic memory regardless of the dose (Morgan et al. 2004a,b).

In another case–control study, Morgan et al. (2010) studied 150 individuals and compared frequent ketamine users who used ketamine more than four times per week (30), infrequent ketamine users who used the drug less than four times per week but at least once a month (30), ex-ketamine users who were abstinent for at least 1 month (30), polydrug users (30), and non–drug users (30). These groups were subjected to a range of neurocognitive and psychological well-being assessments at the start and 1 year later. The frequent ketamine user group had impaired working memory, episodic memory, and executive functioning. The frequent users when followed up for a year showed a decline in cognitive functioning in the domains of spatial working memory, pattern recognition memory, and verbal recognition memory. In users who had increased their ketamine use during this time, there was poorer performance on the spatial memory tasks demonstrating a dose–response effect of ketamine. Frequent ketamine users were also found to have impaired short- and long-term memory (Morgan and Curran 2006; Morgan et al. 2010).

The same study examining cognitive function in infrequent ketamine users showed no long-term cognitive impairment in infrequent users of ketamine. Ex-users showed neither an improvement nor a decline in their cognitive function (Morgan and Curran 2006; Morgan et al. 2010). A similar study comparing 18 frequent (defined as ketamine use more than twice a month) and 19 infrequent (defined as ketamine use twice or less in a month) users of ketamine on day 0 (on the day of drug use) and on day 3 (no ketamine use for 3 days) showed that regular users had significantly impaired episodic and semantic memory as compared to infrequent users who were abstinent for 3 days (Curran and Monaghan 2001).

Another study compared 20 ketamine users with 20 polydrug users at day 0 after acute intake of ketamine and 3 days after being abstinent and performed tests of source memory tasks (Morgan et al. 2004d). They found that ketamine users had persistent impairment in source memory; this study, when compared with the double-blind study involving intravenous ketamine described in more detail above, found that there was no impairment in memory when participants were followed up on day 3 after ketamine use (Morgan et al. 2004a). This confirms that ketamine affects memory because of its chronic effects on cognition and not because of an acute effect.

5.5.4.3.4 Chronic neurological changes related to ketamine use. Over the last decade, more data on the long-term neurological changes caused by ketamine use have emerged. One study recruited 12 chronic ketamine users and performed PET using a dopamine receptor radiotracer and demonstrated increased binding of D_1 receptors in the right dorsolateral prefrontal cortex, suggesting up-regulation of dopaminergic receptors (Narendran et al. 2005). The prefrontal cortex is involved in executive functioning of the brain and deranged dopamine neurotransmission is likely to cause impaired cognition (Moghaddam et al. 1997).

In a Chinese study comparing 44 healthy human volunteers with 41 chronic ketamine users, brain magnetic resonance imaging (MRI) scanning showed that, in chronic ketamine users, there were bilateral frontal and left temporoparietal region abnormalities of the white matter. The findings were dose dependent and provide a microstructural basis for the changes in cognition reported with prolonged ketamine use (Liao et al. 2010). These findings have been further confirmed in a recent study where 16 ketamine users (ketamine use three or more times per week) were compared with 16 polydrug user controls. Microstructural white matter abnormalities were seen on MRI scanning in the right hemisphere mostly predominantly underlying the prefrontal cortex of ketamine users (Edward Roberts et al. 2014).

References

Archer, J.R.H., Dargan, P.I., Wood, D.M. and Winstock, A.R. 2013. Hospital and prehospital emergency service utilisation as an impact of acute recreational drug and ethanol toxicity. *J Subst Use* 18:129–37.

Asia & Pacific Amphetamine-Type Stimulants Information Centre, APAIC. Available at: http://www.apaic.org (accessed on June 11, 2014).

Baduni, N., Sanwal, A.R., Jain, A. and Kachru, N. 2010. Recurrent episodes of intractable laryngospasm followed by laryngeal and pulmonary oedema during dissociative anaesthesia with intravenous ketamine. *Indian J Anaesth* 54:364–5.

Bell, R.F., Eccleston, C. and Kalso, E.A. 2012. Ketamine as an adjuvant to opioids for cancer pain. *Cochrane Database Syst Rev* 14(11):CD003351.

Bettschart-Wolfensberger, R., Bowen, I.M., Freeman, S.L., Weller, R. and Clarke, K.W. 2003. Medetomidine-ketamine anaesthesia induction followed by medetomidine-propofol in ponies: Infusion rates and cardiopulmonary side effects. *Equine Vet J* 35:308–13.

Breier, A., Malhotra, A.K., Pinals, D.A., Weisenfeld, N.I. and Pickar, D. 1997. Association of ketamine-induced psychosis with focal activation of the prefrontal cortex in healthy volunteers. *Am J Psych* 154:805–11.

Canadian Alcohol and Drug Use Monitoring Survey. 2012. Available at: http://www.hc-sc.gc.ca/hc-ps/drugs-drogues/stat/_2012/tables-tableaux-eng (accessed on June 11, 2014).

CEM (College of Emergency Medicine). 2009. Clinical Effectiveness Committee: Guidelines for ketamine sedation in children in Emergency Departments. Available at: https://secure.collemergencymed.ac.uk/code/document.asp?ID=4880.

Central Registry of Drug Abuse Sixty Second Report. 2013. Drug Abuse Trends for 2003–2012. Available at: http://www.nd.gov.hk/en/crda_62nd_report.htm (accessed on June 11, 2014).

Chan, Y.C., Tse, M.L. and Lau, F.L. 2012. Hong Kong Poison Information Centre: Annual report 2010. *Hong Kong J Emerg Med* 19:110–20.

Chang, T., Lin, C.C., Lin, A.T., Fan, Y.H. and Chen, K.K. 2012. Ketamine-induced uropathy: A new clinical entity causing lower urinary tract symptoms. *LUTS* 2(4):19–24.

Chen, Y.C., Chen, Y.L., Huang, G.S. and Wu, C.J. 2012. Ketamine-associated vesicopathy. *QJM* 105:1023–4.

Cheng, J.Y., Chan, K.K. and Mok, V.K. 2005. An epidemiological study on alcohol/drugs related fatal traffic crash cases of deceased drivers in Hong Kong between 1996–2000. *Forensic Sci Int* 153:196–201.

Chiang, G., Patra, P., Letourneau, R. et al. 2000. Pentosanpolysulfate inhibits mast cell histamine secretion and intracellular calcium ion levels: An alternative explanation of its beneficial effect in interstitial cystitis. *J Urol* 164:2119–25.

Chiew, Y.W. and Yang, S. 2009. Disabling frequent urination in a young adult. Ketamine-associated ulcerative cystitis. *Kidney Int* 76:123–4.

Chu, P.S., Kwok, S.C., Lam, K.M. et al. 2007. 'Street ketamine'–associated bladder dysfunction: A report of ten cases. *Hong Kong Med J* 13:311–3.

Chu, P.S., Ma, W.K., Wong, S.C. et al. 2008. The destruction of the lower urinary tract by ketamine abuse: A new syndrome? *Br J Urol Int* 102:1616–22.

Chung, S.D., Wang, C.C. and Kuo, H.C. 2013. Augmentation enterocystoplasty is effective in relieving refractory ketamine-related bladder pain. *Neurourol Urodyn* 10:1002.

Colebunders, B. and Van Erps, P. 2008. Cystitis due to the use of ketamine as a recreational drug: A case report. *J Med Case Rep* 2:219.

Copeland, J. and Dillon, P. 2005. Health and psycho social consequences of ketamine use. *Int J Drug Policy* 16:122–31.

Corssen, G. and Domino, E.F. 1966. Dissociative anaesthesia: Further pharmaco-
     logic studies and first clinical experience with the phencyclidine derivative
     CI-581. *Anesth Analg* 45:29–40.
Cottrell, A.K., Warren, R., Ayres, P., Weinstock, P., Kumar, V. and Gillatt, D. 2008.
     The destruction of the lower urinary tract by ketamine abuse: A new syn-
     drome? *Br J Urol Int* 102:1178–9.
Cottrell, A.M. and Gillatt, D.A. 2008. Ketamine-associated urinary tract pathology:
     The tip of the iceberg for urologists. *Br J Med Surg Urol* 1:136–8.
CSEW (Crime Survey for England and Wales). 2012. Drug Misuse Declared:
     Findings from the 2011 to 2012 (Second Edition). Available at: https://www
     .gov.uk/government/publications/drug-misuse-declared-findings-from
     -the-2011-to-2012-crime-survey-for-england-and-wales-csew-second-edition
     /drug-misuse-declared-findings-from-the-2011-to-2012-crime-survey-for
     -england-and-wales-csew-second-edition (accessed on June 11, 2014).
Cui, Y., Nie, H., Ma, H. et al. 2011. Ketamine inhibits lung fluid clearance through
     reducing alveolar sodium transport. *J Biomed Biotech* 460596:6.
Curran, H.V. and Monaghan, L. 2001. In and out of the K-hole: A comparison of the
     acute and residual effects of ketamine in frequent and infrequent ketamine
     users. *Addiction* 96:749–60.
Curran, H.V. and Morgan, C. 2000. Cognitive, dissociative and psychotogenic
     effects of ketamine in recreational users on the night of drug use and 3 days
     later. *Addiction* 95:575–90.
Curran, H.V. and Morgan, C. 2011. Ketamine use a review. *Addiction* 107:27–38.
Dalgarno, P.J. and Shewan, D. 1996. Illicit use of ketamine in Scotland. *J Psychoactive
     Drugs* 28:191–9.
Dhillon, B.S., Nuttall, M.C., Coull, N. and O'Brien, T.S. 2008. Minerva. *Br Med J*
     336:898.
Dillon, P., Copeland, J. and Jansen, K.L.R. 2003. Patterns of use and harms associ-
     ated with non-medical ketamine use. *Drug Alcohol Dep* 69:23–8.
Dinis-Oliveira, R.J., Carvalho, F., Duarte, J.A., Dias, R., Magalhães, T. and Santos,
     A. 2010. Suicide by hanging under the influence of ketamine and ethanol.
     Forensic Sci Int 202:23–7.
Du Mont, J., Macdonald, S. and Rotbard, N. 2010. Drug-facilitated sexual assault
     in Ontario, Canada: Toxicological and DNA findings. *J Forensic Leg Med*
     17:333–8.
Dundee, J.W., Fee, J.P., Moore, J., McIlroy, P.D. and Wilson, D.B. 1980. Changes in
     serum enzyme levels following ketamine infusions. *Anaesthesia* 35:12–6.
Edward Roberts, R., Curran, H.V., Friston, K.J. and Morgan, C.J. 2014. Abnormalities
     in white matter microstructure associated with chronic ketamine use.
     Neuropsychopharmacology 39:329–38.
Elia, N. and Tramer, M.R. 2005. Ketamine and post-operative pain—A quantitative
     systemic review of randomised controlled trials. *Pain* 113:61–70.
Erowid Ketamine Vault. Available at: https://erowid.org/chemicals/ketaine/keta
     mine.shtml (accessed on June 11, 2014).
García-Larrosa, A., Castillo, C., Ventura, M., Lorente, J.A., Bielsa, O. and Arango,
     O. 2012. Cystitis and ketamine associated bladder dysfunction. *Actas Urol
     Esp.* 36(1):60–4.
Ghodse, H., Corkery, J., Oyefeso, A., Schifano, F., Tonia, T. and Annan, J. 2013. Drug-
     Related Deaths in the UK National Programme on Substance Abuse Deaths.
     Available at: http://www.sgul.ac.uk/research/projects/icdp/our-work

-programmes/pdfs/National%20Programme%20on%20Substance%20
Abuse%20Deaths%20-%20Annual%20Report%202013%20on%20Drug
-related%20Deaths%20in%20the%20UK%20January-December%202012%20
PDF.pdf (accessed on June 11, 2014).

Gill, J.R. and Stajíc, M.J. 2000. Ketamine in non-hospital and hospital deaths in
New York City. *For Sci* 45:655–8.

Green, S.M., Rothrock, S.G., Lynch, E.L. et al. 1998. Intramuscular ketamine for
pediatric sedation in the emergency department: Safety profile in 1,022 cases.
Ann Emerg Med 31:688–97.

Gregoire, M.C., MacLellan, D.L. and Finley, G.A. 2008. A pediatric case of
ketamine-associated cystitis. *Urology* 71:1232–3.

Haas, D.A. and Harper, D.G. 1992. Ketamine: A review of its pharmacologic properties and use in ambulatory anesthesia. *Anesth Prog* 39:61–6.

Hemmingsen, C., Nielsen, P.K. and Odorico, J. 1994. Ketamine in the treatment
of bronchospasm during mechanical ventilation. *Am J Emerg Med* 12:417–20.

Ho, C.C., Pezhman, H., Praveen, S. et al. 2010. Ketamine-associated ulcerative cystitis: A case report and literature review. *Malays J Med Sci.* 17(2):61–5.

Home Office National Statistics on Drug Misuse. 2012. Available at: https://www
.gov.uk/government/statistics/drug-misuse-findings-from-the-2012-to-
2013-csew (accessed on June 11, 2014).

Home Office Statistical Bulletin. 2010. Crime in England and Wales 2008/09
Volume 1. Findings from the British Crime Survey and Police Recorded Crime.
Home Office, UK. Available at: https://www.gov.uk/government/publica
tions/drug-misuse-declared-findings-from-the-2009-10-british-crime-survey
-england-and-wales (accessed on June 11, 2014).

Home Office Statistical Bulletin. 2012. Available at: https: www.gov.uk/government
/statistics/drug-misuse-findings-from-the-2012-to-2013-csew   (accessed on
June 11, 2014).

Hopcroft, S.A., Cottrell, A.M., Mason, K., Abrams, P. and Oxley, J.D. 2011. Ureteric
intestinal metaplasia in association with chronic recreational ketamine abuse.
J Clin Pathol 64:551–2.

Huang, L.K., Wang, J.H., Shen, S.H., Lin, A.T. and Chang, C.Y. 2014. Evaluation of
the extent of ketamine-induced uropathy: The role of CT urography. *Postgrad
Med J* 90:185–90.

Jacobson, J.D. and Hartsfield, S.M. 1993. Cardiorespiratory effects of intravenous
bolus administration and infusion of ketamine-midazolam in dogs. *Am J Vet
Res* 54:1710–4.

Jansen, K.L.R. 1993. Non–medical use of ketamine. *Br Med J* 306:601–2.

Jansen, K.L.R. 2001a. A review of the non-medical use of ketamine: Part 1: Use,
users and consequences. *J Psychoactive Drugs* 32:419–33.

Jansen, K.L.R. 2001b. *Ketamine: Dreams and Realities*. Published by Multidisciplinary
Association for Psychedelic Studies [MAPS], Sarasota, FL.

Jennings, P.A., Cameron, P. and Bernard, S. 2011. Ketamine as an analgesic in the
pre-hospital setting: A systematic review. *Acta Anaesthesiol Scand* 55:638–43.

Jensen, A.G., Callesen, T., Hagemo, J.S. et al. 2010. Scandinavian clinical practice
guidelines on general anaesthesia for emergency situations. *Acta Anaesthesiol
Scand* 54:922–50.

Kallestrup, E.B., Jorgensen, S.S., Nording, J. and Hald, T. 2005. Treatment of interstitial cystitis with Cystistat: A hyaluronic acid product. *Scand J Urol Nephrol*
39:143–7.

Kalsi, S., Wood, D.M. and Dargan, P.I. 2011. Epidemiology and patterns of acute and chronic toxicity. *Emerg Health Threat J* 4:7107.

Krystal, J.H., Karper, L.P., Seibyl, J.P. et al. 1994. Subanesthetic effects of the non-competitive NMDA antagonist, ketamine, in humans. Psychotomimetic, perceptual, cognitive, and neuroendocrine responses. *Arch Gen Psych* 51:199–214.

Lai, Y., Wu, S., Ni, L. et al. 2012. Ketamine-associated urinary tract dysfunction: An underrecognized clinical entity. *Urol Int* 89:93–6.

Lee, S.T., Wu, T.T., Yu, P.Y. and Chen, R.M. 2009. Apoptotic insults to human HepG2 cells induced by S-(+)-ketamine occurs through activation of a Bax-mitochondria-caspase protease pathway. *Br J Anaesth* 102:80–9.

Li, J.H., Vicknasingam, B., Cheung, Y.W. et al. 2011. To use or not to use: An update on licit and illicit ketamine use. *Subst Abuse Rehab* 2:11–20.

Liao, Y., Tang, J., Ma, M. et al. 2010. Frontal white matter abnormalities following chronic ketamine use: A diffusion tensor imaging study. *Brain* 133:2115–22.

Lieb, M., Bader, M., Palm, U., Stief, C.G. and Baghai, T.C. 2012. Ketamine-induced vesicopathy. *Psychiatr Prax.* 39(1):43–5.

Mason, K., Cottrell, A.M., Corrigan, A.G., Gillatt, D.A. and Mitchelmore, A.E. 2010. Ketamine-associated lower urinary tract destruction: A new radiological challenge. *Clin Radiol* 65:795–800.

Measham, F., Moore, K. and Welch, Z. 2012. Emerging Drug Trends in Lancashire: Nightclub Surveys. Phase Three Report. Lancaster University and Lancashire Drug and Alcohol Action Team, Lancaster, pp. 1–92. Available at: http://www.ldaat.org (accessed on June 11, 2014).

Measham, F., Wood, D., Dargan, P. and Moore, K. 2011. The rise in legal highs: Prevalence and patterns in the use of illegal drugs and first and second generation 'legal highs' in south London gay dance clubs. *J Subst Use* 16:263–72.

Mellor, A.J. 2005. Anaesthesia in austere environments. *J R Army Med Corp* 151:272–6.

Mercadante, S., Arcuri, E., Tirelli, W. and Casuccio, A. 2000. Analgesic effect of intravenous ketamine in cancer patients on morphine therapy: A randomized, controlled, double-blind, crossover, double-dose study. *J Pain Symptom Manage* 20:246–52.

Middela, S. and Pearce, I. 2011. Ketamine-induced vesicopathy: A literature review *Int J Clin Pract* 65:27–30.

Miller, A.C., Jamin, C.T. and Elamin, E.M. 2011. Continuous intravenous infusion of ketamine for maintenance sedation. *Minerva Anestesiol* 77:812–20.

MixMag—Global Drug Survey Results. 2014. Available at: http://www.global drugsurvey.com/wp-content/uploads/2014/04/last-12-months-drug -prevalence.pdf (accessed on June 11, 2014).

Moghaddam, B., Adams, B., Verma, A. and Daly, D. 1997. Activation of glutama-tergic neurotransmission by ketamine: A novel step in the pathway from NMDA receptor blockade to dopaminergic and cognitive disruptions associated with the prefrontal cortex. *J Neurosci* 17:2921–7.

Monitoring the Future, National Survey Results. 2012. Available at: http://www.monitoringthefuture.org//pubs/monographs/mtf-vol2_2012.pdf (accessed on June 11, 2014).

Moore, K.A., Sklerov, J., Levine, B. and Jacobs, A.J. 2001. Urine concentrations of ketamine and norketamine following illegal consumption. *JAT* 25:583–8.

Morgan, C.J. and Curran, H.V. 2006. Acute and chronic effects of ketamine upon human memory: A review. *Psychopharmacology* 188:408–24.

Morgan, C.J., Curran, H.V. and Independent Scientific Committee on Drugs. 2012. Ketamine use: A review. *Addiction*. 107(1):27–38.

Morgan, C.J., Mofeez, A., Brandner, B., Bromley, L. and Curran, H.V. 2004a. Ketamine impairs response inhibition and is positively reinforcing in healthy volunteers: A dose response study. *Psychopharmacology* 172:298–308.

Morgan, C.J., Mofeez, A., Brandner, B., Bromley, L. and Curran, H.V. 2004b. Acute effects of ketamine on memory systems and psychotic symptoms in healthy volunteers. *Neuropsychopharmacology* 29:208–18.

Morgan, C.J., Monaghan, L. and Curran, H.V. 2004c. Beyond the K-hole: A 3-year longitudinal investigation of the cognitive and subjective effects of ketamine in recreational users who have substantially reduced their use of the drug. *Addiction* 99:1450–61.

Morgan, C.J., Ricelli, M., Maitland, C.H. and Curran, H.V. 2004d. Long-term effects of ketamine: Evidence for a persisting impairment of source memory in recreational users. *Drug Alcohol Dep* 75:301–8.

Morgan, C.J., Muetzelfeldt, L. and Curran, H.V. 2010. Consequences of chronic ketamine self-administration upon neurocognitive function and psychological wellbeing: A 1-year longitudinal study. *Addiction* 105:121–33.

Mowry, J.B., Spyker, D.A., Cantilena Jr., L.R., Bailey, J.E. and Ford, M. 2013. 2012 Annual Report of the American Association of Poison Control Centers' National Poison Data System (NPDS): 30th Annual Report. *Clin Toxicol (Phila)* 51:949–1229.

Muetzelfeldt, L., Kamboj, S.K., Rees, H., Taylor, J., Morgan, C.J. and Curran, H.V. 2008. Journey through the K-hole: Phenomenological aspects of ketamine use. *Drug Alcohol Dep* 95:219–29.

Narendran, R., Frankle, W.G., Keefe, R. et al. 2005. Altered prefrontal dopaminergic function in chronic recreational ketamine users. *Am J Psych* 162:2352–9.

National Drug Strategy Household Survey Report. 2010. The Australian Institute of Health and Welfare. Available at: http://www.aihw.gov.au/WorkArea/DownloadAsset.aspx?id=10737421314 (accessed on June 11, 2014).

National Poisons Information Service Annual Report 2012–13. 2013. Figure 6.3 Drugs Most Commonly Involved in Telephone Enquiries and TOXBASE Accesses during 2012/13. Available at: http://www.hpa.org.uk/webc/HPAwebFile/HPAweb_C/1317139808601 (accessed on June 11, 2014).

New Zealand Alcohol and Drug Use 2007/2008 Survey. 2010. Table 70 Used Ketamine for Recreational Purposes in the Last 12 Months, among Total Population Aged 16–64 Years, by Ethnic Group. Available at: http://www.health.govt.nz/system/files/documents/publications/drug-use-in-nz-v2-jan2010.pdf (accessed on June 11, 2014).

Newcomer, J.W., Farber, N.B. and Olney, J.W. 2000. NMDA receptor function, memory, and brain aging. *Dial Clin Neuro Sci* 2:219–32.

Ng, S.H., Lee, H.K., Chan, Y.C. and Lau, F.L. 2009. Dilated common bile ducts in ketamine abusers. *Hong Kong Med J* 15:157.

Ng, S.H., Tse, M.L., Ng, H.W. and Lau, F.L. 2010. Emergency department presentation of ketamine abusers in Hong Kong: A review of 233 cases. *Hong Kong Med J* 16:6–11.

Noorzurani, R., Vicknasingam, B. and Narayanan S. 2010. Illicit ketamine induced frequency of micturition in a young Malay woman. *Drug Alcohol Rev*. 29(3):334–6.

Noppers, I.M., Niesters, M., Aarts, L.P. et al. 2011. Drug-induced liver injury following a repeated course of ketamine treatment for chronic pain in CRPS type 1 patients: A report of 3 cases. *Pain* 152:2173–8.

Oxley, J.D., Cottrell, A.M., Adams, S. and Gillatt, D. 2009. Ketamine cystitis as a mimic of carcinoma in situ. *Histopathology* 55:705–8.

Pandey, C.K., Mathur, N., Singh, N. and Chandola, H.C. 2000. Fulminant pulmonary edema after intramuscular ketamine. *Can J Anaesth* 47:894–6.

Parthasarathy, S.M., Ravishankar, S., Selvarajan, S. and Anbalagan, T. 2009. Ketamine and pulmonary oedema-report of two cases. *Ind J Anaesth* 53:486–8.

Patterns and Trends of Amphetamine-Type Stimulants and Other Drugs: Asia and the Pacific. 2013. Available at: http://www.apaic.org/images/stories/publi cations/2013_regional_ats_report_web.pdf (accessed on June 11, 2014).

Petrillo, T.M., Fortenberry, J.D., Linzer, J.F. and Simon, H.K. 2001. Emergency department use of ketamine in pediatric status asthmaticus. *J Asthma* 38:657–64.

Pomarol-Clotet, E., Honey, G.D., Murray, G.K. et al. 2006. Psychological effects of ketamine in healthy volunteers. Phenomenological study. *Br J Psych* 189: 173–9.

Poon, T.L., Wong, K.F., Chan, M.Y. et al. 2010. Upper gastrointestinal problems in inhalational ketamine abusers. *J Digest Dis* 11:106–10.

RCoA/CEM (Royal College of Anaesthetists and College of Emergency Medicine) Report and Recommendations. 2012. Safe Sedation of Adults in the Emergency Department. Available at: http://secure.collemergencymed.ac.uk/code/docu ment.asp?ID=6691 (accessed on June 11, 2014).

Reich, D.L. and Silvay, G. 1989. Ketamine: An update on the first twenty-five years of clinical experience. *Can J Anaesth* 36:186–97.

SALSUS (Scottish Schools Adolescent Lifestyle and Substance Use Survey) National Report. 2011. The Scottish Government Edinburgh. Smoking, Drinking and Drug Use among 13 and 15 Year Olds in Scotland in 2010. Available at: http://www.drugmisuse.isdscotland.org/publications/local /SALSUS_2010.pdf (accessed on June 11, 2014).

Scottish Crime and Justice Survey: Drug Use. 2012. Available at: http://www .scotland.gov.uk/Publications/2014/03/9823 (accessed on June 11, 2014).

Scott-Ham, M. and Burton, F.C. 2005. Toxicological findings in cases of alleged drug-facilitated sexual assault in the United Kingdom over a 3-year period. *J Clin Forensic Med* 12:175–86.

Shahani, R., Streutker, C., Dickson, B. and Stewart, R.J. 2007. Ketamine-associated ulcerative cystitis: A new clinical entity. *Urology* 69:810–2.

Shahzad, K. and Svec, A. 2012. Analgesic ketamine use leading to cystectomy: A case report. *Br J Med Surg Urol* 5:188–91.

Sinner, B. and Graf, B.M. 2008. Ketamine. *Hand Exp Pharmacol* 182:313–33.

Smoking, Drinking and Drug Use Survey 2012. 2013. Health and Social Care Information Centre. Available at: http://www.hscic.gov.uk/catalogue /PUB11334/smok-drin-drug-youn-peop-eng-2012-repo.pdf (accessed on June 11, 2014).

Storr, T.M. and Quibell, R. 2009. Can ketamine prescribed for pain cause damage to the urinary tract? *Palliat Med* 23:670.

Strayer, R.J. and Nelson, L.S. 2008. Adverse events associated with ketamine for procedural sedation in adults. *Am J Emerg Med* 26:985–1028.

Tan, S., Chan, W.M., Wai, M.S. et al. 2011. Ketamine effects on the urogenital system—Changes in the urinary bladder and sperm motility. *Microsc Res Tech* 74(12):1192–8.

Tarnow, J. and Hess, W. 1978. Pulmonary hypertension and edema after intravenous ketamine. *Anaesthesist* 27:486–7.

The Ketamine Discussion Thread. Available at: http://www.bluelight.org/vb /threads/606919-The-Ketamine-Discussion-thread (accessed on June 11, 2014).

Tsai, J.H., Tsai, K.B. and Jang, M.Y. 2008. Ulcerative cystitis associated with ketamine. *Am J Addict*. 17(5):453.

Tsai, T.H., Cha, T.L. and Lin, C.M. 2009. Ketamine-associated bladder dysfunction. International *J Urol* 16:826–9.

Tweed, W.A., Minuck, M. and Mymin, D. 1972. Circulatory responses to ketamine anaesthesia. *Anesthesiology* 37:613–9.

UK Office for National Statistics. 2012. Available at: http://www.ons.gov.uk/ons /rel/subnational-health3/deaths-related-to-drug-poisoning/2012/stb ---deaths-related-to-drug-poisoning-2012.html (accessed on June 11, 2014).

Visser, E. and Schug, S.A. 2006. The role of ketamine in pain management. *Biomed Pharmacother* 60:341–8.

Vollenweider, F.X., Leenders, K.L., Scharfetter, C. et al. 1997. Metabolic hyperfrontality and psychopathology in the ketamine model of psychosis using positron emission tomography (PET) and [18F]fluorodeoxyglucose (FDG). *Eur Neuropsychopharmacol* 7:9–24.

Wai, M.S., Chan, W.M., Zhang, A.Q., Wu, Y. and Yew, D.T. 2012. Long-term ketamine and ketamine plus alcohol treatments produced damages in liver and kidney. *Hum Exp Toxicol* 31:877–86.

Wei, Y.B., Yang, J.R., Yang, Z. et al. 2013. Genitourinary toxicity of ketamine. *Hong Kong Med J* 19:341–8.

Weiner, A.L., Vieira, L., McKay Jr., C.A. and Bayer, M.J. 2000. Ketamine abusers presenting to the emergency department: A case series. *J Emerg Med* 18(4):447–51.

Weinstock, P., Ayres, R., Deans, K., Cottrell, A.M. and Gillatt, D. 2013. Ketamine associated ulcerative cystitis and other adverse effects reported by ketamine users in Bristol—Review and results of a questionnaire survey. Submitted for publication. *Br J Urol* 18:643–7.

Winstock, A.R., Mitcheson, L., Gillatt, D.A. and Cottrell, A.M. 2012. The prevalence and natural history of urinary symptoms among recreational ketamine users. *Br J Urol Int* 110:1762–6.

Wong, S.W., Lee, K.F., Wong, J. et al. 2009. Dilated common bile ducts mimicking choledochal cysts in ketamine abusers. *Hong Kong Med J* 15:53–6.

Wood, D.M., Bishop, C.R., Greene, S.L. and Dargan, P.I. 2008. Ketamine-related toxicology presentations to the ED. *Clin Toxicol (Phila)* 46:630.

Wood, D., Cottrell, A., Baker, S.C. et al. 2011. Recreational ketamine: From pleasure to pain. *Br J Urol Int* 107:1881–4.

Wood D.M., Hunter L., Measham F. and Dargan P.I. 2012a. Limited use of novel psychoactive substances in South London nightclubs. *QJM*. 105(10):959–64.

Wood, D.M., Measham, F. and Dargan, P.I. 2012b. 'Our favourite drug': Prevalence of use and preference for mephedrone in the London night time economy one year after control. *J Subst Use* 17:91–7.

Wood, D.M., Nicolaou, M. and Dargan, P.I. 2009. Epidemiology of recreational drug toxicity in a nightclub environment. *Subst Use Misuse* 44:1495–502.

World Drug Report. 2013. United Nations Office on Drugs and Crime. Available at: http://www.unodc.org/unodc/secured/wdr/wdr2013/World_Drug_Report _2013.pdf (accessed on June 11, 2014).

Wu, S., Lai, Y., He, Y., Li, X., Guan, Z. and Cai, Z. 2012. Lower urinary tract destruction due to ketamine: A report of 4 cases and review of literature. *J Addict Med* 6:85–8.

Yeung, L.Y., Rudd, J.A., Lam, W.P., Mak, Y.T. and Yew, D.T. 2009. Mice are prone to kidney pathology after prolonged ketamine addiction. *Toxicol Lett* 91:275–8.

Yiu-Cheung, C. 2012. Acute and chronic toxicity pattern in ketamine abusers in Hong Kong. *J Med Toxicol* 8:267–70.

chapter six

Imaging the effects of ketamine use and abuse in the brain

Qi Li, Sherry K.W. Chan, Lin Sun, Wai-Chi Chan,
Eric Y.H. Chen, and Pak C. Sham

Contents

6.1 Introduction

The use of ketamine as a recreational drug is spreading rapidly among young people across the world. Ketamine is classified as an *N*-methyl-D-aspartate (NMDA) receptor antagonist like phencyclidine (PCP) and MK801, which can produce schizophrenia-like psychosis in humans (Javitt et al. 2012). Neurochemical models based on the actions of PCP and ketamine have become progressively well established, with increased focus on glutamatergic dysfunction as a basis for both symptoms and cognitive dysfunction in schizophrenia. These same effects have contributed to it becoming used as a recreational drug. The reported negative consequences of chronic ketamine use include cognitive impairment, deterioration in psychological well-being, psychotic symptoms, bladder dysfunction, and, in some cases, death. Moreover, chronic and prolonged use of ketamine has led to damage of the nervous system, including neuronal loss, synaptic changes, white matter integrity, and formation of mutated tau protein in neurons, as described in ketamine abuse models of humans, primates, and rodents. Given that the use of ketamine as a "club drug" has increased dramatically in recent years, the side effects are of increasing global concern. In addition, accumulating evidence suggests that ketamine has rapid and potent antidepressant effects in treatment-resistant major depressive disorder and bipolar depression.

In vivo brain imaging enables the systematic examination of trait and state variables that contribute to the etiology of human diseases. There has been an explosion of interest in the development and use of brain imaging techniques for the scientific study of ketamine use. This chapter provides a critical overview of functional and structural neuroimaging studies of ketamine users/abusers in humans, primates, and rodents, including functional magnetic resonance imaging (fMRI)/pharmacological fMRI (phMRI), positron emission tomography (PET), magnetic resonance spectroscopy (MRS), and diffusion tensor imaging (DTI). The effect of short- and long-term use of ketamine on the molecular mechanisms in the central nervous system (CNS) and a comparison of ketamine use/abuse and schizophrenia as determined by imaging are also reviewed within the context of these studies.

6.2 Functional magnetic resonance imaging/pharmacological fMRI

fMRI is a functional neuroimaging procedure using MRI technology to measure brain activity by detecting associated changes in blood flow. Blood oxygen level–dependent (BOLD) imaging can be exploited as an indicator of cerebral blood flow. BOLD signal changes are assumed to reflect changes in neuronal activity with the systematic variation of a stimulus or task performance (Willson et al. 2004). Most fMRI studies have used task activation such as photic stimulation, finger movements, or a cognitive challenge to elicit neuronal activity. One approach to simplifying fMRI might be to study the functioning of the human brain during rest. Recent advances in resting-state functional connectivity neuroimaging techniques can be used to evaluate regional interactions that occur when a subject is not performing any explicit tasks (Biswal 2012). The fMRI images obtained using BOLD contrast show signal fluctuations at rest that occur at low frequencies (0.01–0.05 Hz), with coherent changes between widely separated brain regions. The characterization of changes in functional connectivity between brain networks serving to promote distinct psychophysiological functions might explain how various psychiatric symptoms arise from disrupted connectivities between distinct functional networks. The default mode network (DMN) is a network of brain regions that are active when an individual is awake and at rest; this is one of the most well-studied networks present during the resting state and is one of the most easy to visualize (Buckner 2012).

The phMRI technique is being used increasingly in both preclinical and clinical models to investigate pharmacological effects on task-free brain function. Similar to conventional fMRI, with phMRI, it is also possible to elicit neuronal activity using various pharmacological agents as stimuli or as a means of modifying the response to a cognitive challenge. However, unlike in conventional task-related fMRI studies where the time course of the stimuli can be controlled at will, in phMRI, the time course is determined by the pharmacokinetic and pharmacodynamic profile of the drug administered to induce the signal changes. phMRI experiments commonly use BOLD contrast to detect signal changes in a T2-weighted MRI time series covering both pre-compound (i.e., baseline) and post-compound administration conditions with continuous acquisition (Shah and Marsden 2004). The phMRI technique is also used to test novel drugs, targeted toward receptors in the brain to determine if they indeed hit their targets; however, this application is still used relatively little. Without the need for a radiolabel, phMRI is a very useful tool for performing drug discovery (Jenkins 2012).

The NMDA receptor hypofunction model, which is now proposed to explain schizophrenia, is supported by remarkably schizophrenia-like psychotic and cognitive symptoms that occur after administration of PCP and ketamine (De Simoni et al. 2013; Fletcher and Honey 2006; Javitt and Zukin 1991; Javitt et al. 2012). Schizophrenia is characterized by widespread cognitive impairment (Strauss 1993), in particular in the working memory, and prefrontal cortex dysfunction is a core feature of this mental disorder (Potkin et al. 2009; Thormodsen et al. 2011). A longer and less accurate reaction time, especially as the memory load increases, is one of the major working memory features in schizophrenia patients (Potkin et al. 2009). However, the brain activation patterns that are responsible for the poor performance remain controversial (Altamura et al. 2007; Karlsgodt et al. 2007; Manoach et al. 1999). Manoach et al. (1999) observed greater activation in the dorsal lateral prefrontal cortex (DLPFC) in schizophrenia patients, as well as positive correlation between activation and task performance. In addition, DLPFC activation appears to be strongly affected by memory load (Manoach et al. 1999). In healthy volunteers, the BOLD signal in the DLPFC region appears to increase with increasing memory load (Altamura et al. 2007; Jaeggi et al. 2003), although with some of the highest load level tasks, activation may decline. On the other hand, schizophrenia patients reach a peak activation of the working memory system at a lower processing load than the healthy controls (Callicott et al. 1999). Therefore, at least on certain cognitive tasks, with tasks of a low level of difficulty, schizophrenia patients may use greater prefrontal resources yet achieve lower accuracy, compared with healthy controls. With tasks at higher levels of difficulty, patients may fail to sustain the prefrontal network that processes the information, achieving an even lower accuracy as a result (Callicott et al. 2003; Cannon et al. 2005; Perlstein et al. 2001). Esslinger et al. (2007) demonstrated that in a mental maze task, activation was decreased in patients. Meanwhile, hypofrontality is frequently reported in schizophrenia, particularly in chronic schizophrenia and with negative symptoms of schizophrenia (Andreasen et al. 1992; Carter et al. 1998; Gur and Gur 1995). The reported findings of both hypo- and hyperactivation in schizophrenic patients may be related to task, load, performance relative to capacity, ability to compensate, and perhaps temporal variation over time. Thus, a clearer understanding of the influence of ketamine upon working memory and prefrontal activation is critical in assessing its validity as a model of schizophrenia. This may also support a role for glutamatergic dysfunction in the pathophysiology of schizophrenia. Most studies have demonstrated that ketamine induces symptoms in healthy individuals comparable to an acute psychotic state. For example, an increase in score of the clinician-administered dissociative states scale (CADSS), brief psychiatric rating scale (BPRS), positive and negative syndrome scale (PANSS), altered states of consciousness rating scale (5D-ASC), and visual analogue scale (VAS), which are all scales used for measuring the severity of a psychotic state, was observed in healthy volunteers with ketamine administration (Abel et al. 2003a,b; Corlett et al. 2013; De Simoni et al. 2013; Deakin et al. 2008; Doyle et al. 2013; Driesen et al. 2013a,b; Fu et al. 2005; Musso et al. 2011; Nagels et al. 2012; Scheidegger et al. 2012). Although a number of studies have indicated that ketamine does not significantly impair task performance, these studies have shown an increased ketamine-induced activation in the frontal areas involved in episodic memory (Honey et al. 2005), working memory (Honey et al. 2004), verbal fluency (Fu et al. 2005; Nagels et al. 2012), and attention tasks (Daumann et al. 2008). For example, Fu et al. (2005) found no significant deficits in task performance during verbal fluency. The absence of a significant effect of ketamine being observed for this type of task performance may have been attributed to the paced presentation of the letter cues, which required subjects to generate a single word for each cue, rather than as many words as possible such as in a classical (unpaced) verbal fluency test. Likewise, studies of patients with schizophrenia using a paced test have not documented impaired performance, while it is evident with unpaced tests (Fletcher et al. 1996; Frith et al. 1995). The hyperactivity of prefrontal areas during ketamine administration may indicate compensatory mechanisms in order to maintain constant cognitive performance. The task-specific effects of ketamine on brain activation may result in the fMRI alterations observed even though overt performance is unimpaired. Moreover, some studies have reported deficits of working memory with ketamine administration (Anticevic et al. 2012; Corlett et al. 2013; Daumann et al. 2010; Driesen et al. 2013b; Honey et al. 2008; Northoff et al. 2005). However, the findings for the prefrontal activities reported in these studies are not consistent. For example, hypofrontality is involved in episodic retrieval tasks (Northoff et al. 2005), working memory (Anticevic et al. 2012), and spatial working memory (Driesen et al. 2013b), whereas hyperfrontality is involved in target detection (Daumann et al. 2010), and hyperactivity in the striatum and thalamus plays a role in verbal working memory tasks (Honey et al. 2008). Similar to the hypo- and hyperactivation fMRI findings in schizophrenic patients, the acute effects of ketamine on brain function may also relate to task, load, performance relative to capacity, ability to compensate, and perhaps temporal variation over time. The first fMRI evidence for the long-term effects of ketamine on brain function in primates was reported by Yu et al. (2012). The results showed that repeated exposure to ketamine markedly reduced neural activities in the ventral tegmental area, the substantia nigra in the midbrain, the posterior cingulate cortex, and the visual cortex, while increased neural activities were observed in the striatum and entorhinal cortex (Yu et al. 2012).

Furthermore, resting-state research directly shows the functional connectivity in response to ketamine administration (Driesen et al. 2013a; Scheidegger et al. 2012). Scheidegger et al. (2012) reported a marked reduction of resting-state functional connectivity between functional nodes of the DMN via the dorsal nexus (DN), pregenual anterior cingulate (PACC), and medioprefrontal cortex after ketamine administration. However, another recent resting-state fMRI study showed enhanced global functional connectivity immediately after ketamine administration and after 45 min of ketamine infusion (Driesen et al. 2013a). The effect of chronic ketamine use on resting-state functional connectivity was determined using the regional homogeneity method and showed increased activity in the left precentral frontal gyrus and decreased activity in the right anterior cingulate cortex (ACC) (Liao et al. 2012) in ketamine users. Most recently, the effect of acute ketamine administration on resting-state BOLD functional connectivity was investigated in the rat. Gass et al. (2014) focused on the hippocampus–prefrontal cortex network (HC–PFC) as a key substrate of cognitive and emotional deficits in psychiatric disorders such as schizophrenia and demonstrated a dose- and exposure-dependent increase in the HC–PFC system in the rat brain with acute ketamine challenge.

The effect of ketamine on the regional BOLD signal was investigated by infusing the drug and performing a resting-state phMRI time series. Deakin et al. (2008) performed two separate experiments. In Experiment (Expt.) 1, ketamine or a placebo was given to healthy volunteers to determine the effects of this drug on the regional BOLD signal. In Expt. 2, volunteers were pretreated with lamotrigine (an inhibitor of glutamate release) before ketamine administration, to identify the regional components of ketamine treatment that are mediated by enhanced glutamate release. In Expt. 1, ketamine induced increased activity in the mid-posterior cingulate, thalamus, and temporal cortical regions, along with decreased activity in the ventromedial frontal cortex, including the orbitofrontal cortex and subgenual cingulate. Manifestations of dissociative and psychotic subjective effects, assessed by CADSS and BPRS, respectively, were found to be correlated with changes in BOLD activation, supporting the connection between NMDA receptor blockade and psychotomimetic symptoms. Expt. 2 demonstrated that a single dose of lamotrigine before ketamine administration prevented many of the BOLD signal changes induced by ketamine. Given that lamotrigine inhibits the release of glutamate, the results indicate that the ketamine-induced changes in both symptoms and BOLD signal are attributed to an increase in glutamate release (Deakin et al. 2008). This is consistent with rodent studies showing increased glutamate efflux in the prefrontal cortex upon acute administration of ketamine (Lorrain et al. 2003). More recently, Doyle et al. (2013) confirmed the results obtained by Deakin et al (2008) and demonstrated the effect of lamotrigine on ketamine-induced changes in the BOLD signal. Doyle et al. (2013) also investigated the effect of pretreatment with risperidone on the ketamine-induced BOLD signal and showed that risperidone attenuates the ketamine effect across most brain regions, including the prefrontal and cingulate regions as well as the thalamus. As risperidone is an atypical antipsychotic with high affinities for the dopamine D_2 and serotonin 2A receptors, Doyle et al.'s (2013) study suggests that serotonergic and dopaminergic mechanisms might play a role in the ketamine-induced prefrontal changes. This provides a way to investigate the mechanistic action of novel compounds relevant for psychiatric disorders.

A detailed comparison of behavioral characteristics and fMRI/phMRI imaging results after ketamine administration is shown in Table 6.1.

6.3 Positron emission tomography/single-photon emission computed tomography

PET/single-photon emission computed tomography (SPECT) is another important functional imaging technique that can be used to investigate the localization of neurotransmitter receptors and transporters directly by mapping human brain function (Krause 2008). PET measures the emissions from radioactively labeled metabolically active chemicals that have been injected into the bloodstream, and the emission data are computer processed to produce multidimensional images of the distribution of the chemicals throughout the brain. The greatest benefit of PET scanning is that the different compounds can show blood flow, oxygen levels, and glucose metabolism in regions of a working brain and thus provide information about how the brain works. Compared with other functional imaging techniques, PET scans are fast and the images generated are high resolution. However, the biggest drawback of this scanning method is that the

Table 6.1 Effect of Ketamine on Behavior and fMRI/phMRI Imaging

References	Number of subjects	Task	Behavior	Imaging results
Abel et al. (2013a,b)	HC = 8 All M	Face discrimination	↑CADSS score, BPRS total, depersonalization score with Ket.	↓Middle occipital gyrus, precentral gyrus (Pla > Ket). ↑Right cerebellum (Pla > Ket) for fearful faces; right precuneus and bilateral caudate nuclei (Ket > Pla) for neutral faces.
Honey et al. (2004)	HC = 12 M = 6 F = 6	Verbal working memory	No effect on task performance.	↑Bilateral dorsolateral prefrontal regions, ventrolateral prefrontal regions, parietal areas, anterior cingulate gyrus, caudate nucleus, and putamen (Ket > Pla) for manipulation compared with maintenance contrast.
Fu et al. (2005)	HC = 10 All M	Verbal fluency task	↑CADSS score and total BPRS with Ket. No effect on task performance.	↑Anterior cingulate gyrus, medial prefrontal cortices, middle and inferior frontal gyri, ventrolateral prefrontal cortices, occipital and cerebellar cortices, caudate, putamen, and nucleus accumbens.
Honey et al. (2005)	HC = 12 M = 6 F = 6	Episodic memory task	No effect on task performance.	↑Bilateral prefrontal cortex, left hippocampus.

(Continued)

Table 6.1 (Continued) Effect of Ketamine on Behavior and fMRI/phMRI Imaging

References	Number of subjects	Task	Behavior	Imaging results
Northoff et al. (2005)	HC = 16 M = 8 F = 8	Episodic memory retrieval task	Psychological effects: ↑Auditory perceptual alterations, oceanic boundlessness. ↓Episodic memory retrieval in "remember"/"know."	↓Posterior cingulate, precuneus, and anterior cingulate cortex.
Corlett et al. (2006)	HC = 15 M = 8 F = 7	Causal associative learning task	↓Error-dependent learning activity with low-dose Ket. ↓Perceptual aberrations with high-dose Ket.	↑Frontal activation
Honey et al. (2008)	HC = 15 M = 8 F = 7	Verbal working memory task, sustained attention, semantic generation, verbal self-monitoring	Psychological effects: ↑Negative symptoms of schizophrenia, thought disorder, and auditory illusions. ↓RTs in overall behavioral tasks.	↑Basal ganglia and thalamus (working memory). Positive correlation between activation of left middle temporal gyrus, anterior cingulate cortex, and right inferior frontal gyri with PIa, with some auditory illusions experienced with Ket.
Daumann et al. (2008)	HC = 14 M = 8 F = 6	Covert orienting of attention task (COVAT)	Psychological effects: ↑Negative symptoms of schizophrenia, body perception disruption, anxiety, and affective blunting. No effect on task performance.	↑Right superior frontal gyrus, left superior temporal gyrus, and right midfrontal gyrus.

Deakin et al. (2008)	*Expt. 1:* HC = 12 All M	All phMRI *Expt. 1:* Ket infusion	*Expt. 1:* ↑CADSS score and total BPRS with Ket.	*Expt. 1:* ↑Precuneus, mid-posterior cingulate gyrus, motor cortex, superior frontal gyrus, inferior temporal gyrus, hippocampus and bilaterally superior temporal gyrus. ↓Bilaterally in medial orbitofrontal cortex, and temporal pole.
	Expt. 2: HC = 19 All M	*Expt. 2:* Lamotrigine pretreatment before Ket infusion	*Expt. 2:* ↓BPRS total and CADSS after lamotrigine pretreatment.	*Expt. 2:* Lamotrigine inhibited most of the effects of Ket.
Daumann et al. (2010)	HC = 14 M = 8 F = 6	A target detection task	↓RTs	↑Cortical activation in the left insula and precentral gyrus in the auditory modality.
Musso et al. (2011)	HC = 24 All M	Visual oddball task	Psychological effects: ↑PANSS and 5D-ASC scale. Reaction hit and false alarm rate separates the two drug conditions.	↓Visual cortex, the anterior cingulate cortex, and temporoparietal cortex.

(Continued)

Table 6.1 (Continued) Effect of Ketamine on Behavior and fMRI/phMRI Imaging

References	Number of subjects	Task	Behavior	Imaging results
Nagels et al. (2012)	HC = 15 All M	Phonological verbal fluency task	Psychological effects: ↑PANSS. No effect on task performance.	↑Frontotemporal language network: left superior temporal and inferior parietal lobe. PANSS score positively correlated with left superior temporal gyrus, the right middle and inferior frontal gyrus, and precuneus for formal thought disorder positive; abstract thinking correlated with prefrontal and anterior cingulated regions; lack of spontaneity and flow of conversation correlated with hyperactivity in the left superior temporal gyrus.
Scheidegger et al. (2012)	HC = 19 M = 9 F = 10	Resting-state fMRI	Psychological effects: ↑5D-ASC scale.	↓DMN to the DN and to PACC and medioprefrontal cortex via the posterior cingulate cortex.
Liao et al. (2012)	HC = 44 M = 34 F = 10 Ket = 41 M = 33 F = 8	Resting-state fMRI; long-term effects of Ket	NA	↑Left precentral frontal gyrus. ↓Right anterior cingulate cortex.

Anticevic et al. (2012)	HC = 19 M = 10 F = 9	Working memory	↓Task performance accuracy.	↓The dorsolateral prefrontal cortex and precuneus.
Driesen et al. (2013a)	Primary HC = 22 M = 14 F = 8 Additional HC = 12	Resting-state fMRI	Psychological effects: ↑PANSS.	↑Global-based connectivity. ↑Right insula, right planum temporale, bilateral pulvinar nuclei, left lingual gyrus and anterior cerebellar vermis with ↑ positive symptoms. ↑Dorsal and medial anterior striatum and the thalamus with ↓ negative symptoms.
Driesen et al. (2013b)	HC = 22 M = 14 F = 8	Spatial working memory task	Psychological effects: ↑PANSS. ↓Task performance accuracy.	↓Left prefrontal cortex. ↓Connectivity between the R-DLPFC seed and many brain areas.
Corlett et al. (2013)	HC = 18 M = 10 F = 8	Fear memory task	Psychological effects: ↑CADSS. ↑The cue reactivated under Ket. ↑Subsequent memory of representation of the CS under Ket.	Inappropriate DLPFC response to the violation of blocking with Ket.

(Continued)

Table 6.1 (Continued) Effect of Ketamine on Behavior and fMRI/phMRI Imaging

References	Number of subjects	Task	Behavior	Imaging results
De Simoni et al. (2013)	HC = 10 All M	phMRI	Psychological effects: ↑PSI, CADSS, VAS.	↑Anterior cingulate cortex, supragenual paracingulate cortex, thalamus, posterior cingulate cortex, supragenual paracingulate motor area, left anterior insula, right anterior insula, left operculum, right operculum, precuneus, medial occipital lobes. ↓Subgenual cingulate cortex.
Doyle et al. (2013)	HC = 16 All M	phMRI	Psychological effects: ↑VAS.	↑Frontal cortex and thalamic regions with Ket. Lamotrigine and risperidone attenuated the Ket-induced increases. ↓Subgenual cingulate cortex with Ket.

Note: 5D-ASC, altered states of consciousness rating scale; BPRS, brief psychiatric rating scale; CADSS, clinician-administered dissociative states scale; CS, conditioned stimulus; DMN, default mode network; DN, dorsal nexus; Expt., experiment; F, females; fMRI, functional magnetic resonance imaging; HC, the number of healthy controls; Ket, ketamine; M, males; NA, not applicable; PACC, pregenual anterior cingulate cortex; PANSS, positive and negative syndrome scale; phMRI, pharmacological MRI; Pla, placebo; PSI, psychotomimetic states inventory; R-DLPFC, right dorsal lateral prefrontal cortex; RTs, reaction times; VAS, visual analogue scale.

radioactivity decays rapidly and subjects are exposed to ionizing radiation. In recent years, PET scanning has been used largely to determine the distribution of important neurotransmitters in the human brain and to advance our understanding of the pathophysiology of schizophrenia (Patel et al. 2010).

PET studies have shown that schizophrenia is associated with the enhanced synthesis and storage of presynaptic striatal dopamine (McGowan et al. 2004; Meyer-Lindenberg et al. 2002). The administration of amphetamine-like drugs induced an increase in striatal dopamine in schizophrenia patients, leading to a change in the specific binding of the dopamine D_2/D_3 receptor (as determined by specific PET radioligands), in drug-free and drug-naive patients with acute symptoms (Breier et al. 1997; Laruelle et al. 1996). Similarly, when healthy volunteers were treated with amphetamine alone, there was an increase in the release of dopamine in the striatum, which was measured by SPECT as a decrease in binding of the D_2/D_3 receptor PET/SPECT ligands. When infused with ketamine first, amphetamine treatment induced a twofold greater release of dopamine, as shown by the significantly lower binding of the D_2/D_3 receptor PET/SPECT ligands (Kegeles et al. 2000). This finding validates the NMDA model of schizophrenia, as the enhanced response to amphetamine resulting from ketamine treatment in healthy subjects mimics the exaggerated response to amphetamine seen in schizophrenia. Meanwhile, several studies have used D_2/D_3 receptor PET/SPECT ligands to measure the effects of ketamine infusion on D_2/D_3 receptor availability in the striatum, as a model of schizophrenia. Half of these studies have shown increased dopamine transmission in the striatum (Breier et al. 1998; Smith et al. 1998), ventral striatum, left caudate, and right putamen (Vollenweider et al. 2000). However, the remaining studies have reported that ketamine has no observable effects on dopamine release in the striatum (Aalto et al. 2002; Kegeles et al. 2002). In addition, Narendran et al. (2005) reported that chronic ketamine users show higher dopamine 1 (D_1) receptor availability in the dorsolateral prefrontal cortex, similar to what has been observed in patients with schizophrenia (Abi-Dargham et al. 2002, 2012). Furthermore, direct in vivo measurements of glutamatergic indices are necessary to translate clinical findings of schizophrenia into effective therapies. However, the approaches used to measure glutamatergic indices are currently limited by a lack of availability of suitable PET radiotracers. In a SPECT study on schizophrenia, the availability of the NMDA receptor in the hippocampus was negatively associated with the severity of symptoms, especially negative symptoms (Pilowsky et al. 2006). Stone et al. (2008) used the NMDA receptor radiotracer ^{123}I-CNS-1261 to study the effect of ketamine binding to NMDA receptors and its relationship to the induction of positive and negative psychotic symptoms. They demonstrated that ketamine induces negative symptoms via direct inhibition of the NMDA receptor, although positive symptoms may arise though a different neurochemical pathway. Whereas the development of PET and SPECT imaging of the glutamate system has lagged behind that of the dopamine system, MRS technology (described in Section 6.4) has more effectively been utilized to measure glutamatergic indices in vivo. Recently, ketamine has been reported to elicit a long-lasting antidepressant effect in patients with major depression (Berman et al. 2000; Zarate et al. 2006). Yamanaka et al. (2014) conducted a PET study with two serotonin (5-HT)-related PET radioligands, ^{11}CAZ10419369 and ^{11}CDASB, to elucidate the involvement of the serotonergic system in the actions of ketamine in the living primate brain. They observed that ketamine administration significantly increased 5-HT1B receptor binding in the nucleus accumbens, ventral global pallidus, and thalamus—all regions that may be critically involved in the antidepressant action of ketamine. PET imaging studies of the serotoninergic system may prove to be useful in the diagnosis of major depression as well as in the development of novel antidepressants.

6.4 *Magnetic resonance spectroscopy*

MRS is a noninvasive, ionizing radiation–free analytical technique, which has been used to study metabolic changes in the brain. It uses proton signals (^{1}H-MRS) to determine the relative or absolute concentrations of target brain metabolites. Compared with PET/SPECT, this is an alternative way to estimate the concentration of glutamatergic metabolites in vivo. However, individual MRS glutamate studies have produced some inconsistent findings for schizophrenia. In general, studies conducted at higher field strengths indicate that frontal glutamine function is elevated in the early stages of psychosis (Bustillo et al. 2010; Théberge et al. 2002), whereas the findings in chronic schizophrenia are more variable (Reid et al. 2010; Shirayama et al. 2010; Tayoshi et al. 2009; Théberge et al. 2003; Wood et al. 2007).

To further test the glutamate/NMDA hypofunction theory of schizophrenia, several MRS studies have been performed to examine the effects of ketamine on glutamatergic levels in the brain. Stone et al. (2012) reported that ketamine increases the glutamate levels in the ACC, which correlates with the degree of severity of positive psychotic symptoms. On the other hand, ketamine does not affect the levels of subcortical gamma-aminobutyric acid. Rowland et al. (2005) observed an increase in glutamine, a putative marker of glutamate neurotransmitter release, in the ACC. However, this increase was not related to schizophrenia-like positive or negative symptoms. Taylor et al. (2012) reported no difference in either glutamate or glutamine (Glx) in the ACC after low-dose ketamine infusion. Overall, these findings are consistent with the findings of increased glutamatergic indices in the prefrontal cortex of unmedicated patients with schizophrenia.

6.5 *Magnetic resonance imaging*

Structural MRI is an imaging technique used primarily in a medical setting to produce high-quality images to investigate the anatomy of the body. During the past decade, the application of MRI techniques in psychiatric disorders has seen a rapid increase. Schizophrenia is one of the most studied psychiatric disorders using the MRI technique.

Voxel-based morphometry (VBM) is a popular tool used for analyzing MRI data. VBM has been used to map the neuroanatomy of schizophrenia. Several recent meta-analyses of various VBM studies published in the literature have focused on identifying which brain regions most consistently show significant differences. Consistent with the functional findings, the structural imaging results suggest that schizophrenia might involve a reduced global integration of structural brain network and a reduced role for key frontal and parietal hubs in the overall network architecture, which in turn lead toward a diminished capacity to integrate information across different regions of the brain (Filippi et al. 2013; Fornito et al. 2012; van den Heuvel et al. 2010). A meta-analysis on VBM studies showed that schizophrenia is characterized mainly by deficits in the anterior cingulate thalamus, frontal lobe, hippocampal–amygdala region, and superior temporal gyrus, as well as volume reduction in the left medial temporal lobe gray matter (Honea et al. 2005). A more recent meta-analysis and meta-regression of a longitudinal MRI study of schizophrenia demonstrated that the underlying pathological process appears to be especially active in the first stages of the disease; it affects the left hemisphere and the superior temporal structures more and is at least partly moderated by the type of antipsychotic treatment received (Vita et al. 2012).

Liao et al. (2011) observed significant decreases in gray matter volume in the bilateral frontal cortex including the left superior frontal gyrus and the right middle frontal gyrus of chronic ketamine users when compared with controls by using VBM analysis of structural MRI. In this study, they also demonstrated a correlation between a longer duration of ketamine use and lower gray matter volume, which is also consistent with the findings from studies of heroin addiction (Yuan et al. 2009). Another recent MRI study (Wang et al. 2013) revealed lesions in many brain regions of long-term ketamine addicts. These lesions appear as minute patches in the first year and become larger sites of atrophy by 4 years of addiction. The brain regions affected include the prefrontal, parietal, occipital, and limbic lobes; brain stem; and corpus striatum. The results obtained in these MRI studies also correlated with the work by Morgan and Curran, suggesting a long-lasting impairment of episodic memory and attentional functioning in chronic ketamine abusers (Morgan and Curran 2006; Morgan et al. 2004, 2009). In addition to structural changes, fMRI and functional studies also confirm functional and cognitive derangements in ketamine addicts

(Driesen et al. 2013a; Morgan and Curran 2006; Morgan et al. 2009; Yu et al. 2012).

Ketamine is reported to cause neurotoxicity (Olney et al. 1991; Vranken et al. 2006), as well as changes in synaptogenesis (Thomson et al. 1985) and apoptosis (Sun et al. 2014), which may contribute to the morphological changes observed in the brain in ketamine addicts via MRI. Chronic NMDA blockade actually sensitizes the dopamine system (Jentsch and Roth 1999); however, prolonged exposure to ketamine is followed by a depression in brain activity and a reduction in the volume of the prefrontal cortex (Liao et al. 2010; Roberts et al. 2014). Evidence indicates that the availability of the D_1 receptor increases in chronic ketamine users, which is compatible with the hypothesis that chronic NMDA antagonist leads to D_1 receptor up-regulation via reduced dopamine levels in the prefrontal cortex (Narendran et al. 2005). On the other hand, the dysfunction of a subset of cortical fast-spiking inhibitory interneurons might explain the brain morphological alterations that occur after repetitive exposure to ketamine (Behrens et al. 2007). In the rat, the transient blockade of NMDA receptors during late fetal or early neonatal life was shown to trigger widespread apoptotic neurodegeneration in which large numbers of neurons were deleted from the developing brain (Ikonomidou et al. 1999). In another, more recent study, ketamine was shown to significantly increase the translocation of Bax from the cytoplasm to the mitochondria, as well as enhancing the activity of caspases 3, 6, and 9 and increasing the level of cytochrome c in human hepatoma HepG2 cells, which resulted in an increase in DNA fragmentation and apoptosis (Lee et al. 2009). Other studies also demonstrated that ketamine increases the level of apoptosis in neurons with increased levels of Bax protein in forebrain cultures from neonatal rats (Wang et al. 2005, 2008) and postnatal monkeys (Wang et al. 2006). In addition, repeated ketamine administration resulted in significant numbers of apoptotic neuronal cells in the frontal cortex as well as in other regions of the brain (Zou et al. 2009). Furthermore, use of MRI in ketamine-treated animal models revealed that the molecular mechanisms responsible for structural changes in the brain are similar to those observed in human ketamine addicts. These include prefrontal cortex apoptosis (Mak et al. 2010; Sun et al. 2014), mutated tau aggregation (Yeung et al. 2010), prefrontal/brain stem chemical changes (Tan et al. 2011, 2012; Yu et al. 2012), and cerebellar apoptosis (Chan et al. 2012).

6.6 Diffusion tensor imaging

Gascon et al. (2007) reported that in a rodent model of ketamine addiction, both neurons and fibers (white matter) in the CNS are affected. They suggested that long-term administration of ketamine may impair the neuronal network by interfering with the fundamental contacts between the neurons. Given the suggestion that chronic NMDA blockade exerts an effect on synaptic plasticity, an evaluation of connectivity-related changes may prove to be useful. White matter organization and integrity can be evaluated with DTI. This is a novel, noninvasive magnetic resonance technique that is capable of providing quantitative investigation of the microstructural organization of white matter based on patterns of water diffusion in the neural tissue (Le Bihan et al. 2001). White matter integrity can be studied by examining the degree of fractional anisotropy (FA), which qualifies the restriction (anisotropy) of water diffusion by tissue microstructure in each image voxel. FA is thought to reflect properties of the white matter microstructure such as axonal size, density, and the level of myelination (Beaulieu 2002). A high FA indicates a nonspherical tensor with preferential orientation in a particular direction, while a lower FA indicates more isotropic diffusion, which is more characteristic of disrupted or damaged white matter (Beaulieu 2002). Recently, DTI has been used to identify white matter abnormalities in schizophrenia patients (Kyriakopoulos et al. 2008a), even at early stages of the disorder (Kyriakopoulos and Frangou 2009; Kyriakopoulos et al. 2008b). Even though abnormalities in the white matter are not found consistently, the frontal white matter seems to be more commonly affected. Moreover, the deleterious effect of schizophrenia on network performance appears to be localized as reduced regional efficiency in hubs such as the frontal, temporal, paralimbic/limbic, and putamen regions. This shows diminished support for the dysconnectivity hypothesis of schizophrenia, suggesting a deficiency of processing integration that matches well with the cognitive impairment underlying the pathology (Wang et al. 2012). Zalesky et al. (2011) showed that a frontoparietal/occipital network may represent the key macro-circuit affected in schizophrenia. An aberrant frontal–temporal complex network structure, especially of the bilateral inferior/superior frontal cortex and temporal brain regions, has been reported in schizophrenia patients (van den Heuvel et al. 2010). Together, these findings suggest that schizophrenia patients have a less integrated brain network with a reduced central role for key frontal hubs, resulting in a limited structural capacity to integrate information across the other brain regions.

There is a compelling similarity in the white matter changes that occur between chronic ketamine use and schizophrenia. Liao et al. (2010) measured white matter integrity using in vivo DTI in 41 ketamine-dependent subjects and 44 healthy controls, and they demonstrated white matter changes with reduced FA in the bilateral frontal and left temporo-parietal cortices associated with chronic ketamine use. They also reported a correlation between the severity of drug use and the extent of white matter disruption in the bilateral frontal cortex. The brain regions shown to be affected by ketamine (Liao et al. 2010) are remarkably consistent with those affected in schizophrenia (Pomarol-Clotet et al. 2010; van den

Heuvel et al. 2010). In addition, there is striking overlap in the symptoms of schizophrenia, such as perceptual changes and hallucinations, and those described for ketamine users (Liao et al. 2010). Most recently, Roberts et al. (2014) partially replicated the findings reported by Liao et al. (2010), describing increased dissociative and schizotypal symptoms in ketamine users as well as a widespread reduction in axial diffusivity (which reflects the packing density and diameter of axons; Song et al. 2003). Roberts et al. (2014) measured white matter integrity and connectivity in the brain of 16 ketamine users and 16 polydrug-using controls and found a reduction of white matter integrity in the right hemisphere network. Within the ketamine user group, there was a significant positive association in the connectivity profile between the caudate nucleus and the lateral prefrontal cortex pathway, and dissociative experiences were reported. These findings suggest that chronic ketamine use is associated with significant changes to the microstructure of the white matter in the brain, and this may contribute to the prediction of the individual differences in symptomatology in chronic ketamine users. It is noteworthy that frontal white matter abnormalities may also occur after the chronic application of other forms of drug (Goldstein and Volkow 2002), such as alcohol (Rosenbloom et al. 2003), amphetamine (Thompson et al. 2004), methamphetamine (Alicata et al. 2009), cannabis (Arnone et al. 2008), heroin (Liu et al. 2008), and nicotine (Wang et al. 2009).

6.7 *Summary*

The recent and rapid advances in MRS and PET/SPECT techniques capable of imaging different components of the neurotransmitter system have paved the way for a greater, more nuanced understanding of the glutamatergic, dopaminergic, and serotonergic systems in the NMDA receptor hypofunction model of schizophrenia. Further investigations are critical and will facilitate a greater understanding of schizophrenia as well as the successful and efficient development of new treatments. fMRI has also improved tremendously in recent years, with attention now focusing on the resting-state functional connectivity of the brain. In addition, phMRI is starting to yield new information regarding selective receptors and neurotransmitter pathways; without the need of a radiolabel, phMRI is a very useful tool for drug discovery studies.

Recent advances in the development of neuroimaging hardware now make translational studies, from clinical to preclinical, possible. This offers new opportunities in translational neuroscience research, accelerates drug discovery programs, and reduces the cost of drug development. Though no one animal model can fully recapitulate neuropsychiatric disorders, the aspects of the disorder being modeled help us not only to understand the neurobiology of those disorders but also to identify and validate molecular targets, which can then be manipulated pharmacologically. In the future, more studies combining neuroimaging techniques with nonhuman primate or rodent NMDA receptor hypofunction models of schizophrenia are necessary to facilitate our longtitudinal understanding of how drugs, disease development, and even social situations might affect neural structure and function.

References

Aalto, S., Hirvonen, J., Kajander, J., Scheinin, H., Någren, K., Vilkman, H., Gustafsson, L., Syvälahti, E., and Hietala, J. (2002). Ketamine does not decrease striatal dopamine D_2 receptor binding in man. *Psychopharmacology (Berl.)* 164:401–406.

Abel, K.M., Allin, M.P.G., Kucharska-Pietura, K., Andrew, C., Williams, S., David, A.S., and Phillips, M.L. (2003a). Ketamine and fMRI BOLD signal: Distinguishing between effects mediated by change in blood flow versus change in cognitive state. *Hum. Brain Mapp.* 18:135–145.

Abel, K.M., Allin, M.P.G., Kucharska-Pietura, K., David, A., Andrew, C., Williams, S., Brammer, M.J., and Phillips, M.L. (2003b). Ketamine alters neural processing of facial emotion recognition in healthy men: An fMRI study. *Neuroreport* 14:387–391.

Abi-Dargham, A., Mawlawi, O., Lombardo, I., Gil, R., Martinez, D., Huang, Y., Hwang, D.R., Keilp, J., Kochan, L., Van Heertum, R., Gorman, J.M., and Laruelle, M. (2002). Prefrontal dopamine D_1 receptors and working memory in schizophrenia. *J. Neurosci.* 22:3708–3719.

Abi-Dargham, A., Xu, X., Thompson, J.L., Gil, R., Kegeles, L.S., Urban, N., Narendran, R., Hwang, D.R., Laruelle, M., and Slifstein, M. (2012). Increased prefrontal cortical D_1 receptors in drug naive patients with schizophrenia: A PET study with [^{11}C]NNC112. *J. Psychopharmacol.* 26:794–805.

Alicata, D., Chang, L., Cloak, C., Abe, K., and Ernst, T. (2009). Higher diffusion in striatum and lower fractional anisotropy in white matter of methamphetamine users. *Psychiatry Res.* 174:1–8.

Altamura, M., Elvevåg, B., Blasi, G., Bertolino, A., Callicott, J.H., Weinberger, D.R., Mattay, V.S., and Goldberg, T.E. (2007). Dissociating the effects of Sternberg working memory demands in prefrontal cortex. *Psychiatry Res.* 154:103–114.

Andreasen, N.C., Rezai, K., Alliger, R., Swayze, V.W. 2nd, Flaum, M., Kirchner, P., Cohen, G., and O'Leary, D.S. (1992). Hypofrontality in neuroleptic-naive patients and in patients with chronic schizophrenia. Assessment with xenon 133 single-photon emission computed tomography and the Tower of London. *Arch. Gen. Psychiatry* 49:943–958.

Anticevic, A., Gancsos, M., Murray, J.D., Repovs, G., Driesen, N.R., Ennis, D.J., Niciu, M.J., Morgan, P.T., Surti, T.S., Bloch, M.H., Ramani, R., Smith, M.A., Wang, X.J., Krystal, J.H., and Corlett, P.R. (2012). NMDA receptor function in large-scale anticorrelated neural systems with implications for cognition and schizophrenia. *Proc. Natl. Acad. Sci. U.S.A.* 109:16720–16725.

Arnone, D., Barrick, T.R., Chengappa, S., Mackay, C.E., Clark, C.A., and Abou-Saleh, M.T. (2008). Corpus callosum damage in heavy marijuana use: Preliminary evidence from diffusion tensor tractography and tract-based spatial statistics. *NeuroImage* 41:1067–1074.

Beaulieu, C. (2002). The basis of anisotropic water diffusion in the nervous system—A technical review. *NMR Biomed.* 15:435–455.

Behrens, M.M., Ali, S.S., Dao, D.N., Lucero, J., Shekhtman, G., Quick, K.L., and Dugan, L.L. (2007). Ketamine-induced loss of phenotype of fast-spiking interneurons is mediated by NADPH-oxidase. *Science* 318:1645–1647.

Berman, R.M., Cappiello, A., Anand, A., Oren, D.A., Heninger, G.R., Charney, D.S., and Krystal, J.H. (2000). Antidepressant effects of ketamine in depressed patients. *Biol. Psychiatry* 47:351–354.

Biswal, B.B. (2012). Resting state fMRI: A personal history. *NeuroImage* 62:938–944.

Breier, A., Adler, C.M., Weisenfeld, N., Su, T.P., Elman, I., Picken, L., Malhotra, A.K., and Pickar, D. (1998). Effects of NMDA antagonism on striatal dopamine release in healthy subjects: Application of a novel PET approach. *Synapse* 29:142–147.

Breier, A., Su, T.P., Saunders, R., Carson, R.E., Kolachana, B.S., De Bartolomeis, A., Weinberger, D.R., Weisenfeld, N., Malhotra, A.K., Eckelman, W.C., and Pickar, D. (1997). Schizophrenia is associated with elevated amphetamine-induced synaptic dopamine concentrations: Evidence from a novel positron emission tomography method. *Proc. Natl. Acad. Sci. U.S.A.* 94:2569–2574.

Buckner, R.L. (2012). The serendipitous discovery of the brain's default network. *NeuroImage* 62:1137–1145.

Bustillo, J.R., Rowland, L.M., Mullins, P., Jung, R., Chen, H., Qualls, C., Hammond, R., Brooks, W.M., and Lauriello, J. (2010). ^{1}H-MRS at 4 Tesla in minimally treated early schizophrenia. *Mol. Psychiatry* 15:629–636.

Callicott, J.H., Mattay, V.S., Bertolino, A., Finn, K., Coppola, R., Frank, J.A., Goldberg, T.E., and Weinberger, D.R. (1999). Physiological characteristics of capacity constraints in working memory as revealed by functional MRI. *Cereb. Cortex* 9:20–26.

Callicott, J.H., Mattay, V.S., Verchinski, B.A., Marenco, S., Egan, M.F., and Weinberger, D.R. (2003). Complexity of prefrontal cortical dysfunction in schizophrenia: More than up or down. *Am. J. Psychiatry* 160:2209–2215.

Cannon, T.D., Glahn, D.C., Kim, J., Van Erp, T.G.M., Karlsgodt, K., Cohen, M.S., Nuechterlein, K.H., Bava, S., and Shirinyan, D. (2005). Dorsolateral prefrontal cortex activity during maintenance and manipulation of information in working memory, in patients with schizophrenia. *Arch. Gen. Psychiatry* 62:1071–1080.

Carter, C.S., Perlstein, W., Ganguli, R., Brar, J., Mintun, M., and Cohen, J.D. (1998). Functional hypofrontality and working memory dysfunction in schizophrenia. *Am. J. Psychiatry* 155:1285–1287.

Chan, W.M., Xu, J., Fan, M., Jiang, Y., Tsui, T.Y., Wai, M.S., Lam, W.P., and Yew, D.T. (2012). Downregulation in the human and mice cerebella after ketamine versus ketamine plus ethanol treatment. *Microsc. Res. Tech.* 75:258–264.

Corlett, P.R., Cambridge, V., Gardner, J.M., Piggot, J.S., Turner, D.C., Everitt, J.C., Arana, F.S., Morgan, H.L., Milton, A.L., Lee, J.L., Aitken, M.R.F., Dickinson, A., Everitt, B.J., Absalom, A.R., Adapa, R., Subramanian, N., Taylor, J.R., Krystal, J.H., and Fletcher, P.C. (2013). Ketamine effects on memory reconsolidation favor a learning model of delusions. *PLoS One* 8(6):e6588. doi:10.1371/journal.pone.0065088.

Corlett, P.R., Honey, G.D., Aitken, M.R., Dickinson, A., Shanks, D.R., Absalom, A.R., Lee, M., Pomarol-Clotet, E., Murray, G.K., Mckenna, P.J., Robbins, T.W., Bullmore, E.T., and Fletcher, P.C. (2006). Frontal responses during learning predict vulnerability to the psychotogenic effects of ketamine: Linking cognition, brain activity, and psychosis. *Arch. Gen. Psychiatry* 63:611–621.

Daumann, J., Heekeren, K., Neukirch, A., Thiel, C.M., Moller-Hartmann, W., and Gouzoulis-Mayfrank, E. (2008). Pharmacological modulation of the neural basis underlying inhibition of return (IOR) in the human 5-HT2A agonist and NMDA antagonist model of psychosis. *Psychopharmacology (Berl.)* 200:573–583.

Daumann, J., Wagner, D., Heekeren, K., Neukirch, A., Thiel, C.M., and Gouzoulis-Mayfrank, E. (2010). Neuronal correlates of visual and auditory alertness in the DMT and ketamine model of psychosis. *J. Psychopharmacol.* 24:1515–1524.

De Simoni, S., Schwarz, A.J., O'Daly, O.G., Marquand, A.F., Brittain, C., Gonzales, C., Stephenson, S., Williams, S.C.R., and Mehta, M.A. (2013). Test-retest reliability of the BOLD pharmacological MRI response to ketamine in healthy volunteers. *NeuroImage* 64:75–90.

Deakin, J.F.W., Lees, J., Mckie, S., Hallak, J.E.C., Williams, S.R., and Dursun, S.M. (2008). Glutamate and the neural basis of the subjective effects of ketamine. *Arch. Gen. Psychiatry* 65:154–164.

Doyle, O.M., De Simoni, S., Schwarz, A.J., Brittain, C., O'Daly, O.G., Williams, S.C.R., and Mehta, M.A. (2013). Quantifying the attenuation of the ketamine pharmacological magnetic resonance imaging response in humans: A validation using antipsychotic and glutamatergic agents. *J. Pharmacol. Exp. Ther.* 345:151–160.

Driesen, N.R., Mccarthy, G., Bhagwagar, Z., Bloch, M., Calhoun, V., D'Souza, D.C., Gueorguieva, R., He, G., Ramachandran, R., Suckow, R.F., Anticevic, A., Morgan, P.T., and Krystal, J.H. (2013a). Relationship of resting brain hyperconnectivity and schizophrenia-like symptoms produced by the NMDA receptor antagonist ketamine in humans. *Mol. Psychiatry* 18:1199–1204.

Driesen, N.R., Mccarthy, G., Bhagwagar, Z., Bloch, M.H., Calhoun, V.D., D'Souza, D.C., Gueorguieva, R., He, G., Leung, H.C., Ramani, R., Anticevic, A., Suckow, R.F., Morgan, P.T., and Krystal, J.H. (2013b). The impact of NMDA receptor blockade on human working memory-related prefrontal function and connectivity. *Neuropsychopharmacology* 38:2613–2622.

Esslinger, C., Gruppe, H., Danos, P., Lis, S., Broll, J., Wiltink, J., Gallhofer, B., and Kirsch, P. (2007). Influence of vigilance and learning on prefrontal activation in schizophrenia. *Neuropsychobiology* 55:194–202.

Filippi, M., van den Heuvel, M.P., Fornito, A., He, Y., Hulshoff Pol, H.E., Agosta, F., Comi, G., and Rocca, M.A. (2013). Assessment of system dysfunction in the brain through MRI-based connectomics. *Lancet Neurol.* 12:1189–1199.

Fletcher, P.C., Frith, C.D., Grasby, P.M., Friston, K.J., and Dolan, R.J. (1996). Local and distributed effects of apomorphine on fronto-temporal function in acute unmedicated schizophrenia. *J. Neurosci.* 16:7055–7062.

Fletcher, P.C., and Honey, G.D. (2006). Schizophrenia, ketamine and cannabis: Evidence of overlapping memory deficits. *Trends Cogn. Sci.* 10:167–174.

Fornito, A., Zalesky, A., Pantelis, C., and Bullmore, E.T. (2012). Schizophrenia, neuroimaging and connectomics. *NeuroImage* 62:2296–2314.

Frith, C.D., Friston, K.J., Herold, S., Silbersweig, D., Fletcher, P., Cahill, C., Dolan, R.J., Frackowiak, R.S., and Liddle, P.F. (1995). Regional brain activity in chronic schizophrenic patients during the performance of a verbal fluency task. *Br. J. Psychiatry* 167:343–349.

Fu, C.H., Abel, K.M., Allin, M.P., Gasston, D., Costafreda, S.G., Suckling, J., Williams, S.C., and McGuire, P.K. (2005). Effects of ketamine on prefrontal and striatal regions in an overt verbal fluency task: A functional magnetic resonance imaging study. *Psychopharmacology (Berl.)* 183:92–102.

Gascon, E., Klauser, P., Kiss, J.Z., and Vutskits, L. (2007). Potentially toxic effects of anaesthetics on the developing central nervous system. *Eur. J. Anaesthesiol.* 24:213–224.

Gass, N., Schwarz, A.J., Sartorius, A., Schenker, E., Risterucci, C., Spedding, M., Zheng, L., Meyer-Lindenberg, A., and Weber-Fahr, W. (2014). Sub-anesthetic ketamine modulates intrinsic BOLD connectivity within the hippocampal-prefrontal circuit in the rat. *Neuropsychopharmacology* 39:895–906.

Goldstein, R.Z., and Volkow, N.D. (2002). Drug addiction and its underlying neurobiological basis: Neuroimaging evidence for the involvement of the frontal cortex. *Am. J. Psychiatry* 159:1642–1652.

Gur, R.C., and Gur, R.E. (1995). Hypofrontality in schizophrenia: RIP. *Lancet* 345: 1383–1384.

Honea, R., Crow, T.J., Passingham, D., and Mackay, C.E. (2005). Regional deficits in brain volume in schizophrenia: A meta-analysis of voxel-based morphometry studies. *Am. J. Psychiatry* 162:2233–2245.

Honey, G.D., Corlett, P.R., Absalom, A.R., Lee, M., Pomarol-Clotet, E., Murray, G.K., McKenna, P.J., Bullmore, E.T., Menon, D.K., and Fletcher, P.C. (2008). Individual differences in psychotic effects of ketamine are predicted by brain function measured under placebo. *J. Neurosci.* 28:6295–6303.

Honey, G.D., Honey, R.A., O'Loughlin, C., Sharar, S.R., Kumaran, D., Suckling, J., Menon, D.K., Sleator, C., Bullmore, E.T., and Fletcher, P.C. (2005). Ketamine disrupts frontal and hippocampal contribution to encoding and retrieval of episodic memory: An fMRI study. *Cereb. Cortex* 15:749–759.

Honey, R.A., Honey, G.D., O'Loughlin, C., Sharar, S.R., Kumaran, D., Bullmore, E.T., Menon, D.K., Donovan, T., Lupson, V.C., Bisbrown-Chippendale, R., and Fletcher, P.C. (2004). Acute ketamine administration alters the brain responses to executive demands in a verbal working memory task: An fMRI study. *Neuropsychopharmacology* 29:1203–1214.

Ikonomidou, C., Bosch, F., Miksa, M., Bittigau, P., Vockler, J., Dikranian, K., Tenkova, T.I., Stefovska, V., Turski, L., and Olney, J.W. (1999). Blockade of NMDA receptors and apoptotic neurodegeneration in the developing brain. *Science* 283:70–74.

Jaeggi, S.M., Seewer, R., Nirkko, A.C., Eckstein, D., Schroth, G., Groner, R., and Gutbrod, K. (2003). Does excessive memory load attenuate activation in the prefrontal cortex? Load-dependent processing in single and dual tasks: Functional magnetic resonance imaging study. *NeuroImage* 19:210–225.

Javitt, D.C., and Zukin, S.R. (1991). Recent advances in the phencyclidine model of schizophrenia. *Am. J. Psychiatry* 148:1301–1308.

Javitt, D.C., Zukin, S.R., Heresco-Levy, U., and Umbricht, D. (2012). Has an angel shown the way? Etiological and therapeutic implications of the PCP/NMDA model of schizophrenia. *Schizophr. Bull.* 38:958–966.

Jenkins, B.G. (2012). Pharmacologic magnetic resonance imaging (phMRI): Imaging drug action in the brain. *NeuroImage* 62:1072–1085.

Jentsch, J.D., and Roth, R.H. (1999). The neuropsychopharmacology of phencyclidine: From NMDA receptor hypofunction to the dopamine hypothesis of schizophrenia. *Neuropsychopharmacology* 20:201–225.

Karlsgodt, K.H., Glahn, D.C., Van Erp, T.G.M., Therman, S., Huttunen, M., Manninen, M., Kaprio, J., Cohen, M.S., Lonnqvist, J., and Cannon, T.D. (2007). The relationship between performance and fMRI signal during working memory in patients with schizophrenia, unaffected co-twins, and control subjects. *Schizophr. Res.* 89:191–197.

Kegeles, L.S., Abi-Dargham, A., Zea-Ponce, Y., Rodenhiser-Hill, J., Mann, J.J., Van Heertum, R.L., Cooper, T.B., Carlsson, A., and Laruelle, M. (2000). Modulation of amphetamine-induced striatal dopamine release by ketamine in humans: Implications for schizophrenia. *Biol. Psychiatry* 48:627–640.

Kegeles, L.S., Martinez, D., Kochan, L.D., Hwang, D.R., Huang, Y.Y., Mawlawi, O., Suckow, R.F., Van Heertum, R.L., and Laruelle, M. (2002). NMDA antagonist effects on striatal dopamine release: Positron emission tomography studies in humans. *Synapse* 43:19–29.

Krause, J. (2008). SPECT and PET of the dopamine transporter in attention-deficit/hyperactivity disorder. *Expert Rev. Neurother.* 8:611–625.

Kyriakopoulos, M., and Frangou, S. (2009). Recent diffusion tensor imaging findings in early stages of schizophrenia. *Curr. Opin. Psychiatry* 22:168–176.

Kyriakopoulos, M., Bargiotas, T., Barker, G.J., and Frangou, S. (2008a). Diffusion tensor imaging in schizophrenia. *Eur. Psychiatry* 23:255–273.

Kyriakopoulos, M., Vyas, N.S., Barker, G.J., Chitnis, X.A., and Frangou, S. (2008b). A diffusion tensor imaging study of white matter in early-onset schizophrenia. *Biol. Psychiatry* 63:519–523.

Laruelle, M., Abi-Dargham, A., Van Dyck, C.H., Gil, R., D'Souza, C.D., Erdos, J., Mccance, E., Rosenblatt, W., Fingado, C., Zoghbi, S.S., Baldwin, R.M., Seibyl, J.P., Krystal, J.H., Charney, D.S., and Innis, R.B. (1996). Single photon emission computerized tomography imaging of amphetamine-induced dopamine release in drug-free schizophrenic subjects. *Proc. Natl. Acad. Sci. U.S.A.* 93:9235–9240.

Le Bihan, D., Mangin, J.F., Poupon, C., Clark, C.A., Pappata, S., Molko, N., and Chabriat, H. (2001). Diffusion tensor imaging: Concepts and applications. *J. Magn. Reson. Imaging* 13:534–546.

Lee, S.T., Wu, T.T., Yu, P.Y., and Chen, R.M. (2009). Apoptotic insults to human HepG2 cells induced by S-(+)-ketamine occurs through activation of a Bax-mitochondria-caspase protease pathway. *Br. J. Anaesth.* 102:80–89.

Liao, Y., Tang, J., Corlett, P.R., Wang, X., Yang, M., Chen, H., Liu, T., Chen, X., Hao, W., and Fletcher, P.C. (2011). Reduced dorsal prefrontal gray matter after chronic ketamine use. *Biol. Psychiatry* 69:42–48.

Liao, Y., Tang, J., Fornito, A., Liu, T., Chen, X., Chen, H., Xiang, X., Wang, X., and Hao, W. (2012). Alterations in regional homogeneity of resting-state brain activity in ketamine addicts. *Neurosci. Lett.* 522:36–40.

Liao, Y.H., Tang, J.S., Ma, M.D., Wu, Z.M., Yang, M., Wang, X.Y., Liu, T.Q., Chen, X.G., Fletcher, P.C., and Hao, W. (2010). Frontal white matter abnormalities following chronic ketamine use: A diffusion tensor imaging study. *Brain* 133:2115–2122.

Liu, H., Li, L., Hao, Y., Cao, D., Xu, L., Rohrbaugh, R., Xue, Z., Hao, W., Shan, B., and Liu, Z. (2008). Disrupted white matter integrity in heroin dependence: A controlled study utilizing diffusion tensor imaging. *Am. J. Drug Alcohol Abuse* 34:562–575.

Lorrain, D.S., Baccei, C.S., Bristow, L.J., Anderson, J.J., and Varney, M.A. (2003). Effects of ketamine and N-methyl-D-aspartate on glutamate and dopamine release in the rat prefrontal cortex: Modulation by a group II selective metabotropic glutamate receptor agonist LY379268. *Neuroscience* 117:697–706.

Mak, Y.T., Lam, W.P., Lu, L., Wong, Y.W., and Yew, D.T. (2010). The toxic effect of ketamine on SH-SY5Y neuroblastoma cell line and human neuron. *Microsc. Res. Tech.* 73:195–201.

Manoach, D.S., Press, D.Z., Thangaraj, V., Searl, M.M., Goff, D.C., Halpern, E., Saper, C.B., and Warach, S. (1999). Schizophrenic subjects activate dorsolateral prefrontal cortex during a working memory task, as measured by fMRI. *Biol. Psychiatry* 45:1128–1137.

McGowan, S., Lawrence, A.D., Sales, T., Quested, D., and Grasby, P. (2004). Presynaptic dopaminergic dysfunction in schizophrenia: A positron emission tomographic [^{18}F]fluorodopa study. *Arch. Gen. Psychiatry* 61:134–142.

Meyer-Lindenberg, A., Miletich, R.S., Kohn, P.D., Esposito, G., Carson, R.E., Quarantelli, M., Weinberger, D.R., and Berman, K.F. (2002). Reduced prefrontal activity predicts exaggerated striatal dopaminergic function in schizophrenia. *Nat. Neurosci.* 5:267–271.

Morgan, C.J., and Curran, H.V. (2006). Acute and chronic effects of ketamine upon human memory: A review. *Psychopharmacology (Berl.)* 188:408–424.

Morgan, C.J.A., Huddy, V., Lipton, M., Curran, H.V., and Joyce, E.M. (2009). Is persistent ketamine use a valid model of the cognitive and oculomotor deficits in schizophrenia? *Biol. Psychiatry* 65:1099–1102.

Morgan, C.J., Monaghan, L., and Curran, H.V. (2004). Beyond the K-hole: A 3-year longitudinal investigation of the cognitive and subjective effects of ketamine in recreational users who have substantially reduced their use of the drug. *Addiction* 99:1450–1461.

Musso, F., Brinkmeyer, J., Ecker, D., London, M.K., Thieme, G., Warbrick, T., Wittsack, H.J., Saleh, A., Greb, W., De Boer, P., and Winterer, G. (2011). Ketamine effects on brain function—Simultaneous fMRI/EEG during a visual oddball task. *NeuroImage* 58:508–525.

Nagels, A., Kirner-Veselinovic, A., Wiese, R., Paulus, F.M., Kircher, T., and Krach, S. (2012). Effects of ketamine-induced psychopathological symptoms on continuous overt rhyme fluency. *Eur. Arch. Psychiatry Clin. Neurosci.* 262:403–414.

Narendran, R., Frankle, W.G., Keefe, R., Gil, R., Martinez, D., Slifstein, M., Kegeles, L.S., Talbot, P.S., Huang, Y., Hwang, D.R., Khenissi, L., Cooper, T.B., Laruelle, M., and Abi-Dargham, A. (2005). Altered prefrontal dopaminergic function in chronic recreational ketamine users. *Am. J. Psychiatry* 162:2352–2359.

Northoff, G., Richter, A., Bermpohl, F., Gnimm, S., Martin, E., Marcar, V.L., Wahl, C., Hell, D., and Boeker, H. (2005). NMDA hypofunction in the posterior cingulate as a model for schizophrenia: An exploratory ketamine administration study in fMRI. *Schizophr. Res.* 72:235–248.

Olney, J.W., Labruyere, J., Wang, G., Wozniak, D.F., Price, M.T., and Sesma, M.A. (1991). NMDA antagonist neurotoxicity: Mechanism and prevention. *Science* 254:1515–1518.

Patel, N.H., Vyas, N.S., Puri, B.K., Nijran, K.S., and Al-Nahhas, A. (2010). Positron emission tomography in schizophrenia: A new perspective. *J. Nucl. Med.* 51:511–520.

Perlstein, W.M., Carter, C.S., Noll, D.C., and Cohen, J.D. (2001). Relation of prefrontal cortex dysfunction to working memory and symptoms in schizophrenia. *Am. J. Psychiatry* 158:1105–1113.

Pilowsky, L.S., Bressan, R.A., Stone, J.M., Erlandsson, K., Mulligan, R.S., Krystal, J.H., and Ell, P.J. (2006). First in vivo evidence of an NMDA receptor deficit in medication-free schizophrenic patients. *Mol. Psychiatry* 11:118–119.

Pomarol-Clotet, E., Canales-Rodriguez, E.J., Salvador, R., Sarro, S., Gomar, J.J., Vila, F., Ortiz-Gil, J., Iturria-Medina, Y., Capdevila, A., and Mckenna, P.J. (2010). Medial prefrontal cortex pathology in schizophrenia as revealed by convergent findings from multimodal imaging. *Mol. Psychiatry* 15:823–830.

Potkin, S.G., Turner, J.A., Brown, G.G., McCarthy, G., Greve, D.N., Glover, G.H., Manoach, D.S., Belger, A., Diaz, M., Wible, C.G., Ford, J.M., Mathalon, D.H., Gollub, R., Lauriello, J., O'Leary, D., van Erp, T.G., Toga, A.W., Preda, A., Lim, K.O., and FBIRN (2009). Working memory and DLPFC inefficiency in schizophrenia: The FBIRN study. *Schizophr. Bull.* 35:19–31.

Reid, M.A., Stoeckel, L.E., White, D.M., Avsar, K.B., Bolding, M.S., Akella, N.S., Knowlton, R.C., den Hollander, J.A., and Lahti, A.C. (2010). Assessments of function and biochemistry of the anterior cingulate cortex in schizophrenia. *Biol. Psychiatry* 68:625–633.

Roberts, R.E., Curran, H.V., Friston, K.J., and Morgan, C.J.A. (2014). Abnormalities in white matter microstructure associated with chronic ketamine use. *Neuropsychopharmacology* 39:329–338.

Rosenbloom, M., Sullivan, E.V., and Pfefferbaum, A. (2003). Using magnetic resonance imaging and diffusion tensor imaging to assess brain damage in alcoholics. *Alcohol Res. Health* 27:146–152.

Rowland, L.M., Bustillo, J.R., Mullins, P.G., Jung, R.E., Lenroot, R., Landgraf, E., Barrow, R., Yeo, R., Lauriello, J., and Brooks, W.M. (2005). Effects of ketamine on anterior cingulate glutamate metabolism in healthy humans: A 4-T proton MRS study. *Am. J. Psychiatry* 162:394–396.

Scheidegger, M., Walter, M., Lehmann, M., Metzger, C., Grimm, S., Boeker, H., Boesiger, P., Henning, A., and Seifritz, E. (2012). Ketamine decreases resting state functional network connectivity in healthy subjects: Implications for antidepressant drug action. *PLoS One* 7:e44799. doi:10.1371/journal.pone.0044799.

Shah, Y.B., and Marsden, C.A. (2004). The application of functional magnetic resonance imaging to neuropharmacology. *Curr. Opin. Pharmacol.* 4:517–521.

Shirayama, Y., Obata, T., Matsuzawa, D., Nonaka, H., Kanazawa, Y., Yoshitome, E., Ikehira, H., Hashimoto, K., and Iyo, M. (2010). Specific metabolites in the medial prefrontal cortex are associated with the neurocognitive deficits in schizophrenia: A preliminary study. *NeuroImage* 49:2783–2790.

Smith, G.S., Schloesser, R., Brodie, J.D., Dewey, S.L., Logan, J., Vitkun, S.A., Simkowitz, P., Hurley, A., Cooper, T., Volkow, N.D., and Cancro, R. (1998). Glutamate modulation of dopamine measured in vivo with positron emission tomography (PET) and ^{11}C-raclopride in normal human subjects. *Neuropsychopharmacology* 18:18–25.

Song, S.K., Sun, S.W., Ju, W.K., Lin, S.J., Cross, A.H., and Neufeld, A.H. (2003). Diffusion tensor imaging detects and differentiates axon and myelin degeneration in mouse optic nerve after retinal ischemia. *NeuroImage* 20:1714–1722.

Stone, J.M., Dietrich, C., Edden, R., Mehta, M.A., De Simoni, S., Reed, L.J., Krystal, J.H., Nutt, D., and Barker, G.J. (2012). Ketamine effects on brain GABA and glutamate levels with ^{1}H-MRS: Relationship to ketamine-induced psychopathology. *Mol. Psychiatry* 17:664–665.

Stone, J.M., Erlandsson, K., Arstad, E., Squassante, L., Teneggi, V., Bressan, R.A., Krystal, J.H., Ell, P.J., and Pilowsky, L.S. (2008). Relationship between ketamine-induced psychotic symptoms and NMDA receptor occupancy—A [^{123}I]CNS-1261 SPET study. *Psychopharmacology (Berl.)* 197:401–408.

Strauss, M.E. (1993). Relations of symptoms to cognitive deficits in schizophrenia. *Schizophr. Bull.* 19:215–231.

Sun, L., Li, Q., Li, Q., Zhang, Y., Liu, D., Jiang, H., Pan, F., and Yew, D.T. (2014). Chronic ketamine exposure induces permanent impairment of brain functions in adolescent cynomolgus monkeys. *Addict. Biol.* 19:185–194.

Tan, S., Lam, W.P., Wai, M.S., Yu, W.H., and Yew, D.T. (2012). Chronic ketamine administration modulates midbrain dopamine system in mice. *PLoS One* 7:e43947. doi:10.1371/journal.pone.0043947.

Tan, S., Rudd, J.A., and Yew, D.T. (2011). Gene expression changes in GABA$_A$ receptors and cognition following chronic ketamine administration in mice. *PLoS One* 6:e21328. doi:10.1371/journal.pone.0021328.

Taylor, M.J., Tiangga, E.R., Mhuircheartaigh, R.N., and Cowen, P.J. (2012). Lack of effect of ketamine on cortical glutamate and glutamine in healthy volunteers: A proton magnetic resonance spectroscopy study. *J. Psychopharmacol.* 26:733–737.

Tayoshi, S., Sumitani, S., Taniguchi, K., Shibuya-Tayoshi, S., Numata, S., Iga, J., Nakataki, M., Ueno, S., Harada, M., and Ohmori, T. (2009). Metabolite changes and gender differences in schizophrenia using 3-Tesla proton magnetic resonance spectroscopy (^{1}H-MRS). *Schizophr. Res.* 108:69–77.

Théberge, J., Al-Semaan, Y., Williamson, P.C., Menon, R.S., Neufeld, R.W., Rajakumar, N., Schaefer, B., Densmore, M., and Drost, D.J. (2003). Glutamate and glutamine in the anterior cingulate and thalamus of medicated patients with chronic schizophrenia and healthy comparison subjects measured with 4.0-T proton MRS. *Am. J. Psychiatry* 160:2231–2233.

Théberge, J., Bartha, R., Drost, D.J., Menon, R.S., Malla, A., Takhar, J., Neufeld, R.W., Rogers, J., Pavlosky, W., Schaefer, B., Densmore, M., Al-Semaan, Y., and Williamson, P.C. (2002). Glutamate and glutamine measured with 4.0 T proton MRS in never-treated patients with schizophrenia and healthy volunteers. *Am. J. Psychiatry* 159:1944–1946.

Thompson, P.M., Hayashi, K.M., Simon, S.L., Geaga, J.A., Hong, M.S., Sui, Y., Lee, J.Y., Toga, A.W., Ling, W., and London, E.D. (2004). Structural abnormalities in the brains of human subjects who use methamphetamine. *J. Neurosci.* 24:6028–6036.

Thomson, A.M., West, D.C., and Lodge, D. (1985). An *N*-methylaspartate receptor-mediated synapse in rat cerebral cortex: A site of action of ketamine? *Nature* 313:479–481.

Thormodsen, R., Jensen, J., Holmen, A., Juuhl-Langseth, M., Emblem, K.E., Andreassen, O.A., and Rund, B.R. (2011). Prefrontal hyperactivation during a working memory task in early-onset schizophrenia spectrum disorders: An fMRI study. *Psychiatry Res.: Neuroimaging* 194:257–262.

van den Heuvel, M.P., Mandl, R.C., Stam, C.J., Kahn, R.S., and Hulshoff Pol, H.E. (2010). Aberrant frontal and temporal complex network structure in schizophrenia: A graph theoretical analysis. *J. Neurosci.* 30:15915–15926.

Vita, A., De Peri, L., Deste, G., and Sacchetti, E. (2012). Progressive loss of cortical gray matter in schizophrenia: A meta-analysis and meta-regression of longitudinal MRI studies. *Transl. Psychiatry* 2:e190. doi:10.1038/tp.2012.116.

Vollenweider, F.X., Vontobel, P., Oye, I., Hell, D., and Leenders, K.L. (2000). Effects of (S)-ketamine on striatal dopamine: A [^{11}C]raclopride PET study of a model psychosis in humans. *J. Psychiatr. Res.* 34:35–43.

Vranken, J.H., Troost, D., de Haan, P., Pennings, F.A., van der Vegt, M.H., Dijkgraaf, M.G., and Hollmann, M.W. (2006). Severe toxic damage to the rabbit spinal cord after intrathecal administration of preservative-free S(+)-ketamine. *Anesthesiology* 105:813–818.

Wang, C., Sadovova, N., Fu, X., Schmued, L., Scallet, A., Hanig, J., and Slikker, W. (2005). The role of the N-methyl-D-aspartate receptor in ketamine-induced apoptosis in rat forebrain culture. *Neuroscience* 132:967–977.

Wang, C., Sadovova, N., Hotchkiss, C., Fu, X., Scallet, A.C., Patterson, T.A., Hanig, J., Paule, M.G., and Slikker, W., Jr. (2006). Blockade of N-methyl-D-aspartate receptors by ketamine produces loss of postnatal day 3 monkey frontal cortical neurons in culture. *Toxicol. Sci.* 91:192–201.

Wang, C., Sadovova, N., Patterson, T.A., Zou, X., Fu, X., Hanig, J.P., Paule, M.G., Ali, S.F., Zhang, X., and Slikker, W., Jr. (2008). Protective effects of 7-nitroindazole on ketamine-induced neurotoxicity in rat forebrain culture. *Neurotoxicology* 29:613–620.

Wang, C., Zheng, D., Xu, J., Lam, W., and Yew, D.T. (2013). Brain damages in ketamine addicts as revealed by magnetic resonance imaging. *Front. Neuroanat.* 7:23. doi:10.3389/fnana.2013.00023.

Wang, J.J., Durazzo, T.C., Gazdzinski, S., Yeh, P.H., Mon, A., and Meyerhoff, D.J. (2009). MRSI and DTI: A multimodal approach for improved detection of white matter abnormalities in alcohol and nicotine dependence. *NMR Biomed.* 22:516–522.

Wang, Q.F., Su, T.P., Zhou, Y., Chou, K.H., Chen, I.Y., Jiang, T.Z., and Lin, C.P. (2012). Anatomical insights into disrupted small-world networks in schizophrenia. *NeuroImage* 59:1085–1093.

Willson, M.C., Wilman, A.H., Bell, E.C., Asghar, S.J., and Silverstone, P.H. (2004). Dextroamphetamine causes a change in regional brain activity in vivo during cognitive tasks: A functional magnetic resonance imaging study of blood oxygen level-dependent response. *Biol. Psychiatry* 56:284–291.

Wood, S.J., Yucel, M., Wellard, R.M., Harrison, B.J., Clarke, K., Fornito, A., Velakoulis, D., and Pantelis, C. (2007). Evidence for neuronal dysfunction in the anterior cingulate of patients with schizophrenia: A proton magnetic resonance spectroscopy study at 3 T. *Schizophr. Res.* 94:328–331.

Yamanaka, H., Yokoyama, C., Mizuma, H., Kurai, S., Finnema, S.J., Halldin, C., Doi, H., and Onoe, H. (2014). A possible mechanism of the nucleus accumbens and ventral pallidum 5-HT1B receptors underlying the antidepressant action of ketamine: A PET study with macaques. *Transl. Psychiatry* 4:e342. doi:10.1038/tp.2013.112.

Yeung, L.Y., Wai, M.S., Fan, M., Mak, Y.T., Lam, W.P., Li, Z., Lu, G., and Yew, D.T. (2010). Hyperphosphorylated tau in the brains of mice and monkeys with long-term administration of ketamine. *Toxicol. Lett.* 193:189–193.

Yu, H., Li, Q., Wang, D., Shi, L., Lu, G., Sun, L., Wang, L., Zhu, W., Mak, Y.T., Wong, N., Wang, Y., Pan, F., and Yew, D.T. (2012). Mapping the central effects of chronic ketamine administration in an adolescent primate model by functional magnetic resonance imaging (fMRI). *Neurotoxicology* 33:70–77.

Yuan, Y., Zhu, Z., Shi, J., Zou, Z., Yuan, F., Liu, Y., Lee, T.M., and Weng, X. (2009). Gray matter density negatively correlates with duration of heroin use in young lifetime heroin-dependent individuals. *Brain Cogn.* 71:223–228.

Zalesky, A., Fornito, A., Seal, M.L., Cocchi, L., Westin, C.F., Bullmore, E.T., Egan, G.F., and Pantelis, C. (2011). Disrupted axonal fiber connectivity in schizophrenia. *Biol. Psychiatry* 69:80–89.

Zarate, C.A. Jr., Singh, J.B., Carlson, P.J., Brutsche, N.E., Ameli, R., Luckenbaugh, D.A., Charney, D.S., and Manji, H.K. (2006). A randomized trial of an N-methyl-D-aspartate antagonist in treatment-resistant major depression. *Arch. Gen. Psychiatry* 63:856–864.

Zou, X., Patterson, T.A., Sadovova, N., Twaddle, N.C., Doerge, D.R., Zhang, X., Fu, X., Hanig, J.P., Paule, M.G., Slikker, W., and Wang, C. (2009). Potential neurotoxicity of ketamine in the developing rat brain. *Toxicol. Sci.* 108:149–158.

chapter seven

Does sniffing drugs affect the respiratory system?

An example being ketamine

Maria S.M. Wai, Jacqueline C. Lam,
Lawrence K. Hui, and David T. Yew

Ketamine, a common anesthetic used in both human and animal surgeries, is an *N*-methyl-D-aspartate glutamate receptor antagonist. Even at clinical concentrations, the drug can induce psychotropic and pathological changes in experimental animals (Jevtovic-Todorovic and Carter 2005). Studies in the past decade have indicated that ketamine might also induce neurodegeneration (Beals et al. 2003; Mellon et al. 2007). Interaction of ketamine with nitrous oxide was shown to have a more extensive and significant effect on the central nervous system (Beals et al. 2003; Nakao et al. 2003), and this reaction was particularly significant in older animals (Beals et al. 2003; Jevtovic-Todorovic and Carter 2005). In addition, as ketamine became one of the more popular drugs of abuse, much attention was being centered on its influence on recreational ketamine users and long-term abusers. Narendran et al. (2005) reported that there is an up-regulation of the D_1 dopamine receptor in the brains of chronic ketamine users, especially in the region of the prefrontal cortex. Using magnetic resonance imaging (MRI), gene expression, and immunocytochemical techniques, many regions of the brain and the associated molecular changes that occur therein have been reported in mice and monkey long-term ketamine abuse models (Chan et al. 2011; Mak et al. 2010; Sun et al. 2011; Tan et al. 2011a; Yeung et al. 2010; Yu et al. 2012). Changes were not only seen in the brain after chronic abuse of ketamine but also observed in other organs such as the heart (Chan et al. 2011; Tan et al. 2011a), kidney, and urinary bladder (Tan et al. 2011b; Yeung et al. 2009). Similar to the administration of drugs such as cocaine, many ketamine abusers choose to sniff the powder into their nasal cavity instead of injecting the drug into their body, as it ensures a faster reaction.

Sniffing of drugs such as ketamine is a common practice for abusers around the world, because the drug is absorbed via the nasal vasculature and thus achieves rapid direct contact with the brain. Insufflation is believed to be a more rapid route to trigger the effect of the drug when compared with either oral intake or intravenous injection. Moreover, the restriction of the blood–brain barrier may also be overcome via the intranasal approach. A study led by Sherry Chow et al. (1999) on the nasal transport of cocaine to the brain in rats provided supportive evidence for this more efficient and faster route. For this reason, taking analgesic drugs intranasally is now more commonly used in routine clinical practice, including for pain relief in children (Nielsen et al. 2014) and in adult patients after surgical operations (Costantini et al. 2011).

Despite the benefits of the rapid onset of drug effects via intranasal administration, the adverse consequences of this approach should not be overlooked, especially in long-term users or abusers. Induced pathological changes of the nasal cavity, paranasal sinuses, and pharynx have often been reported in active cocaine (Rachapalli and Kiely 2008; Sittel and Eckel 1998) and narcotic (Yewell et al. 2002) abusers over the past 10 years. Drug abusers who use the nasal route are frequently admitted to the hospital claiming to have nasal obstruction, epistaxis, abnormal nasal discharge, and nasal or facial pain. A detailed medical examination reveals that necrotic damage of the nasal and pharyngeal mucosae and perforation of the nasal septum are common symptoms that occur in these drug abusers. In addition, the growth of numerous fungal flora in the nasal mucosa has been reported in drug abusers, indicating that localized immunosuppression might be triggered by the intranasal administration of narcotics (Yewell et al. 2002). In addition, cocaine-induced midline nasal lesions, including erosion of the nasal septum and palate, and including the presence of anti-neutrophil cytoplasmic antibodies, might be misdiagnosed as granulomatosis (Perez Alamino and Espinoza 2013; Rachapalli and Kiely 2008), which have similar symptoms. The accumulating evidence for the effect of drug sniffing on intranasal destruction has recently provoked more extensive studies on the damage caused by other drugs such as heroin (Peyrière et al. 2013) and methamphetamine (Bakhshaee et al. 2013), the latter being a prescribed psychostimulant, which elevates mood and alertness.

The damage to the nasal cavity caused by sniffing ketamine has yet to be fully determined. However, side effects such as nasal passage irritation and a transient change in taste have been reported in patients who have daily intranasal administration of ketamine for chronic pain relief (Carr et al. 2004). We have performed experiments on 2- and 4-month-old mice to study the effect of ketamine on the nasal structures via histopathology and MRI. As 1-year-old mice are approximately equal to a 40- to 50-year-old human, our 2- and 4-month-old mice are equivalent to 6- to 8-year-old and 13- to 16-year-old humans, respectively. Prominent damage to the nasal epithelium and glands and the formation of polyps and granuloma were observed after just 1 month of ketamine treatment. In addition, the normal pseudostratified respiratory epithelium (Figure 7.1a) was lost in most experimental animals (Figure 7.1b). In some specimens, denuded epithelium on the septum was demonstrated, with only the basal layer of the epithelium retained (Figure 7.2a), and signs of metaplasia on the epithelium of the septum were also observed (Figure 7.2b). In addition, severe hyperemia occurred, in which congested blood vessels developed in the lamina propria of the lateral nasal wall epithelium (Figure 7.3a), including that covering the conchae (Figure 7.3b). The effect of drug sniffing on the blood vessels has also been reported in some cocaine abusers, when inflammation of small blood vessels (i.e., vasculitis) in the nasal cavity was documented (Perez Alamino and Espinoza 2013). This is somewhat similar to a form of cutaneous drug reaction (Verma et al. 2013).

Our results also demonstrated that the severity in the damage caused to the nasal cavity was dependent on the age of the animal. Older animals

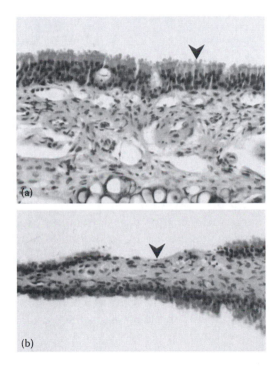

Figure 7.1 Hematoxylin and eosin–labeled sections showing the nasal cavity epithelium in control and ketamine-treated 2-month-old mice. (a) The normal epithelium in the nasal cavity of the control mouse. Note the pseudostratified columnar epithelium (arrowhead) with cilia on top of the dense connective tissue and glands. (b) After administration of ketamine, the nasal epithelium exhibits a denuded surface (arrowhead). Magnification is ×400.

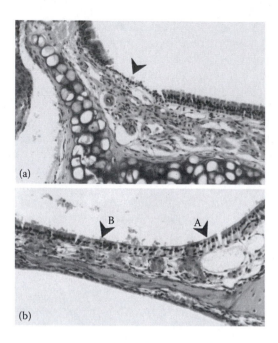

Figure 7.2 Hematoxylin and eosin–labeled sections showing the denuded epithelium of ketamine-treated 2-month-old mice. Images show (a) that the basal layer was retained (arrowhead) and (b) while some epithelia were normal (A), others showed metaplasia (B). Magnification is ×400.

(i.e., 4-month-old mice) were more prone to severe damage than younger animals after the same ketamine treatment regimen. More drastic changes were displayed in the older mice after exposure to ketamine for a month. These included large areas of the nasal epithelium becoming monolayered, and in the lamina propria, the accumulation of exudates and degeneration of glands occurred (Figure 7.4a). In addition, the nasal cavity wall revealed a severe focal loss of epithelium and an accumulation of inflammatory cells (Figure 7.4b). Moreover, large polyps were seen to protrude out from the concha (Figure 7.5a), which had a nonciliated, squamous epithelium (Figure 7.5b) rather than the usual pseudostratified columnar ciliated epithelium. In addition, there were glands and blood vessels inside the polyps, and while some of the polyps that protruded out from the wall of the nasal cavity were round, others were fungiform in appearance (Figure 7.5c). Furthermore, many foci of immune cells with denuded epithelium were clearly noted in the wall of the nasal cavity (Figure 7.5d). In essence, the older animals exhibited more obvious histopathological changes in the nasal cavity after intranasal ketamine treatment. The more severe damage noted in the older animals is consistent with other reports,

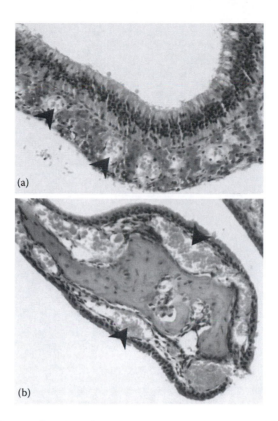

Figure 7.3 In the nasal cavity of the ketamine-treated 2-month-old mouse, hyperemia of tissue was seen in the (a) nasal wall (arrowheads) and (b) nasal conchae (arrowheads). Magnification is ×400.

which demonstrated that older animals suffered more with drug-toxicity effects (Beals et al. 2003; Jevtovic-Todorovic and Carter 2005).

In our ketamine experiments, examination of the nasal cavity with MRI indicated that approximately 70% of 2-month-old mice had high-contrast material in the nasal cavity not seen in the normal controls (Figure 7.6a and b). On the other hand, more severe effects were observed in the older animals, in that degeneration of brain tissue in the temporal lobes of the brain and erosion of the bones in the surrounding skull were observed (Figure 7.6c).

In mice, even a low dose of ketamine (i.e., 0.02 mg per animal) caused apparent adverse changes to the functional structures of the nasal cavity. On the other hand, severe damage to the nasal cavity was observed in long-term ketamine addicts who, in general, use as much as 200 mg of ketamine per dose per day (personal information supplied by local psychiatrists and the law enforcement agency), and thus prominent pathology is

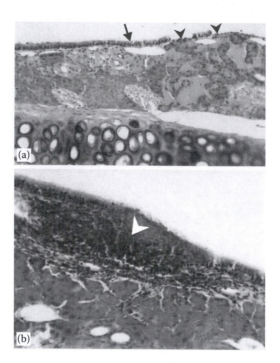

Figure 7.4 Hematoxylin and eosin–labeled sections through the nasal cavity of a representative ketamine-treated 4-month-old mouse. Images show (a) denuded area (arrowheads) and thin epithelium (arrow) in the nasal cavity, with a few glands, and (b) inflammatory changes. Images at (a) ×400 and (b) ×50 magnification.

unavoidable. As far as time required for initiation of damage in the nasal cavity, it is clear that after just a month of sniffing ketamine, significant reactions can already be observed in both young and old animals.

To date, relatively few studies have been reported that employ animal models to investigate the nasal and lung damage that occurs after continuous application of abusive drugs via the intranasal route. Herculiani et al. (2009) exposed mice to a 5-g daily dose of crack cocaine for 2 months and reported a thinning of the epithelium in the mucosae of the nose and in the bronchi, with a concomitant thickening of the pulmonary arteries owing to vasoconstriction. They also reported an elevated level of hemosiderin in the alveoli and an increase in the number of macrophages. In our study, we demonstrated that sniffing ketamine caused the initial severe changes in the nasal cavity. These were mostly granulomatous lesions or polyps, although metaplasia of nasal epithelium was also noted. It appeared that aging led to more vigorous damages. Metaplasia should be of great concern to long-term addicts as it might lead to neoplastic changes. Continuous use of ketamine for long periods, no matter whether via the intranasal application or

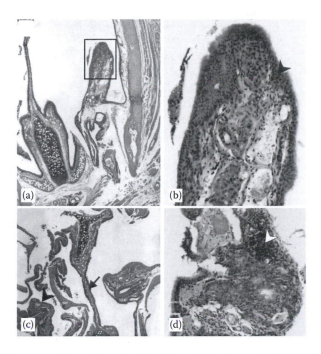

Figure 7.5 Hematoxylin and eosin–labeled sections through the nasal cavity of a representative 4-month-old mouse after intranasal ketamine treatment. Images show (a) a polyp at low magnification (×50). The region bounded by the rectangle in panel (a) is shown at higher magnification (×400) in panel (b), and the arrowhead shows the squamous epithelium. (c) A low magnification view (×50) of a fungiform polyp (arrowhead) and nasal septum (arrow), and (d) a granulomatous lesion-like (magnification: ×50) structure containing immune cells (arrowhead).

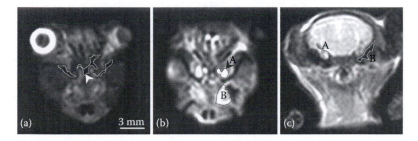

Figure 7.6 Magnetic resonance images of mouse brains. Images show (a) a normal (control) 2-month-old mouse with a clear nasal cavity (white arrowhead), (b) a 2-month-old mouse after ketamine treatment displaying infiltration in both the nasal conchae (A) and maxillary sinus (B), and (c) another ketamine-treated mouse displaying temporal lobe lesion (A) and bone erosion (B).

via system injection or ingestion, can lead to lung fibrosis, with an increase in the number of collagen fibers, first around the bronchi and then around the alveoli (Figure 7.7a and b). We first observed this pathological change in mice after treatment with ketamine for over 3 months. It has long been known that addicts are prone to lung abnormalities and infections; however, the effect of long-term ketamine use on the lung physiology in human addicts has still not been formally investigated. Nonetheless, clinicians often blame these symptoms on the low immune resistance that is acquired because of drug usage.

Although there are relatively few reports that describe drugs of abuse, such as ketamine, causing fibrosis of the lung, far more reports describe drug-induced lesions in the lung. For many years, paraquat, used to treat urinary tract infections, was known to induce changes in the lungs of women during pregnancy (Chomchai and Tiawilai 2007). In this case,

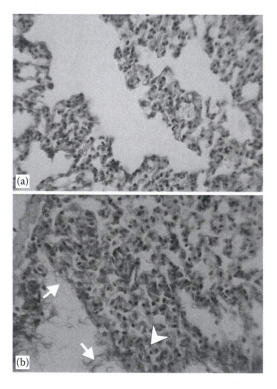

Figure 7.7 Histological images of the lungs of normal and ketamine-treated mice. (a) Lung of a representative normal (control) mouse with no fibrosis (note the clear alveoli). (b) Fibrosis of lung after 1 month of ketamine treatment. Note the formation of the collagen fibers as revealed by Sirius Red staining in the bronchi (white arrows) and alveolar wall (white arrowhead). Magnification is ×400.

pleural effusion and pulmonary hemorrhage occurred along with the infiltration of eosinophils, and this progressed to interstitial pneumonia (Boggess et al. 1996). Other agents, such as nilutamide and bleomycin, have also been reported to lead to interstitial pneumonia, and pathological features such as lymphocytosis and bronchoalveolar fluid collection may result in an immunological episode (van der Drift and Kaajan 2002). Interstitial pneumonia usually leads to fibrosis, via the use of amiodarone, methotrexate, or antiandrogens (van der Drift and Kaajan 2002). In addition, inhalation of inorganic particles such as silica and liquid droplets such as mineral oil may lead to fibrosis of the lung (Dogan et al. 2014; Fazzi et al. 2014; Morfeld et al. 2013). In all of these induced lung diseases, inflammation is a key factor to the fibrosis that results (Rastrick and Birrell 2014). The results we obtained from our study of ketamine inhalation in mice are in accord with the previous reports, as we observed that lymphocytes are present in lung before fibrosis occurs (unpublished data). Other environmental agents including dust after long-term exposure could also be a risk for lung fibrosis in addition to other lung allergic or infectious reactions (Momoh et al. 2013).

The prognosis of idiopathic lung fibrosis and that of either induced fibrosis or that which occurs secondary to another disease differs (Alhamad et al. 2012). Idiopathic disease patients usually have an increased rate of mortality (de Lauretis et al. 2011; Strand et al. 2014; Vij et al. 2011). In fact, the histology of idiopathic lung disease is not the same as that of induced or secondary lung disease (Tansey et al. 2004).

It is known that in humans, heroin causes extensive damage to the central nervous system, especially to the white matter in areas including the cerebellum, posterior internal capsule, frontoparietal cortex, occipital lobe, and brainstem (Jee et al. 2009; Keogh et al. 2003; Tormoehlen 2011). This condition, known as leukoencephalopathy (Tormoehlen 2011), can also occur in patients overdosed with morphine (Salazar and Dubow 2012). With regard to ketamine, brain lesions also initially occurred in the white matter, but they eventually occur in both the white and gray matters (Roberts et al. 2014; Wang et al. 2013). The same diffuse regions of the brain are affected by ketamine as they are by heroin and cocaine. The question here is whether the mode of drug application (i.e., inhalation or injection) might affect the type of lesions that form. Our MRI images of mice brains after 1 month of ketamine delivery via the nose show that both nasal and brain lesions had formed. This appears to signify that the ketamine-induced damage to the brain may be induced by nasal administration as well as by injection.

Indeed, any damage caused to the body and particularly to the respiratory system as a result of the inhalation of toxic materials depends on the total dosage, as well as the size of the particles and their density in the environment, as was clearly demonstrated in the study on cigarette deposits in the alveolus by Gower and Hammond (2007). For our nasal ketamine study, we used ketamine solution, which was applied inside the nasal cavity directly. Our treatment regimen lasted for just 1 month, but in that time, damage to both the nasal cavities and the lung was apparent. The lung damage observed in our animals was consistent with that observed in animals that had been injected with ketamine, where fibrosis of the lung occurred in both treatment groups. As the dose of ketamine we used was lower and the treatment period was shorter for the animals injected with the drug, our results appeared to indicate that the accelerated damage to the lung could manifest after just a short period of nasal sniffing of minute quantities of the drug.

When comparing the mode of ketamine application that we used in our mice experiments with that used by human addicts, there are two major differences. Due to the small nostril size of mice and the fact that ketamine powder is difficult to apply and to ensure that it stays within the nasal cavity, we applied the solution form of ketamine into the mouse nose. Human addicts, on the other hand, insufflate ketamine in its powdered form. In addition, while the purity of ketamine used for the mouse experiments can be closely controlled, the purity of that used by human abusers can never be determined as these drugs are accessed illegally and so different batches might be of differing purities. This was something that always posed a problem. As ketamine is getting to be more popular as an addictive agent worldwide, possibly because of its affordable price, it is important that more studies are conducted to determine the specific damage caused when it is delivered via the nasal route.

References

Alhamad, E.H., Al-Kassimi, F.A., Alboukai, A.A., Raddaoui, E., Al-Hajjaj, M.S., Hajjar, W., and Shaik, S.A. 2012. Comparison of three groups of patients with usual interstitial pneumonia. *Respir. Med.* 106:1575–1585.

Bakhshaee, M., Khadivi, E., Naseri Sadr, M., and Esmatinia, F. 2013. Nasal septum perforation due to methamphetamine abuse. *Iran. J. Otorhinolaryngol.* 25:53–56.

Beals, J.K., Carter, L.B., and Jevtovic-Todorovic, V. 2003. Neurotoxicity of nitrous oxide and ketamine is more severe in aged than in young rat brain. *Ann. N.Y. Acad. Sci.* 993:123–124.

Boggess, K.A., Benedetti, T.J., and Raghu, G. 1996. Nitrofurantoin-induced pulmonary toxicity during pregnancy: A report of a case and review of the literature. *Obstet. Gynecol. Surv.* 51:367–370.

Carr, D.B., Goudas, L.C., Denman, W.T., Brookoff, D., Staats, P.S., Brennen, L., Green, G., Albin, R., Hamilton, D., Rogers, M.C., Firestone, L., Lavin, P.T., and Mermelstein, F. 2004. Safety and efficacy of intranasal ketamine for the treatment of breakthrough pain in patients with chronic pain: A randomized, double-blind, placebo-controlled, crossover study. *Pain* 108:17–27.

Chan, W.M., Liang, Y., Wai, M.S., Hung, A.S., and Yew, D.T. 2011. Cardiotoxicity induced in mice by long term ketamine and ketamine plus alcohol treatment. *Toxicol. Lett.* 207:191–196.

Chomchai, C., and Tiawilai, A. 2007. Fetal poisoning after maternal paraquat ingestion during third trimester of pregnancy: Case report and literature review. *J. Med. Toxicol.* 3:182–186.

Costantini, R., Affaitati, G., Fabrizio, A., and Giamberardino, M.A., 2011. Controlling pain in the post-operative setting. *Int J Clin Pharmacol Ther.* 49(2):116–127.

de Lauretis, A., Veeraraghavan, S., and Renzoni, E. 2011. Review series: Aspects of interstitial lung disease: Connective tissue disease-associated interstitial lung disease: How does it differ from IPF? How should the clinical approach differ? *Chron. Respir. Dis.* 8:53–82.

Dogan, H., Akgun, M., Araz, O., Ucar, E.Y., Yoruk, O., Diyarbakir, E., Atis, O., Akdemir, F., Acemoglu, H., and Pirim, I. 2014. The association of human leukocyte antigen polymorphisms with disease severity and latency period in patients with silicosis. *Multidiscip. Respir. Med.* 9:17. doi: 10.1186/2049-6958-9-17.

Fazzi, F., Njah, J., Di Giuseppe, M., Winnica, D.E., Go, K., Sala, E., St Croix, C.M., Watkins, S.C., Tyurin, V.A., Phinney, D.G., Fattman, C.L., Leikauf, G.D., Kagan, V.E., and Ortiz, L.A. 2014. TNFR1/Phox interaction and TNFR1 mitochondrial translocation thwart silica-induced pulmonary fibrosis. *J. Immunol.* 192:3837–3846.

Gower, S., and Hammond, D. 2007. CSP deposit to the alveolar region of the lung implications of cigarette design. *Risk Anal.* 27:1519–1533.

Herculiani, P.P., Pires-Neto, R.C., Bueno, H.M., Zorzetto, J.C., Silva, L.C., Santos, A.B., Garcia, R.C., Yonamine, M., Detregiachi, C.R., Saldiva, P.H., and Mauad, T. 2009. Effect of chronic exposure to crack cocaine on the respiratory tract of mice. *Toxicol. Pathol.* 37:324–332.

Jee, R.C., Tsao, W.L., Shyu, W.C., Yen, P.S., Hsu, Y.H., and Liu, S.H. 2009. Heroin vapor inhalation-induced spongiform leukoencephalopathy. *J. Formos. Med. Assoc.* 108:518–522.

Jevtovic-Todorovic, V., and Carter, L.B. 2005. The anesthetics nitrous oxide and ketamine are more neurotoxic to old than young rat brain. *Neurobiol. Aging* 26:947–956.

Keogh, C.F., Andrews, G.T., Spacey, S.D., Forkheim, K.E., and Graeb, D.A. 2003. Neuroimaging features of heroin inhalation toxicity "chasing the dragon." *AJR Am. J. Roentgenol.* 80:847–850.

Mak, Y.T., Lam, W.P., Lü, L., Wong, Y.W., and Yew, D.T. 2010. The toxic effect of ketamine on SH-SY5Y neuroblastoma cell line and human neuron. *Microsc. Res. Tech.* 73:195–201.

Mellon, R.D., Simone, A.F., and Rappaport, B.A. 2007. Use of anesthetic agents in neonates and young children. *Anesth. Analg.* 104:509–520.

Momoh, A., Mhlongo, S.E., Abiodun, O., Muzerengi, C., and Mudanalwo, M. 2013. Potential implications of mine dusts on human health: A case study of Mukula Mine, Limpopo Province, South Africa. *Pak. J. Med. Sci.* 29:1444–1446.

Morfeld, P., Mundt, K.A., Taeger, D., Guldner, K., Steinig, O., and Miller, B.G. 2013. Threshold value estimation for respirable quartz dust exposure and silicosis incidence among workers in the German porcelain industry. *J. Occup. Environ. Med.* 56:123–125.

Nakao, S., Nagata, A., Masuzawa, M., Miyamoto, E., Yamada, M., Nishizawa, N., and Shingu, K. 2003. NMDA receptor antagonist neurotoxicity and psychotomimetic activity. *Masui* 52:594–602.

Narendran, R., Frankle, W.G., Keefe, R., Gil, R., Martinez, D., Slifstein, M., Kegeles, L.S., Talbot, P.S., Huang, Y., Hwang, D.R., Khenissi, L., Cooper, T.B., Laruelle, M., and Abi-Darghan, A. 2005. Altered prefrontal dopaminergic function in chronic recreational ketamine users. *Am. J. Psychiatry* 162:2352–2359.

Nielsen, B.N., Friis, S.M., Rømsing, J., Schmiegelow, K., Anderson, B.J., Ferreirós, N., Labocha, S., and Henneberg, S.W. 2014. Intranasal sufentanil/ketamine analgesia in children. *Paediatr Anaesth.* 24(2):170–180.

Perez Alamino, R., and Espinoza, L.R. 2013. Vasculitis mimics: Cocaine-induced midline destructive lesions. *Am. J. Med. Sci.* 346:430–431.

Peyrière, H., Léglise, Y., Rousseau, A., Cartier, C., Gibaja, V., and Galland, P. 2013. Necrosis of the intranasal structures and soft palate as a result of heroin snorting: A case series. *Subst. Abus.* 34:409–414.

Rachapalli, S.M., and Kiely, P.D. 2008. Cocaine-induced midline destructive lesions mimicking ENT-limited Wegener's granulomatosis. *Scand. J. Rheumatol.* 37: 477–480.

Rastrick, J., and Birrell, M. 2014. The role of the inflammasome in fibrotic respiratory diseases. *Minerva Med.* 105:9–23.

Roberts, R.E., Curran, H.V., Friston, K.J., and Morgan, C.J.A. 2014. Abnormalities in white matter microstructure associated with chronic ketamine use. *Neuropsychopharmacology* 39:329–338.

Salazar, R., and Dubow, J. 2012. Delayed posthypoxic leukoencephalopathy following a morphine overdose. *J. Clin. Neurosci.* 19:1060–1062.

Sherry Chow, H.-H., Chen, Z., and Matsuura, G.T. 1999. Direct transport of cocaine from the nasal cavity to the brain following intranasal cocaine administration in rats. *J. Pharm. Sci.* 88:754–758.

Sittel, C., and Eckel, H.E. 1998. Nasal cocaine abuse presenting as a central facial destructive granuloma. *Eur. Arch. Otorhinolaryngol.* 255:446–447.

Strand, M.J., Sprunger, D., Cosgrove, G.P., Fernandez-Perez, E.R., Frankel, S.K., Huie, T.J., Olson, A.L., Solomon, J., Brown, K.K., and Swigris, J.J. 2014. Pulmonary function and survival in idiopathic vs secondary usual interstitial pneumonia. *Chest* 146:775–785. doi: 10.1378/chest.13-2388.

Sun, L., Lam, W.P., Wong, Y.W., Lam, L.H., Tang, H.C., Wai, M.S., Mak, Y.T., Pan, F., and Yew, D.T. 2011. Permanent deficits in brain functions caused by longterm ketamine treatment in mice. *Hum. Exp. Toxicol.* 30:1287–1296.

Tan, S., Rudd, J.A., and Yew, D.T. 2011a. Gene expression changes in $GABA_A$ receptors and cognition following chronic ketamine administration in mice. *PLoS One* 6:e21328. doi: 10.1371/journal.pone.0021328.

Tan, S., Chan, W.M., Wai, M.S., Hui, L.K., Hui, V.W., James, A.E., Yeung, L.Y., and Yew, D.T. 2011b. Ketamine effects on the urogenital system—changes in the urinary bladder and sperm motility. *Microsc. Res. Tech.* 74:1192–1198.

Tansey, D., Wells, A.U., Colby, T.V., Ip, S., Nikolakoupolou, A., du Bois, R.M., Hansell, D.M., and Nicholson, A.G. 2004. Variations in histological patterns of interstitial pneumonia between connective tissue disorders and their relationship to prognosis. *Histopathology* 44:585–596.

Tormoehlen, L.M. 2011. Toxic leukoencephalopathies. *Neurol. Clin.* 29(3):591–605.

van der Drift, M.A., and Kaajan, J.P. 2002. Respiratory side-effects of drugs. *Ned. Tijdschr. Geneeskd.* 146:145–150.

Verma, R., Vasudevan, B., and Pragasam, V. 2013. Severe cutaneous adverse drug reactions. *Med. J. Armed Forces India* 69:375–383.

Vij, R., Noth, I., and Strek, M.E. 2011. Autoimmune-featured interstitial lung disease: A distinct entity. *Chest* 140:1292–1299.

Wang, C., Zheng, D., Xu, J., Lam, W., and Yew, D.T. 2013. Brain damages in ketamine addicts as revealed by magnetic resonance imaging. *Front. Neuroanat.* 7:23. doi: 10.3389/fnana.2013.00023.

Yeung, L.Y., Rudd, J.A., Lam, W.P., Mak, Y.T., and Yew, D.T. 2009. Mice are prone to kidney pathology after prolonged ketamine addiction. *Toxicol. Lett.* 191:275–278.

Yeung, L.Y., Wai, M.S., Fan, M., Mak, Y.T., Lam, W.P., Li, Z., Lu, G., and Yew, D.T. 2010. Hyperphosphorylated tau in the brains of mice and monkeys with long-term administration of ketamine. *Toxicol. Lett.* 193:189–193.

Yewell, J., Haydon, R., Archer, S., and Manaligod, J.M. 2002. Complications of intranasal prescription narcotic abuse. *Ann. Otol. Rhinol. Laryngol.* 111:174–177.

Yu, H., Li, Q., Wang, D., Shi, L., Lu, G., Sun, L., Wang, L., Zhu, W., Mak, Y.T., Wong, N., Wang, Y., Pan, F., and Yew, D.T. 2012. Mapping the central effects of chronic ketamine administration in an adolescent primate model by functional magnetic resonance imaging (fMRI). *Neurotoxicology* 33:70–77.

chapter eight

Long-term ketamine use causes damage to the pancreas and adrenal glands

Wai Ping Lam, Tan Sijie, Lok Hang Lam, Yeak Wan Wong, and Chun-Mei Wang

Contents

8.1 Introduction

The chronic effects observed after long-term ketamine usage, as seen in human addicts, only started to receive attention in the last half decade. For instance, the effect of ketamine on the central nervous system was explored, and areas of ketamine damage in the prefrontal cortex, hippocampus, and striatum were demonstrated by functional magnetic resonance imaging (fMRI), as well as via morphological and immunohistochemical studies (Chan et al. 2012; Mak et al. 2010; Sun et al. 2011; Yeung et al. 2010; Yu et al. 2012). These studies illustrated apoptosis of the neurons (Chan et al. 2012; Mak et al. 2010; Sun et al. 2011), formation of hyperphosphorylated tau (hypertau) in the neurons (Yeung et al. 2010), and changes in fMRI patterns of the brain (Chan et al. 2012; Yu et al. 2012). In the heart, necrosis of cardiac cells and electrocardiogram abnormalities were observed (Chan et al. 2011); in the urinary bladder, loss of muscle cells followed by fibrosis was demonstrated (Tan et al. 2011), while in the kidney, long-term damage by ketamine was shown to lead to glomerular and tubular changes that resulted in proteinuria (Wai et al. 2012).

The effects of ketamine on the pancreas and adrenal glands as vital organs in the body have rarely been addressed in drug addiction studies. The few reports that have been published all describe the effects of ketamine after acute administration but not after long-term ketamine treatment (Changmin et al. 2010; Reyes Toso et al. 1995; Saha et al. 2005). In this study, therefore, we explored the various changes that occur in these organs after long-term chronic ketamine treatment in mice.

8.2 Materials and methods

8.2.1 Grouping of experimental animals and treatments

Four-week-old male Institute of Cancer Research (ICR) mice, raised in the Laboratory Animal Services Centre of the Chinese University of Hong Kong, were kept in cages with water and food pellets in a room that was maintained at 22°C ± 1°C with a 12-h light–dark cycle. The Animal Ethics Committee of the Chinese University of Hong Kong approved this study. Twenty-seven ICR mice were divided randomly into the three groups comprising the saline-treated ($n = 9$), ketamine-treated ($n = 9$), and ketamine plus alcohol–treated ($n = 9$) groups. The ketamine-treated and ketamine plus alcohol–treated mice received daily intraperitoneal injections of ketamine (HK-37715, Alfasan, Holland) at 30 mg/kg for 6 months. This ketamine treatment period was selected based on preliminary results, which showed that pathological changes in the adrenal gland and pancreas were apparent in mice after 6 months of treatment. Mice in the corresponding control group were injected with saline alone. For the ketamine with alcohol group, mice were injected with ketamine for 6 months, and in the final (i.e., 6th) month, they were also given 10% ethanol orally (0.05 ml/day). At the end of the treatment period, all the mice were sacrificed by cervical dislocation. The adrenal glands and pancreas were dissected out and fixed in 4% paraformaldehyde. They were then dehydrated in ethanol and xylene, embedded in paraffin, and sectioned at 5 µm.

8.2.2 Histological studies on the pancreas and adrenal glands

Specimens from each group were deparaffinized, rehydrated, and immersed in Mayer's hematoxylin for 5 min. After rinsing with deionized water, these sections were dipped into 0.1% acid water and then immersed in Scott's tap water for 1 min for the blue color to develop. They were then immersed in 1% eosin for 5–10 min for the development of the red-colored stain. The sections were then run through a graded alcohol series, cleared in xylene, mounted under Permount (USB Corporation, Cleveland, Ohio, USA), and then observed via light microscopy.

8.2.3 *Immunohistochemistry on the pancreas and adrenal glands*

The adrenal gland sections were immunolabeled with antibodies to proliferative cell nuclear antigen (PCNA), lactic acid dehydrogenase (LDH), tyrosine hydroxylase (TH), and dopamine β-hydroxylase (DBH). In addition, the pancreatic sections were immunolabeled with antibodies to PCNA and LDH. The selected sections were dewaxed, rehydrated, and permeabilized for 10 min with 1×phosphate-buffered saline (PBS) supplemented with 0.1% Triton X and 0.05% Tween 20, followed by three rinses for 5 min each in 1×PBS. The endogenous peroxidase activity was blocked with 3% hydrogen peroxidase in methanol for 45 min. After three more rinses in 1×PBS, nonspecific binding was suppressed by incubation for 1 h with either 5% normal rabbit serum (10510, Invitrogen, California, USA) for the LDH and TH labeling or 5% normal goat serum (PCN500, Invitrogen) for the PCNA and DBH labeling. Thereafter, individual sections were incubated overnight at 4°C with one of the following anti-sera: PCNA (dilution: 1:1000; ab29, abcam, Cambridge, UK), LDH (dilution: 1:500; sc27230, Santa Cruz Biotechnology, USA), TH (dilution: 1:1000; sc-7847, Santa Cruz Biotechnology), and DBH (dilution: 1:1000; ab43868, abcam). PCNA is a proliferation marker (Wai et al. 2008) while LDH is a marker of necrosis (Ostrovskiĭ et al. 2011). TH plays an essential role in the synthesis of catecholamines (Bobrovskaya et al. 2010) and DBH plays an essential role in the production of noradrenalin (Kvetnanský et al. 2008). On the following day, sections were rinsed three times with 1×PBS before incubation with the respective diluted secondary antibodies: anti-rabbit goat-biotinylated secondary antibodies for DBH (dilution: 1:500, 656140, Invitrogen), anti-mouse goat-biotinylated secondary antibodies for PCNA (dilution: 1:500, 626520, Invitrogen), and anti-goat rabbit-biotinylated secondary antibodies for TH and LDH (dilution: 1:500, 611640, Invitrogen) in 2% (w/v) bovine serum albumin for 2 h at room temperature. After rinsing with 1×PBS with Tween 20 (PBST) three times, the sections were incubated with Streptavidin-HRP (dilution: 1:500; 43432, Invitrogen) in PBST for 2 h at room temperature. After rinsing with PBST three times again, the sections were then incubated in 5% 3,3-diaminobenzidine (DAB) (Sigma-Aldrich, St. Louis, Missouri, USA) in 1×PBS containing 0.01% hydrogen peroxide (H_2O_2) for 5–10 min until the brown-colored labeling had developed. The sections were then run through an ascending alcohol series, cleared in xylene, mounted under Permount (USB Corporation) and observed with a light microscope.

For quantitation of the level of TH and DBH labeling, five random sections (from anterior to posterior) of each adrenal gland were selected. The positive regions (darkly stained areas) of each section of the adrenal gland were calculated as a percentage of the whole adrenal area in that section. All percentages from the five sections were averaged and represented the percentage of the total area of TH- or DBH-positive sites in the adrenal of that animal. Bearing in mind that each group had 9 animals, we therefore had $n = 9$ adrenal glands for each group. They were then averaged and standard deviations were calculated. Means of the different groups were then compared with the controls using one-way analysis of variance and $p < 0.05$ was considered to be significant.

8.2.4 TUNEL evaluation

Apoptosis in both the pancreas and adrenal glands was detected by in situ terminal transferase mediated dUTP nick end labeling (TUNEL), using an ApopTag kit (s7100, Millipore Corporation, Massachusetts, USA). First, randomly selected sections from each group of animals (five slides for each group) were deparaffinized, rehydrated, and incubated with proteinase-K in PBS at 20 µg/ml for 15 min at room temperature. Sections were rinsed with 1×PBS and then they were quenched in 3.0% H_2O_2 in PBS for 5 min, followed by a rinse with 1×PBS again. Subsequently, the sections were treated with biotin-deoxyuridine triphosphate in the working solution of deoxynucleotidyl transferase for 1 h in a humidified chamber at 37°C, followed by incubation in the working stop/wash buffer for 30 min at 37°C. The sections were then rinsed with 1×PBS three times (5 min each), and three drops of anti-digoxigenin-peroxidase were applied to each section and incubated for 30 min at room temperature. The sections were rinsed with PBS and reacted with DAB and 0.01% H_2O_2 for visualization of apoptotic cells. They were subsequently rinsed with distilled water and dehydrated before being mounted under Permount and stained slides were then observed via light microscopy.

8.3 Results

Histology on the saline-treated adrenal glands revealed a normal arrangement of the three cortical zones (i.e., the glomerulosa, fasciculata, and reticularis) and the medulla (Figure 8.1a). After ketamine treatment alone, the three cortical zones were morphologically the same as the controls (Figure 8.1b), whereas after ketamine plus alcohol treatment, the zona fasciculata demonstrated focal misalignment of fasciculata cells (Figure 8.1c), and in approximately 10% of animals receiving ketamine plus alcohol treatment, a nodular aggregation of cells with dense nuclei was apparent in the cortex (Figure 8.1d).

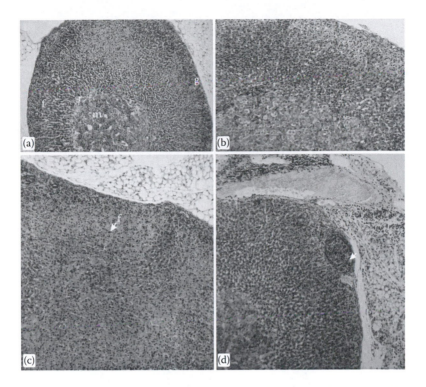

Figure 8.1 Hematoxylin and eosin (HE) staining of the adrenal gland. (a) Normal adrenal gland with zona glomerulosa (g), zona fasciculata (f), zona reticularis (r), and medulla (m). (b) A representative adrenal gland exhibiting normal histology in a ketamine-treated mouse. (c and d) The adrenal gland of a representative ketamine and alcohol–treated mouse showing (c) disorganization of some areas of the zona fasciculata (f and arrow) and (d) nodular infiltration of cells with dense nuclei (arrow) in the cortex. Magnification, ×100.

Immunohistochemistry for LDH revealed no positive cells in the controls (Figure 8.2a), while some LDH-positive cells were apparent in the deep adrenal cortex of the ketamine alone–treated animals (Figure 8.2b). In the ketamine plus alcohol–treated mice, LDH-positive cells were observed in both the cortex and the medulla of the adrenal glands (Figure 8.2c). On the other hand, PCNA-positive nuclei were present in both the cortex and medulla of the controls (Figure 8.3a) but were less obvious in the ketamine and ketamine plus alcohol treatment group (Figure 8.3b and c). No TUNEL-positive sites were seen in any group (Figure 8.3d and e).

Immunocytochemistry for TH showed positive sites in the zona glomerulosa, fasciculata, and the medulla in the controls (Figure 8.4a), whereas after ketamine treatment, there was a significant decrease in the density of

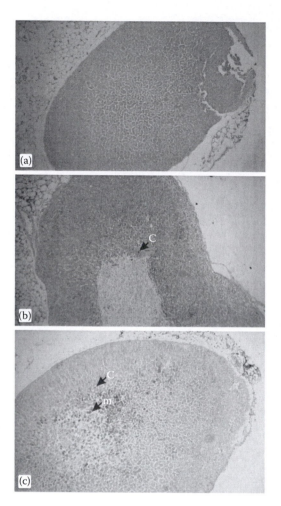

Figure 8.2 Immunochemistry of LDH in the adrenal gland. (a) Normal adrenal gland with few LDH-positive sites. (b) The adrenal gland of a representative ketamine-treated mouse showing some LDH-positive sites (arrow) in the cortex (C). (c) The adrenal gland of a representative ketamine and alcohol–treated mouse showing more LDH-positive sites (arrow) in both the cortex (C) and medulla (m). Magnification, ×100.

these sites in both the medulla and the cortex (Figure 8.4b). Further down-regulation in the number of these positive sites was observed in the keta-mine plus alcohol–treated group (Figure 8.4c). Since catecholamines in the adrenal medulla are of great physiological importance, the optical density of TH was quantified in the adrenal medulla for all three treatment groups with the microphotometer (Minolta III). The optical density of TH

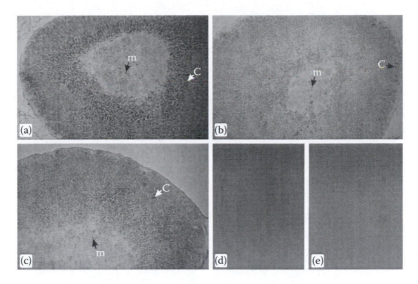

Figure 8.3 Immunochemistry of PCNA, and TUNEL in situ hybridization in the adrenal gland. (a–c) Representative examples of adrenal glands from (a) a normal mouse, (b) a ketamine-treated mouse, and (c) a ketamine plus alcohol–treated mouse. (a) The adrenal gland from the normal mouse has many PCNA-positive cells (arrows) in both the medulla and the cortex, whereas after (b) ketamine or (c) ketamine and alcohol treatment, progressively fewer PCNA-positive cells (arrows) were observed. In panels (a) through (c), C and m are cortex and medulla, respectively. (d and e) TUNEL in situ hybridization of (d) a normal adrenal gland and (e) an adrenal gland from a ketamine and alcohol–treated mouse. No TUNEL-positive sites were seen in either panel (d) or (e). Magnification, ×100.

was determined to be 91.39% ± 3.42% in the controls, 31.6% ± 15.06% in the ketamine-treated group, and 12.15% ± 8.03% in the ketamine plus alcohol–treated animals ($n = 9$ for each group). In addition, immunohistochemistry for DBH confirmed its down-regulation in both the cortex and the medulla of the ketamine-treated and the ketamine plus alcohol–treated group when compared with the controls (Figure 8.5a and b). Interestingly, there was not much difference between the ketamine-treated and ketamine plus alcohol–treated group in this case (Figure 8.5c). It was also interesting that although the down-regulatory trend of TH and DBH was the same after ketamine treatment, the number of DBH-positive sites was much less than that of TH-positive sites.

At first glance, the pancreas of all animals appeared to have a normal histology, with complements of both acinar cells and islets, and some of the islets were huge in size in the normal control mice (Figure 8.6a). After ketamine treatment, the acinar cells were still normal in

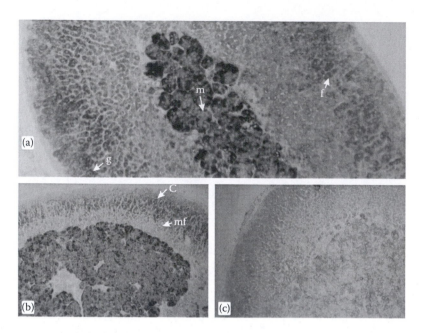

Figure 8.4 Immunochemistry of TH in the adrenal gland. (a) Representative normal adrenal gland with TH-positive signals (arrows) in the cortical glomerular (g) and fasciculata (f) cells as well as in the reactive medulla (m). (b) A representative ketamine-treated mouse showing TH-positive cells in the cortex (C) and in focal areas of the medulla (mf). (c) A representative ketamine and alcohol–treated mouse with few TH-positive sites. Magnification, ×100.

appearance (Figure 8.6b) but many of the islets appeared to be smaller in size, and after the ketamine plus alcohol treatment, some islets were even smaller and appeared to go into atresia (Figure 8.6c and d). As an example, the number of islets above the size of 30 μm^2 in area per 700 μm^2 area ($n = 30$ fields) was 1.67 ± 0.9 ($n = 30$ fields) in the control, 0.67 ± 0.48 in the ketamine-treated group, and 0.32 ± 0.21 in the ketamine plus alcohol–treated animals. In two of the nine animals, mononuclear inflammatory cells infiltrated the pancreas (Figure 8.6e) and some of these inflammatory cells were located around the pancreatic ducts (Figure 8.6f).

PCNA-positive nuclei were found in both the acinar cells and the islets of the normal control mice (Figure 8.7a). After ketamine treatment, PCNA-positive nuclei were still present in the acinar cells and the islets (Figure 8.7b) but were fewer in number (Figure 8.7b) than those in the control mice (Figure 8.7a). After ketamine plus alcohol treatment, PCNA-positive nuclei were present in the acinar cells only (Figure 8.7c). LDH, a

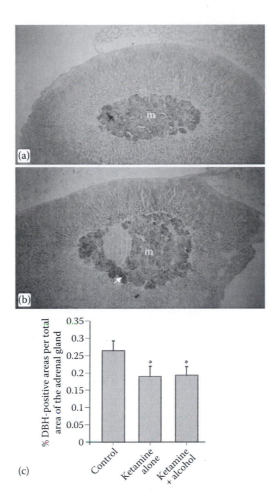

Figure 8.5 Immunochemistry of DBH in the adrenal medulla. (a) A representative ketamine-treated mouse with few DBH-positive cells (arrow) in focal areas of the medulla (m). (b) A representative ketamine and alcohol–treated mouse with DBH-positive cells in focal areas (arrow). (c) The percentage of DBH-positive areas per total area of the adrenal gland in the control, ketamine, and ketamine plus alcohol groups. Bars represent mean ± SD. *Statistically significant data at $p < 0.05$. Magnification, ×100.

marker of necrosis, showed no reaction in the control specimens (Figure 8.8a) while only acinar cells were positive in the ketamine-treated animals (Figure 8.8b), and in the ketamine plus alcohol treatment group, the cells in the islets also showed LDH activity (Figure 8.8c). There were no TUNEL-positive cells in any of the groups (Figure 8.9a and b).

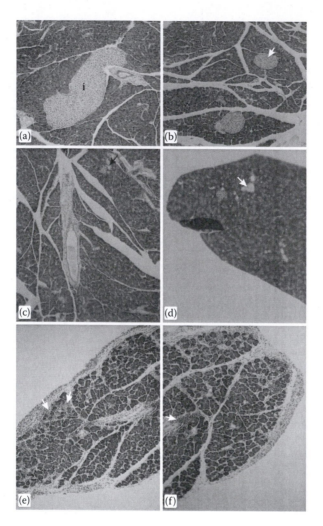

Figure 8.6 Hematoxylin and eosin (HE) staining of an islet in the pancreas. (a) Normal pancreas with a large islet (i). (b) The pancreas of a ketamine-treated mouse with smaller islet (arrow). (c and d) The pancreas of a ketamine and alcohol–treated mouse with degenerating islets (arrows). (e and f) The pancreas of a ketamine and alcohol–treated mouse showing the infiltration of mononuclear cells (e) into the acinar cells (arrows) and (f) around the duct (arrow). Magnification, ×100.

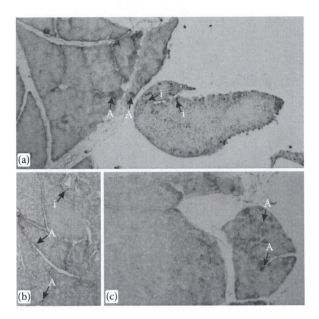

Figure 8.7 Immunochemistry of PCNA in the pancreas. (a) Normal pancreas with PCNA-positive signals (arrows) in the acinar (A) and islet (i) cells. (b) The pancreas from a representative ketamine-treated mouse with relatively few PCNA-positive signals (arrows) in the acinar (A) and islet (i) cells. (c) The pancreas from a representative ketamine and alcohol–treated mouse with PCNA-positive sites (arrows) in the acinar (A) cells alone. Magnification, ×100.

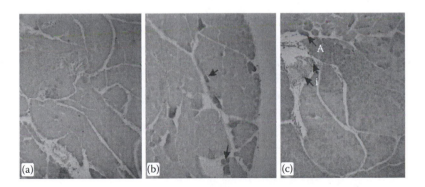

Figure 8.8 Immunochemistry of LDH in the pancreas. (a) Normal pancreas without any LDH-positive cells. (b) The pancreas from a representative ketamine-treated mouse with LDH-positive cells (arrows). (c) The pancreas from a representative ketamine and alcohol–treated mouse with LDH-positive sites (arrows) in both the acinar (A) and islet (i) cells. Magnification, ×100.

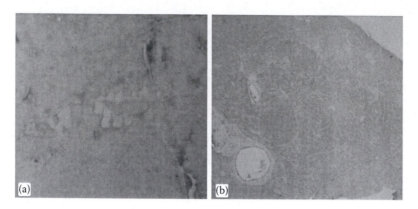

Figure 8.9 TUNEL in situ hybridization in the pancreas. (a) Normal pancreas. (b) The pancreas from a representative ketamine and alcohol–treated mouse. No positive TUNEL nuclei were found in either panel (a) or (b). Magnification, ×100.

8.4 Discussion

Only a small number of animal studies have been conducted to investigate the effect of ketamine on the pancreas and the adrenal glands. Saha et al. (2005) reported that in fasting rats, ketamine at 100 mg/kg produced acute hyperglycemia with 174.8 ± 5.7 mg/dl glucose, reaching a maximum of almost 300 mg/dl within just 2 h after the administration of ketamine. Such a dramatic increase in the level of glucose was not recorded in fed rats. This induced hyperglycemia in fasting rats could be blocked by yohimbine, an α_2-adrenergic receptor antagonist, at 1 to 4 mg/kg (Saha et al. 2005). These results corroborated those from another study conducted 10 years earlier, which suggested that ketamine or barbitone could inhibit insulin secretion mediated by the adrenergic innervation of the pancreas of the rat (Reyes Toso et al. 1995). It was further suggested that after acute ketamine treatment, circulatory changes occur in the catecholamines secreted by the adrenal medulla. More recently, Changmin et al. (2010) confirmed that after short-term catecholamine treatment, there is an increase in the plasma glucose level and a decrease in the level of insulin, as well as in adrenalin and noradrenalin. Furthermore, it was most recently reported that ketamine treatment can deplete the formation of liver glycogen (Wong et al. 2012). The fluctuation in blood glucose level after ketamine treatment must therefore be carefully monitored.

The results recorded in this study are novel for the pancreas and the adrenal glands. We showed that after long-term ketamine treatment, the pancreatic islets appear to be much smaller in size, even though the TUNEL results indicated no apoptosis. In addition, immunolabeling of LDH (a marker for necrosis) only showed a slight increase in the level of this protein in the pancreas. On the other hand, the smaller size of the islets might indicate functional hormonal disturbance, resulting in less secretion of insulin. This, together with liver cirrhosis, which might disrupt the storage of cellular glycogen (Wai et al. 2012) and thus cause a rise in plasma glucose, are just two of the many reasons that might lead to an increase in glucose in the blood. Our findings from these long-term experiments correspond with those reported by Reyes Toso et al. (1995), Saha et al. (2005), and Changmin et al. (2010) in their acute ketamine experiments, as well as with those of Wong et al. (2012) in their chronic-treatment experiments.

Our study also reveals several novel findings in the pancreas and adrenal glands after treatment with ketamine, both with and without the addition of alcohol. First, we showed that there was a decrease in the number of proliferative (i.e., PCNA-positive) cells and an increase in the number of necrotic (i.e., LDH-positive) cells in these organs after ketamine treatment. Perhaps the most significant result was the decrease in the localization of TH- and DBH-expressing cells as both are important productive enzymes in the catecholaminergic pathway in the adrenal glands. Such a reduction indicates a significant disruption of catecholamine production in general, and of noradrenalin specifically, which might prove to be life-threatening. A decrease in noradrenalin or adrenalin might be attributed to too much noradrenalin or adrenalin being released into the blood or a decrease in the production of these chemicals. Our results seemed to favor the latter. Further studies need to be conducted in order to elucidate the effect of ketamine on adrenal gland function.

The presence of catecholamines, especially dopamine and its analogues, in the adrenal glands has been well documented (Bird et al. 1998; Charlton et al. 1992; Choi et al. 1993; Toth et al. 1997). It was suggested that noradrenaline in the fasciculata cells is probably acting on vessels (Bird et al. 1998; Charlton et al. 1992; Choi et al. 1993; Toth et al. 1997) while dopamine itself might stimulate cortisol secretion (Bird et al. 1998). Dopamine analogues in the medulla are, of course, very important for whole-body physiology (Choi et al. 1993). The depletion of catecholamines upon the down-regulation of synthesis induced by long-term ketamine administration thus had a profound effect on the homeostasis of the body both directly and indirectly and is thus worthy of further research. Taken together, disturbances in the adrenal gland might involve both hormones and neurotransmitters.

Our results appeared to show that the toxic effects of ketamine were focused mainly in the adrenal cortex and medulla, both of which are instrumental for vital body functions, with the middle cortex apparently being affected primarily. It is common knowledge that glucocorticoids are secreted by this part of the adrenal cortex. Thus, impairment or degeneration of this area is likely to affect gluconeogenesis in the liver, glucose regulation by cells, and fluctuation in the blood glucose supply.

Our results appear to correspond with those reported previously where there was a lack of glycogen in the liver in chronic ketamine-treated mice (Wong et al. 2012). A disturbance in the level of glucose is an essential parameter that can compromise the health and integrity of living individuals. Fortunately, our study revealed that other areas of the adrenal cortex appeared to be spared from obvious damage at least at the histological level, and thus mineral corticoids and ions such as potassium would likely not be affected.

Our results also showed that the adrenal medulla, and hence the secretion of catecholamines, was also affected by chronic ketamine administration. Maintaining catecholamines at a stable level is of great importance for our survival. In ketamine addicts, the irregular production of catecholamines by the adrenal glands might lead to a sudden drop in blood pressure and, with the added weakening effect of the drug on the hearts, result in sudden death. This has previously been demonstrated to occur in heroin addicts (Li et al. 2005).

We showed that after ketamine treatment, the pancreatic islets in mice appeared to be smaller in size than those in the controls and of particular significance was the decrease in the number of large β islets after ketamine alone and the ketamine–alcohol combined treatment. The decrease in the number of large-sized β islets after ketamine treatment was approximately 40% and the decrease of those after ketamine–alcohol treatment was an additional 30%. The decrease of the large-sized β islets would likely result in a decrease in insulin production. This might be complicated further by the lack of glycogen in the liver, as together they would lead to an increased level of blood sugar in the mouse.

Liver damage after ketamine treatment is to be expected as the drug is detoxified and metabolized in this organ. In our experiment in mice, the effects on the liver were evident even after just 3 months of ketamine treatment, with changes in transaminases, cellular damage, and cellular death being apparent. The combination of ketamine and alcohol is, as expected, an extra burden on the liver. However, the idea that ketamine plus alcohol might double the liver damage was not always valid. In most cases, alcohol added to the damage but the level of damage as reflected in the quantity of transaminases in the ketamine–alcohol-treated mice was not double that of the ketamine alone–treated animals (Wai et al. 2012, 2013).

Although some patients with chronic ketamine addiction complained of an irritable abdomen, subsequent histopathological examination failed to reveal any major findings in the stomach, with only occasional areas of focal atrophy. In the intestine, however, the effects of the drug were more profound, as proliferation of the lymphoid glands and retardation of intestinal movement were noted (Wong et al. 2012). It appears that ketamine might affect the lymphoid system as well as the autonomic balance in the intestine.

The ailments induced in all of these vital organs might potentially be fatal and particular care must be taken to warn people against ketamine addiction.

The pancreatic lesions resulting from alcohol are well documented in the literature, but those resulting from abusive drugs are rarely reported. Chronic pancreatitis resulting from abnormal pancreatic enzymatic action (i.e., initially down-regulation but then up-regulation) and precipitation of protein plugs are features in this disease (Goebell and Singer 1978). This decrease and increase in enzyme action is, however, not ubiquitous. For example, trypsinogen was shown to increase 8 months after alcohol intake but diminished 16 months later in the rat model (Gronroos et al. 1988), which is somewhat contrary to the results reported by Goebell and Singer (1978). It was suggested that many of the defective enzyme action effects observed with alcohol might be caused by the inhibitory effect of alcohol on the cholinergic system via the vagus nerve (Siegmund and Singer 2005; Singer 1985). In our study, we recorded the cellular and immunohistochemical changes in the pancreas after ketamine treatment and confirmed the occurrence of chronic pancreatitis. Occasionally, we also observed precipitation of particles similar to the protein precipitation mentioned in other drug studies (Goebell and Singer 1978). Although we did not study either enzymatic secretion or the cholinergic system after ketamine treatment on the pancreas, the effect of ketamine on the urinary system seemed to indicate modes of nerve involvement (Dargan et al. 2014). Pancreatic lesions were also often associated with liver cirrhosis, as well as stomach and esophagus mucosal lesions (Hastier et al. 1999; Siegmund and Singer 2005). Further detailed ketamine studies on these organs are thus warranted.

While we document the effect of ketamine on the adrenal glands here, the effect of alcohol and other drugs on these glands has been previously described. The effect of alcohol, for example, has been studied in chronically fed rats, and it was shown that an increase of adrenomedullary TH, DBH, and phenylethanolamine N-methyltransferase mRNA was observed, which was associated with increased plasma catecholamine level (Patterson-Buckendahl et al. 2005). In addition, reserpine and tetrabenazine were shown to inhibit chromaffin cell catecholamine uptake, which led to the depletion of catecholamine and to an increase of opioid peptides produced by these cells (Wilson et al. 1981). On the other hand, the sympatho-reaction caused by cocaine had a biphasic response with an initial cardiovascular excitation and subsequent inhibition, probably via the adrenal glands and the rest of the sympathetic system (Knuepfer and Branch 1992). Cocaine also induces an increase of blood pressure and an increase in the local cortical blood flow, which might increase the probability of hemorrhagic stroke (Kelley et al. 1993; Muir and Ellis 1993). Amphetamine had a similar effect, raising the ACTH level initially in rat followed by a decline 30 min later (Swerdlow et al. 1993). Long-term abusive drug addiction, however, had more profound effect on the adrenal glands, and there was a report from a postmortem case of a heroin addict where a depletion of adrenal medulla cells was described (Li et al. 2005).

Drug toxicity on the liver is, of course, not solely indicated by ketamine toxicity. A wide spectrum of drugs can lead to the generation of reactive metabolites via the mitochondrial cytochrome system, which results in the overproduction of reactive oxygen species by the damaged mitochondria, along with changes in fatty acid oxidation, lipid deposition, and steatosis (Begriche et al. 2011). Compared with the above, there are often alanine aminotransferase and aspartate aminotransferase changes, glutathione depletion, and cellular degeneration (Kumar et al. 2011).

Ecstasy (3,4-methylenedioxymethamphetamine) is a drug that is highly toxic to the liver, which can result in hyperthermia, hyperexcitation, hypertension, jaundice, cholestasis, and hepatomegaly, which are often associated with changes in the kidney similar to those that occur after ketamine treatment (Ellis et al. 1996; Garbino et al. 2001). Cocaine hepatotoxicity is often aggravated by being used in combination with barbital (Charles and Powell 1992; Roth et al. 1992). Liver enzymes might increase by 15 times in a mere 2 h after an acute overdose (Charles and Powell 1992), and perilobular hepatocytes degenerate at first and then migrate centrally (Roth et al. 1992). In our chronic models with ketamine, fibrosis was a feature after 3 months of treatment. The nodule of fibrosis also occurred peripherally and went hand in hand with a moderate increase in enzymes (Wai et al. 2012). More interesting was the down-regulation of glycogen storage in the liver (Wong et al. 2012). With regard to the latter, we have yet to evaluate whether these changes are attributed to altered glycogen synthesis or glycogen degradation. The influence of ketamine on glucose metabolism is therefore instrumental for an individual's survival and will open a new chapter in the control of, and changes in, this important metabolism.

Acknowledgment

This study was funded by a grant from the Beat Drugs Fund Association, Hong Kong Government (Project Ref. No. BDF100052).

References

Begriche, K., Massart, J., Robin, M.A., Borgne-Sanchez, A., and Fromenty, B. 2011. Drug-induced toxicity on mitochondria and lipid metabolism: Mechanistic diversity and deleterious consequences for the liver. *J. Hepatol.* 54:773–794.

Bird, I.M., Lightly, E.R., Nicol, M., Williams, B.C., and Walker, S.W. 1998. Dopaminergic stimulation of cortisol secretion from bovine zfr cells occurs through nonspecific stimulation of adrenergic beta-receptors. *Endocr. Res.* 24:769–772.

Bobrovskaya, L., Damanhuri, H.A., Ong, L.K., Schneider, J.J., Dickson, P.W., Dunkley, P.R., Goodchild, A.K. 2010. Signal transduction pathways and tyrosine hydroxylase regulation in the adrenal medulla following glucoprivation: An in vivo analysis. *Neurochem. Int.* 57(2):162–167.

Chan, W.M., Liang, Y., Wai, M.S., Hung, A.S., and Yew, D.T. 2011. Cardiotoxicity induced in mice by long term ketamine and ketamine plus alcohol treatment. *Toxicol. Lett.* 207:191–196.

Chan, W.M., Xu, J., Fan, M., Jiang, Y., Tsui, T.Y., Wai, M.S., Lam, W.P., and Yew, D.T. 2012. Downregulation in the human and mice cerebella after ketamine versus ketamine plus ethanol treatment. *Microsc. Res. Tech.* 75:258–264.

Changmin, H., Jianguo, C., Dongming, L., Guohong, L., and Mingxing, D. 2010. Effects of xylazole alone and in combination with ketamine on the metabolic and neurohumoral responses in healthy dogs. *Vet. Anaesth. Analg.* 37:322–328.

Charles, S.J., and Powell, C.J. 1992. Rapidly developing cocaine-induced peripheral portal liver damage. *Toxicol. Lett.* 64–65 Spec No:729–737.

Charlton, B.G., McGadey, J., Russell, D., and Neal, D.E. 1992. Noradrenergic innervation of the human adrenal cortex as revealed by dopamine-beta-hydroxylase immunohistochemistry. *J. Anat.* 180:501–506.

Choi, W.S., Kim, M.O., and Ramirez, V.D. 1993. Immunocytochemical localization of dopamine-releasing protein in the rat adrenal. *Neuroendocrinology* 58:440–447.

Dargan, P.L., Tang, H.C., Liang, W., Wood, D.M., and Yew, D.T. 2014. Three months of methoxetamine administration is associated with significant bladder and renal toxicity in mice. *Clin Toxicol (Phila).* 52(3):176–180.

Ellis, A.J., Wendon, J.A., Portmann, B., and Williams, R. 1996. Acute liver damage and ecstasy ingestion. *Gut* 38:454–458.

Garbino, J., Henry, J.A., Mentha, G., and Romand, J.A. 2001. Ecstasy ingestion and fulminant hepatic failure: Liver transplantation to be considered as a last therapeutic option. *Vet. Hum. Toxicol.* 43:99–102.

Goebell, H., and Singer, M.V. 1978. Effect of alcohol on the human and animal pancreas. *Leber. Magen. Darm.* 8:304–314.

Gronroos, J.M., Aho, H.J., Meklin, S.S., Hakala, J., and Nevalainen, T.J. 1988. Pancreatic digestive enzymes and ultrastructure after chronic alcohol intake in the rat. *Exp. Pathol.* 35:197–208.

Hastier, P., Buckley, M.J., Francois, E., Peten, E.P., Dumas, R., Caroli-Bosc, F.X., and Delmont, J.P. 1999. A prospective study of pancreatic disease in patients with alcoholic cirrhosis: Comparative diagnostic value of ERCP and EUS and long-term significance of isolated parenchymal abnormalities. *Gastrointest. Endosc.* 49:705–709.

Kelley, P.A., Sharkey, J., Philip, R., and Ritchie, I.M. 1993. Acute cocaine alters cerebrovascular autoregulation in the rat neocortex. *Brain Res. Bull.* 31:581–585.

Knuepfer, M.M., and Branch, C.A. 1992. Cardiovascular responses to cocaine are initially mediated by the central nervous system in rats. *J. Pharmacol. Exp. Ther.* 263:734–741.

Kumar, K.J., Chu, F.H., Hsieh, H.W., Liao, J.W., Li, W.H., Lin, J.C., Shaw, J.F., and Wang, S.Y. 2011. Antroquinonol from ethanolic extract of mycelium of *Antrodia cinnamomea* protects hepatic cells from ethanol-induced oxidative stress through Nrf-2 activation. *J. Ethnopharmacol.* 136:168–177.

Kvetnanský, R., Krizanova, O., Tillinger, A., Sabban, E.L., Thomas, S.A., and Kubovcakova, L. 2008. Regulation of gene expression of catecholamine biosynthetic enzymes in dopamine-beta-hydroxylase- and CRH-knockout mice exposed to stress. *Ann. NY Acad. Sci.* 1148:257–268.

Li, L., Lu, G., Yao, H., Zhao, Y., Feng, Z., and Yew, D.T. 2005. Postmortem changes in the central nervous system and adrenal medulla of the heroin addicts. *Int. J. Neurosci.* 115:1443–1449.

Mak, Y.T., Lam, W.P., Lu, L., Wong, Y.W., and Yew, D.T. 2010. The toxic effect of ketamine on SH-SY5Y neuroblastoma cell line and human neuron. *Microsc. Res. Tech.* 73:195–201.

Muir, J.K., and Ellis, E.F. 1993. Cocaine potentiates the blood pressure and cerebral blood flow response to norepinephrine in rats. *Eur. J. Pharmacol.* 249:287–292.

Ostrovskiĭ, V.K., Makarov, S.V., Rodionov, P.N., and Kochetkov, L.N. 2011. Indices of tissue necrosis markers in acute pyo-destructive diseases of organs of the abdominal cavity. *Vestn Khir Im I I Grek.* 170(5):28–30.

Patterson-Buckendahl, P., Kubovcakova, L., Krizanova, O., Pohorecky, L.A., and Kvetnansky, R. 2005. Ethanol consumption increases rat stress hormones and adrenomedullary gene expression. *Alcohol* 37:157–166.

Reyes Toso, C.F., Linares, L.M., and Rodriguez, R.R. 1995. Blood sugar concentrations during ketamine or pentobarbitone anesthesia in rats with or without alpha and beta adrenergic blockade. *Medicina (B Aires)* 55:311–316.

Roth, L., Harbison, R.D., James, R.C., Tobin, T., and Roberts, S.M. 1992. Cocaine hepatotoxicity: Influence of hepatic enzyme inducing and inhibiting agents on the site of necrosis. *Hepatology* 15:934–940.

Saha, J.K., Xia, J., Grondin, J.M., Engle, S.K., and Jakubowski, J.A. 2005. Acute hyperglycemia induced by ketamine/xylazine anesthesia in rats: Mechanisms and implications for preclinical models. *Exp. Biol. Med. (Maywood)* 230:777–784.

Siegmund, S.V., and Singer, M.V. 2005. Effects of alcohol on the upper gastrointestinal tract and the pancreas—An up-to-date overview. *Z. Gastroenterol.* 43:723–736.

Singer, M.V. 1985. The pancreas and alcohol. *Schweiz. Med. Wochenschr.* 115:973–987.

Sun, L., Lam, W.P., Wong, Y.W., Lam, L.H., Tang, H.C., Wai, M.S., Mak, Y.T., Pan, F., and Yew, D.T. 2011. Permanent deficits in brain functions caused by long-term ketamine treatment in mice. *Hum. Exp. Toxicol.* 30:1287–1296.

Swerdlow, N.R., Koob, G.F., Cador, M., Lorang, M., and Hauger, R.L. 1993. Pituitary-adrenal axis responses to acute amphetamine in the rat. *Pharmacol. Biochem. Behav.* 45:629–637.

Tan, S., Chan, W.M., Wai, M.S., Hui, L.K., Hui, V.W., James, A.E., Yeung, L.Y., and Yew, D.T. 2011. Ketamine effects on the urogenital system—Changes in the urinary bladder and sperm motility. *Microsc. Res. Tech.* 74:1192–1198.

Toth, I.E., Vizi, E.S., Hinson, J.P., and Vinson, G.P. 1997. Innervation of the adrenal cortex, its physiological relevance, with primary focus on the noradrenergic transmission. *Microsc. Res. Tech.* 36:534–545.

Wai, M.S., Chan, W.M., Zhang, A.Q., Wu, Y., and Yew, D.T. 2012. Long-term ketamine and ketamine plus alcohol treatments produced damages in liver and kidney. *Hum. Exp. Toxicol.* 31:877–886.

Wai, M.S., Luan, P., Jiang, Y., Chan, W.M., Tsui, T.Y., Tang, H.C., Lam, W.P., Fan, M., and Yew, D.T. 2013. Long term ketamine and ketamine plus alcohol toxicity—What can we learn from animal models? *Mini Rev. Med. Chem.* 13:273–279.

Wai, M.S., Shi, C., Kwong, W.H., Zhang, L., Lam, W.P., and Yew, D.T. 2008. Development of the human insular cortex: Differentiation, proliferation, cell death, and appearance of 5HT-2A receptors. *Histochem. Cell Biol.* 130:1199–1204.

Wilson, S.P., Chang, K.J., and Viveros, O.H. 1981. Opioid peptide synthesis in bovine and human adrenal chromaffin cells. *Peptides* 2 Suppl 1:83–88.

Wong, Y.W., Lam, L.H., Tang, H.C., Liang, Y., Tan, S., and Yew, D.T. 2012. Intestinal and liver changes after chronic ketamine and ketamine plus alcohol treatment. *Microsc. Res. Tech.* 75:1170–1175.

Yeung, L.Y., Wai, M.S., Fan, M., Mak, Y.T., Lam, W.P., Li, Z., Lu, G., and Yew, D.T. 2010. Hyperphosphorylated tau in the brains of mice and monkeys with long-term administration of ketamine. *Toxicol. Lett.* 193:189–193.

Yu, H., Li, Q., Wang, D., Shi, L., Lu, G., Sun, L., Wang, L., Zhu, W., Mak, Y.T., Wong, N., Wang, Y., Pan, F., and Yew, D.T. 2012. Mapping the central effects of chronic ketamine administration in an adolescent primate model by functional magnetic resonance imaging (fMRI). *Neurotoxicology* 33:70–77.

chapter nine

Ketamine uropathy
Hong Kong experience

Peggy Sau Kwan Chu, Chi Fai Ng, and Wai Kit Ma

Contents

9.1 Introduction

Ketamine is a nonbarbituate phencyclidine derivative that has been used as a dissociative anesthetic in both human and veterinary medicine since 1971. Ketamine was first synthesized in the United States in 1962 (Domino et al. 1965). Due to its relative hemodynamic stability, ketamine was generally accepted in the 1980s as a safe anesthetic without long-term adverse effects (Reich and Silvay 1989; Shorn and Whitwam 1980). It was considered an ideal "battlefield anesthetic," popularized during the Vietnam War (Malchow and Black 2008) and is authorized as a medicine in at least 60 countries including the United States and the European Union (World Health Organization 2012).

Ketamine is both water and lipid soluble and allows administration by many routes. Intramuscular, intravenous, subcutaneous, oral, nasal, and rectal administration are prescribed therapeutically and are utilized for both recreational and nonmedical ketamine misuse (Jansen 2000; Reich and Silvay 1989; Sinner and Graf 2008). It is rapidly absorbed when administered through the intramuscular (1–5 min), nasal (5–10 min), and oral (15–20 min) routes. Extensive first-pass metabolism in the liver and intestine is largely accountable for the low bioavailability of ketamine when given orally (17%) or rectally (25%). The bioavailability after nasal, intravenous, and intramuscular administration is approximately 50%, 100%, and 93%, respectively (Malinovsky et al. 1996). Upon administration, ketamine is converted to norketamine via biotransformation in the liver, which is then eliminated via the hepatic route into conjugated hydroxyl metabolites to be excreted renally. Ketamine (5–11 days), norketamine (6–14 days), and dehydronorketamine (10 days) can be detected in the urine within a fortnight after administration (Adamowicz and Kala 2005; Parkin et al. 2008).

9.2   Epidemiology of ketamine misuse and ketamine-associated cystitis

Ketamine has been misused as a hallucinogen for almost 30 years owing to it producing an effect similar to that of phencyclidine but with a much shorter duration (Siegel 1978). Ketamine affects the perception of body, time, surroundings, and reality, producing a "psychedelic" state of mind that resembles schizophrenic psychosis. It can cause a dose-related high and a biphasic effect on anxiety, but tolerance to the effects of ketamine develops (World Health Organization 2012).

Ketamine has a more prominent place as an anesthetic in human medicine in developing countries such as Ethiopia, Nigeria, Tanzania, and Benin (World Health Organization 2012), where the facilities are much poorer, leading to an increase in the prevalence of ketamine abuse. As this problem may differ from region to region and can be underestimated, ketamine has not been classified as a scheduled drug in the Conventions of United Nations. On the basis of the 2008 World Health Organization questionnaire for the preparation of the 35th Expert Committee on Drug Dependence, 16 out of the 64 countries that responded to the questionnaire, with an emphasis on East and Southeast Asia and South America, reported harmful ketamine use. In Australia, 1.1% of the general population have used ketamine at least once in their lifetime (World Health Organization 2012). Recent data from Monitoring the Future Study (2011) from the United States reported an annual prevalence of 0.8%, 1.2%, and 1.7% for school students in the 8th, 10th, and 12th grade, respectively (Johnston et al. 2011).

In Hong Kong, ketamine is the most common substance of abuse among teenagers since 2001 (Narcotics Division, Security Bureau, the Government of Hong Kong Special Administrative Region of the People's Republic of China 2014). The national survey data in Taiwan (2004–2006) revealed ketamine as the second commonly used illegal drug among middle and high school students (Chen et al. 2009).

Cystitis, bladder dysfunction, and, in severe cases, secondary renal damage have been reported in chronic ketamine abusers since 2007 (Chu et al. 2007, 2008; Shahani et al. 2007).

In a large case study in Hong Kong involving 284 ketamine abusers, 188 of the subjects had been admitted acutely to the Accident and Emergency Department (acute cases), while 96 presented chronically to the outpatient clinics (chronic cases). While urinary symptoms were only present in 60 of 188 acute cases (32%), 88 of the 96 chronic cases (92%) displayed urinary symptoms such as dysuria, urgency, frequency, urge incontinence, decreased bladder volume, and painful hematuria (Chan 2012). Similar findings were also reported by a group of toxicologists in the United Kingdom (Kalsi et al. 2011).

According to a recent systemic review, 110 documented reports of irritative urinary tract symptoms from ketamine dependence exist. Urinary symptoms have been collectively referred to as "ketamine-induced ulcerative cystitis" or "ketamine-induced vesiculopathy" (Middela and Pearce 2011).

9.3   Clinical symptomatology and urological damages

The detrimental effects of ketamine on the urinary tract are acknowledged as a clinical syndrome being that of a small, painful bladder, incontinence, and upper tract obstruction with or without papillary necrosis (Chu et al. 2007; Shahani et al. 2007; Wood et al. 2011). The first index case in the literature was a 28-year-old Canadian man who presented with a 6-month history of painful hematuria, dysuria, urgency, and post-micturition pain (Shahani et al. 2007). Since then, numerous case series have been published worldwide on this new clinical entity, including Hong Kong (Chu et al. 2008; Mak et al. 2011; Ng et al. 2010; Tam et al. 2014), Taiwan (Chiew and Yang 2009; Huang et al. 2008; Tsai et al. 2009), United Kingdom (Selby et al. 2008), Belgium (Colebunders and Van Erps 2008), Malaysia (Ho et al. 2010), and Spain (Garcia-Larrosa et al. 2012). Termed "ketamine-associated cystitis," "ketamine-induced ulcerative cystitis," or "ketamine-induced vesiculopathy," this syndrome is characterized primarily by symptoms of lower urinary tract irritation related to ketamine use among young adults. Further studies of this entity confirmed the involvement of organs beyond the bladder, where the symptoms are composed of a spectrum of urinary tract damage ranging from mild cystitis changes on endoscopy (Chu et al. 2008) to obstructive uropathy and kidney injury (Selby et al. 2008). This condition is considered to be a classical bladder pain and LUTS (lower urinary tract symptoms) syndrome (frequency, urgency, nocturia, dysuria, or hematuria) with cystitis and a contracted bladder that is associated with ketamine abuse, which ensues without other known causes such as bacterial infection, stone disease, or neurogenic problems. Painful frequent small-volume voids is the classical chief complaint of the affected ketamine abusers.

A total of around 500 cases have been reported in the literature to date, with the largest series reported from Hong Kong (Chu et al. 2008; Ng et al. 2010; Tam et al. 2014). The true prevalence of ketamine-associated vesiculopathy or uropathy, however, is underdetermined because of underreporting and reluctance of ketamine abusers to seek medical consultation. In an online self-reporting cross-sectional survey carried out in the United Kingdom, among the 1285 people reporting ketamine use within the last 12 months, 340 (26.6%) participants reported experiencing at least one urinary symptom (Winstock et al. 2012). In 2012/2013, it was estimated that around 120,000 individuals had misused ketamine in the United Kingdom (Advisory Council on the Misuse of Drugs 2013), indicating that the true number of ketamine abusers affected by various degrees of urinary symptoms stretches beyond the reported series.

A comprehensive investigation pathway has been established in urology centers pioneering in the study of this syndrome (Chu et al. 2008). This includes symptom documentation and quantification by questionnaires, blood tests (routine renal and liver function), urine tests (culture and toxicology), urinary system ultrasonography, uroflowmetry study, flexible cystoscopy, video urodynamic study, and computed tomography for severe cases with possible upper tract involvement. Symptoms were documented with standardized frequency/voiding charts and the pelvic pain and urgency/frequency (PUF) symptom scale. This questionnaire comprises seven questions concerning daytime and nighttime frequency, pelvic or urological pain and its severity, and urgency and its degree of severity, and has been validated and used in screening and diagnosing patients with interstitial cystitis (Parsons et al. 2002). It includes a symptom score and a bother score, totaling a maximum of 35 points. In view of the clinical and histopathological resemblance between interstitial cystitis and ketamine-associated cystitis (Ma et al. 2008), centers in Hong Kong adopted the PUF symptom scale as an assessment tool for symptom quantification in ketamine-associated cystitis patients. The Chinese version of the PUF symptom scale has been validated, and correlations with the symptomatology and investigation results were evaluated (Ng et al. 2012). In a series of 50 patients with a mean age of 24 years and a mean duration of ketamine abuse of 4.7 ± 2.8 years, the prevalence of urinary symptoms was as follows: urinary urgency, 92%; frequency, 84%; nocturia, 88%; dysuria, 86%;

and hematuria, 68%. The same study suggested that higher mean PUF total scores were noted in patients with positive cystoscopic, urodynamic, and ultrasonographic investigation results, and a higher PUF score was associated with smaller bladder capacity. The cutoff value of 17 is suggestive of more serious urological sequelae: endoscopically confirmed cystitis (83% vs. 47%), detrusor instability (48% vs. 0%), vesicoureteric reflux (14% vs. 0%), poor bladder compliance (48% vs. 0%), and hydronephrosis (37% vs. 0%). In a more recent Hong Kong series involving 318 ketamine abusers with a mean duration of 81 months of ketamine use, the mean voided volume was 111 mL and the mean bladder capacity was 152 mL, with a mean bladder emptying efficiency of 73% (Tam et al. 2014). In more severe patients, the typical voided volume can be less than 50 mL, and they are napkin dependent because of severe urge incontinence. In an earlier large series in which videocystometrogram had been performed in 47 patients (Chu et al. 2008), the mean cystometric bladder capacity was 154.5 mL (range, 14–600 mL), with 51% (24) of the patients having a bladder capacity of ≤100 mL. Most of the patients showed a decreased bladder compliance or detrusor overactivity of different magnitudes at a very low bladder infusion volume (as low as 14 mL). Thirteen percent (6) of the patients showed vesicoureteral reflux as a secondary event to the severely contracted bladder with high detrusor pressure. This finding correlated well with the symptoms of these patients, in that both the functional and cystometric bladder capacities were markedly decreased, causing them to have very frequent small voids.

Cystoscopy examination aims to reveal if there is any inflammation and cystitis change endoscopically, and the cause of hematuria, if any (Figure 9.1). However, because of the small, painful bladder, this procedure is not well tolerated by patients with severe symptoms if performed under local anesthesia. To date, the largest series on cystoscopic findings on ketamine abusers is from Hong Kong involving 42 patients, in which

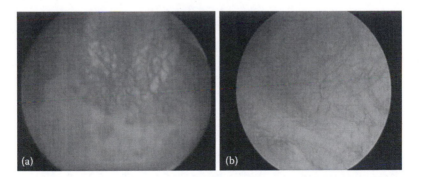

Figure 9.1 Cystoscopic views showing severe cystitis with erythematous and irregular inflamed bladder mucosa in a 26-year-old patient abusing ketamine for 7 years (a) compared to normal healthy mucosa (b).

30 had cystoscopy under local anesthesia while the others had cystoscopy and transurethral resection biopsy under regional or general anesthesia (Chu et al. 2008). All patients showed various degrees of epithelial inflammation of the bladder and neovascularization. Severe cases showed petechial hemorrhages, as classically described in patients with interstitial cystitis. Histological evidence indicated mucosal ulceration, striking urothelial reactive atypia, lamina propria inflammation with predominant lymphocyte infiltration, and a variable number of eosinophil cells. Ultrastructural examination by electron microscopy revealed the existence of querciphylloid muscle cells (vacuoles at the periphery of muscle cells). This feature has also been found in interstitial cystitis. In another UK series involving 17 patients with cystoscopy and bladder biopsies performed, a significant number with urothelial atypia mimicking carcinoma in situ were identified (Oxley et al. 2009). Marked urothelial atypia was seen in the biopsy specimens of 12 patients with nuclear enlargement and loss of polarity. Immunohistochemistry for CK20, p53, and Ki67 was performed in 10 cases, in which high expression of p53 was present in 9 cases and that of Ki67 in 6 cases. None of the biopsy specimens, however, showed CK20 expression in the atypical urothelium, which would be against carcinoma in situ. In addition, squamous metaplasia, nephrogenic metaplasia, and calcification have been described. The histopathological findings have great implications to its underlying pathophysiology, as the mechanism of chronic interstitial inflammation by ketamine metabolites is the main area of ongoing research.

Imaging of the urinary tract in severe ketamine cystitis patients typically reveals a small, contracted bladder with a thickened wall, with or without secondary upper tract changes like ureteric narrowing and hydroureteronephrosis. In a radiological review of 36 patients, 8% and 19% had unilateral and bilateral hydronephrosis, respectively (Yeung et al. 2012). Bilateral ureteral narrowing was demonstrated in 14% of the patients. A small number of patients had renal stones and bladder stones detected; 11% of patients had episodic or continuously elevated serum creatinine levels. Eleven percent had a percutaneous nephrostomy for acute renal failure or tight ureteric strictures. In the series by Chu and Ma, four patients were found to have ultrasonographic evidence of papillary necrosis while two of them had acute papillary necrosis with para-aortic lymphadenopathy and a thickened ureteric wall, suggestive of marked ongoing transmural inflammatory changes. The radiological investigation protocol for such young patients should emphasize the need to minimize radiation doses. A stepwise approach may begin with an ultrasonography to look for evidence of urinary tract obstruction, renal parenchymal disease, and thickened bladder wall. Depending on features of significant obstruction and presence of renal impairment, further investigation may include an intravenous urogram, as it involves a lower radiation dose by computed tomography (McTavish et al. 2002), as the holding up of contrast can provide information on presence of radio-opaque renal stones, hydrone-phrosis, papillary necrosis, ureteric stricture, and significant obstruction.

The dose–symptom relationship between ketamine use and symp-tomatology occurrence has been established. In a large United Kingdom online questionnaire series, 1285 people reporting use of ketamine within the last 12 months were studied. It was found that higher typical doses (>1 g) were associated with significantly higher rates of experiencing burning or stinging when passing urine ($\chi^2 = 37.95$, $P < 0.001$), frequency of urination ($\chi^2 = 78.29$, $P < 0.001$), lower abdominal pain ($\chi^2 = 39.0$, $P < 0.001$), blood in the urine ($\chi^2 = 6.52$, $P = 0.038$), and leakage of urine ($\chi^2 = 7.44$, $P = 0.024$) (Winstock et al. 2012).

9.3.1 Pathophysiology of bladder damage

The exact pathophysiology that leads to all the symptoms and clinical findings in the urinary tract is hitherto unknown. However, existing evi-dence points to the following mechanism of pathogenesis—persistent high concentration of ketamine metabolites in the urine causes chemi-cal irritation to the urothelium, which then stimulates an inflammatory response. The urinary bladder, being the storage organ for urine before it is eliminated, has the longest contact time with the "contaminated" urine and is consequently the first organ in the urinary system to be affected. Thus, most of the patients present with lower urinary tract symptoms suggestive of cystitis: frequency (95.2%), urgency (82.7%), dysuria (63.5%), hematuria (53.8%), and nocturia (87.5%) (Chu et al. 2008). If the chemical irritation is severe, denudation of the urothelium may occur, leading to transmural inflammation, loss of muscle thickness, and fibrosis of the detrusor muscle. This causes poor bladder compliance leading to vesico-ureteric reflux or urinary stasis in the ureter and in turn results in chronic ureteric inflammation and subsequently ureteric stricture.

9.3.1.1 Urine cytology findings

In a series of 104 otherwise healthy ketamine abusers in Hong Kong pre-senting with lower urinary tract symptoms, 51.9% and 70.2% of patients had leucocytes and erythrocytes detected in their urine microscopy examination, respectively (Figure 9.2) (Chu et al. 2008). However, only 12 (14.1%) out of 85 midstream urine cultures were found to be positive for bacteria. This further reinforced the postulation that most patients were having a form of nonbacterial cystitis with the inflammatory process elic-ited by ketamine or its metabolites in urine.

9.3.1.2 Urinary bladder histology

Shahani first reported on the presence of histological eosinophilic cystitis in four of his ketamine abuse patients where he described an epithelial

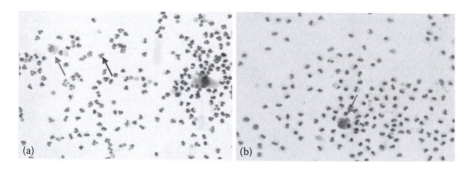

Figure 9.2 Urine cytology (pap stain, 400×) from a 26-year-old ketamine abuser, showing the presence of numerous neutrophils (dark arrow) (a). Histiocytes (gray arrow) are also observed (a and b).

denudation and inflammation with a mild eosinophilic infiltrate (Shahani et al. 2007). In the bladder biopsies of 42 patients in our series, similar histology features could be demonstrated, with mixed acute and chronic inflammatory infiltrates observed in some chronic abusers (Figures 9.3 and 9.4). In severe cases, multiple ulcerations could be seen during cystoscopic examination (Chung et al. 2007). The severity of cystitis both cystoscopically and histologically is apparently dose related. Animal studies also reported dose-related toxic effects of ketamine on cultures of marmoset bladder epithelial cells and whole rat bladders in vitro (Nemitz et al. 2002). In rare cases, cystitis glandularis and intestinal metaplasia may be found in daily ketamine abusers (Figure 9.5). Researchers in the United Kingdom recently demonstrated histologically that severe bladder pain in

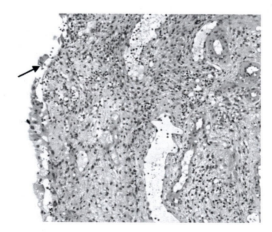

Figure 9.3 Bladder biopsy (hematoxylin and eosin stain, 100×) from a 26-year-old female ketamine abuser. The urothelium (arrow) is denudated with abundant mixed inflammatory infiltrates in the lamina propria.

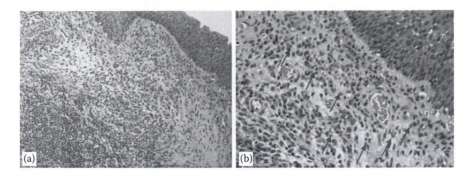

Figure 9.4 Urinary bladder mucosa histology obtained from a 28-year-old man abusing ketamine for 5 years. (a) Mixed acute and chronic inflammatory infiltrates are seen in the lamina propria (hematoxylin and eosin [H&E] stain, 200×). Lymphoid follicle is present in the left lower corner. (b) High-power view shows numerous neutrophils (arrows) in the lamina propria (H&E stain, 400×).

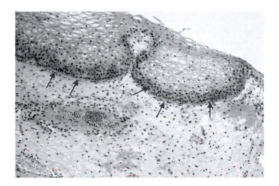

Figure 9.5 Squamous metaplasia (arrows) of the urothelial layer is demonstrated in the bladder biopsy of a 29-year-old female ketamine abuser (hematoxylin and eosin stain, 200×).

chronic ketamine abusers may be related to nerve hyperplasia resembling that of Morton's neuroma (Baker et al. 2013). When compared to the control groups of patients with benign bladder conditions, including interstitial cystitis, stress urinary incontinence, idiopathic neurogenic overactivity, or prostatic cancer undergoing radical prostatectomy, the ketamine abuser's bladder biopsy specimen were found to have prominent peripheral nerve fascicle hyperplasia with increased nerve growth factor expression.

Researchers from Taiwan demonstrated that the histology of bladder biopsy specimen from chronic ketamine abuse patients had a markedly decreased expression of E-cadherin and increased apoptosis when compared to patients with interstitial cystitis or painful bladder syndrome (IC/PBS) (Lee et al. 2013). E-cadherin is a calcium-dependent glycoprotein, which plays a critical role in cell-to-cell adhesion, and previous studies have demonstrated an association between reduced E-cadherin expression and an increase in visual analogue pain scores in patients with IC/BPS. The author thus suggested that E-cadherin is associated with bladder sensation and barrier function. Apoptosis is a stepwise process characterized by a series of stereotypical morphological changes that eventually lead to cell death, and the authors observed that the apoptotic process was highly activated in the urothelial cells of the ketamine cystitis specimens. Moreover, the percentages of apoptotic endothelial cells in these patients were negatively associated with maximum bladder capacity. The degree of apoptosis was also related to the symptom severity of ketamine cystitis.

9.3.1.3 Animal studies on effect of ketamine on bladder

Ketamine is known to be absorbable through intraperitoneal injection in laboratory animals (Flecknell 1998). The Chinese University of Hong Kong had recently reported the adverse effect of ketamine injected intravenously into monkey and intraperitoneally into mouse (Yew 2010). In the study, young adult (2-month-old) mice were injected intraperitoneally with ketamine daily for 3 and 6 months. The controls received a normal saline dose. Apoptosis in the bladder epithelium was observed initially in the 3-month group of mice injected with ketamine, where the bladder showed low mucosal infolding and thinning of muscle in all regions, and these changes were more apparent in the 6-month group (Tan et al. 2011). Extensive infiltration of connective tissue was seen in the whole bladder from lamina propria to the muscle layer. Meng from Taiwan demonstrated that mice treated daily with intraperitoneal injection of ketamine for 16 weeks demonstrated enhanced noncholinergic contractions and P2X1 receptor expression in the detrusor strip, indicating that the dysregulation of purinergic neurotransmission may underlie detrusor overactivity in cases of ketamine-induced bladder dysfunction (Meng et al. 2011).

9.3.2 Ketamine damage on the kidney and related animal studies

Papillary necrosis of the kidney as evident by ultrasound and computerized tomogram was found in some ketamine abusers (Figure 9.6) (Chu et al. 2008). The postulated pathophysiology is acute tubular injury caused by the toxic concentration of ketamine. Chronic exposure of the kidney ketamine of higher than therapeutic anesthetic dose leads to chronic tubulo-interstitial damage in the inner medulla, causing tubular atrophy. This may result in largely irreversible interstitial fibrosis (Braden et al. 2005). In the study by Yeung et al. in 2009, young adult mice were subjected to daily intraperitoneal ketamine injection for 1, 3, or 6 months, and the kidneys were examined histologically. Foci of infiltration of mononuclear white cells were observed in all the ketamine-injected mice near the glomeruli

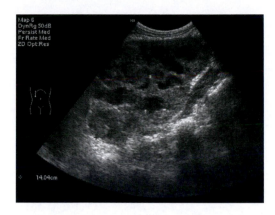

Figure 9.6 Ultrasonography showing papillary necrosis of a chronic ketamine abuser.

or blood vessels (arteries or veins), forming perivascular cuffing (Yeung et al. 2009). In addition, these mononuclear infiltrations were demonstrated in the distal and collecting tubules inside the renal medulla in all addicted cases. Severe degeneration in parts of the kidney was evident in more than 15% of injected mice. On the contrary, none of the controls (1, 3, and 6 months) had significant infiltration of mononuclear white cells. These findings also point to the theory of chronic inflammatory changes of the renal tubules as the cause of subsequent scarring and renal damages.

9.3.3   Ketamine damage on ureter

Fifteen percent of patients in the Hong Kong series of 104 ketamine abusers had either unilateral or bilateral hydronephrosis on ultrasonography of the kidneys (Chu et al. 2011). In some of the chronic ketamine abusers requiring percutaneous nephrostomy drainage for deranged renal function, unilateral and bilateral ureteric strictures were found (Figure 9.7) (Chu et al. 2007, 2008), along with the occasional occurence of a thickened ureteric wall and para-aortic lymphadenopathy. Histology of the resected ureteric stricture revealed transmural edema and inflammation (Figure 9.8).

9.3.4   Blood inflammatory biochemical markers

In the Hong Kong series of 104 patients, 28% of the male ketamine abusers had a raised erythrocyte sedimentation rate (ESR, defined as >15 mm/h) while the ESR of 43.9% of their female counterparts was above normal (>20 mm/h). The C-reactive protein was raised (>3.6 mg/L) in 39.6% of the authors' series (Chu et al. 2011). All these findings further concur with the

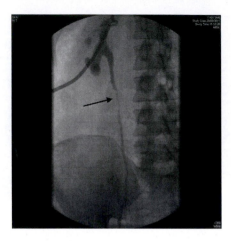

Figure 9.7 Antegrade nephrostogram of a ketamine abuser patient showing a segment of ureteric stricture at the L3/4 level.

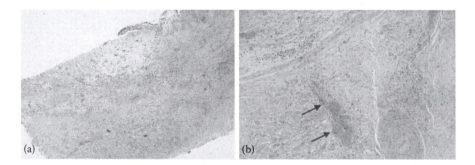

Figure 9.8 Histology of left ureter from a 31-year-old man with ketamine abuse resulting in ureteric stricture (a) with a diffuse denudation of the epithelium (hematoxylin and eosin [H&E] stain, 40×). The lamina propria is edematous and congested. (b) The muscularis propria layer shows mild fibrosis (H&E stain, 100×).

idea of intense inflammatory reaction triggered by the presence of ketamine or its metabolites in contact with the urothelium.

9.4   *Pathophysiology of street ketamine uropathy: Discussion*

Although the exact pathogenesis of urogenital system damage by ketamine abuse is still unknown, contemporary clinical, biochemical, and histological evidence, together with findings from laboratory animal studies, all point to an initial acute inflammatory reaction of the urothelium to toxic concentration of ketamine and its metabolites. In the initial series of nine Canadian patients described by Shahani et al. (2007), bladder biopsies obtained from four patients showed histologic features primarily characterized by marked changes in the epithelium and lamina propria with granulation tissue formation and fibrosis, which appeared to be distinct from what has been described as classic eosinophilic cystitis attributed to other causes. These features were echoed in the subsequent series of 59 patients from Hong Kong in which 42 patients had bladder biopsies showing various degrees of epithelial inflammation of the bladder and infiltrates that are composed predominantly of lymphocytes and a variable number of eosinophil cells (Chu et al. 2011). Case reports with similar histologic findings have been described in different countries since then. Another recent series of 17 ketamine abusers from the United Kingdom had bladder biopsies done showing urothelial atypia so marked as to mimic carcinoma in situ (Oxley et al. 2009). All these histopathological features suggested that the primary event occurring in ketamine-associated cystitis is the action of a metabolite of ketamine at high doses on the urothelium, leading to inflammatory changes. Upon chronic exposure, interstitial fibrosis sets in. The urinary bladder, being the storage organ for urine, is usually the first to be damaged by ketamine. This results in bladder contracture with poor compliance displaying all the clinical symptoms of a diminished functional bladder capacity. Due to the difficulty of obtaining a transmural thickened sample of the bladder for histological examination in human patients, animal studies are the best way to prove the above postulation. From Yew's study of mice urinary bladder after intraperitoneal injection of ketamine for 6 months, severe fibrosis and a comparatively small amount of detrusor muscle fibers were reported (Yew 2010). The clinical finding of a thickened bladder wall but a shrunken bladder on computed tomography examination of these patients is the result of the thinning of the detrusor muscle and its replacement by a severely fibrotic suburothelial layer. Due to the presence of a poor compliant bladder, urinary stasis occurs in the ureters, causing transmural inflammatory changes that result in ureteric stricture. A case report from the United Kingdom demonstrated the presence of ureteric intestinal metaplasia in a patient with 12 years of daily ketamine abuse presenting with bilateral ureteric stricture (Hopcroft et al. 2011). In addition to the obstructive uropathy caused by ureteric stricture, ketamine affects the kidneys directly, causing acute or chronic tubulo-interstitial damage to a different extent that results in nephron loss. Another earlier case report revealed reversible hydronephrosis owing to precipitation of ketamine metabolites in the ureters and acute renal failure requiring dialysis (Selby et al. 2008). Ketamine does not usually precipitate in the pelvicalyceal systems, and the presence of a higher cumulative dose may therefore be crucial. It is important to note that the hydronephrosis resolved during the patient's hospital stay when ketamine administration was terminated.

Recently, a rare case of germ cell tumor of the urinary bladder in a 31-year-old female chronic ketamine abuser was reported (Mui et al. 2014). However, the correlation between ketamine abuse and the formation of a urinary bladder malignancy had not been previously documented (Hopcroft et al. 2011).

9.5 Management of street ketamine uropathy

The presence of predominant irritative symptoms in young adults or adolescents associated with chronic ketamine abuse, with or without hematuria, formed the basis for the diagnosis of ketamine-associated cystitis. Therefore, diagnosis of ketamine-associated cystitis is usually quite straightforward and is mainly focused on assessing the severity of symptom and the degree of urinary tract damage. An initial noninvasive investigation approach may help improve the compliance of patients and build up rapport between patients and the health care provider (Tam et al. 2014). However, the management of the condition remains a great challenge to urologists. The difficulty in management is attributed not only to the relatively unclear pathophysiology but also to complex psychosocial factors in patients on a case-by-case basis. Therefore, the simple management of the condition as a pure physical disease will usually result in an unsatisfactory outcome or relapse of the condition.

A full explanation about the condition, including the symptomatology, the association with ketamine usage, and potential serious irreversible consequences, will help patients understand their situation and hence increase their motivation to cooperate and comply with condition management. While the underlying pathophysiology is still under investigation, the clear linkage with ketamine usage made abstinence the cornerstone in the management. Many reports have shown that ketamine abstinence led to symptom improvement (Chen et al. 2011; Shahani et al. 2007). Unfortunately, complicated psychosocial backgrounds and poor personal coping skills have greatly affected the success of detoxification in many patients. Therefore, a joint approach of management involving family members, health care providers, social workers, and even government input will help improve the success of detoxification (Tam et al. 2014; Wood et al. 2011).

As pathological findings from many series suggested the inflammatory basis of the condition (Chen et al. 2009; Shahani et al. 2007), anti-inflammatory agents including both nonsteroidal anti-inflammation drugs and COX-II inhibitors are frequently used in the initial management of these patients (Tam et al. 2014; Tsai et al. 2009). Depending on the degree of dysuria and pelvic pain, additional analgesics such as phenazo-pyridine, paracetamol, and narcotic analgesics may lead to better pain control. As ketamine itself is a type of analgesic and sedative agent, some patients will experience an increase in pain during initial detoxification, driving them back to ketamine consumption. Therefore, a short period of intense analgesia would help overcome the discomfort related to detoxification and increase the chance of success.

A majority of patients are reported to experience marked irritative urinary symptoms. For patients with a positive urine culture, a course of appropriate antibiotics may help control some of the symptoms. While anti-inflammatory agents and analgesics may relieve some of the irritative symptoms secondary to cystitis, anticholinergic agents may also be used to help symptom control. However, some of the patients were found to have diminished bladder-emptying efficiencies as suggested by relatively large post-void residual volume (Tam et al. 2014), and the use of anticholinergics may lead to difficulty in passing urine and worsening of the symptom. Therefore, we strongly recommend assessing post-void residual urine before the commencement of anticholinergic treatment. Moreover, as some patients try to decrease fluid intake in order to decrease urinary frequency, advice on the possibility of constipation after treatment should be given.

As the damage of the glycosaminoglycan layer was one of the proposed pathophysiology of ketamine-associated cystitis (Middela and Pearce 2011), an attempt to repair this protective layer was proposed as a treatment option, along with oral pentosan polysulfate (Elmiron) and intravesical instillation of hyaluronic acid (Cystitstat). All reports suggested symptom improvement after usage of these agents (Chen et al. 2011; Shahani et al. 2007; Tsai et al. 2009).

For patients with hydronephrosis, invasive intervention may be needed to preserve renal function. The causes of hydronephrosis include ureteric reflux secondary to detrusor overactivity or a poor compliance bladder, and ureteric stricture related to ketamine usage (Chu et al. 2007, 2008). Intravenous urogram and computerized tomogram would help identify the level of ureteric obstruction. A urodynamic study would provide information about the function of the lower urinary tract and help determine whether the hydronephrosis is secondary to urinary reflux or high voiding pressure. For patients with hydronephrosis secondary to reflux, urethral catheter placement may help alleviate the problem, though this is not well tolerated in all patients. For patients with hydronephrosis secondary to ureteric stricture, temporary bilateral nephrostomy insertion would be needed to protect kidney function till definitive management, but the presence of cystitis may cause difficulty in ureteric stenting as well as low tolerance in patients. Though the definitive management plan for ureteric stricture would follow the same approach for other ureteric strictures, the frequent association of small contracted bladder may limit the choice of reconstruction.

While abstinence and medical therapy may help the majority of patients improve their urinary symptom, some patients will still have significant residual symptoms (Cheung et al. 2011). In severe cases, the residual symptom is probably related to fibrotic changes in the bladder secondary to prolonged inflammation. For these patients with a contracted bladder, augmentation cystoplasty may be needed to counteract this irreversible damage. While the surgical principle would be the same as for the augmentation of contracted bladders of other etiologies, one should bear in mind potential problems related to ketamine usage during the counseling. Ng et al. (2013) report the outcome of four patients with augmentation cystoplasty performed for contracted bladders secondary to ketamine usage. While the patients abstained from ketamine usage before surgery, all of them resumed usage at different stages after surgery. Three of them developed serious complications, including renal failure, convulsion, and ureteric strictures. One of the possible causes for the rapid development of complications was the recirculation of ketamine and its metabolites in the body in relation to the intestine used for augmentation. Therefore, careful selection of patients and repeated emphasis on the maintenance of abstinence are important steps before surgery.

References

Adamowicz, P, and M Kala. 2005. "Urine excretion rates of ketamine and norketamine following therapeutic ketamine administration: Method and detection window considerations." *J Anal Toxicol* 29: 376–82.

Advisory Council on the Misuse of Drugs. 2013. "Ketamine: A review of use and harm." London. Available at https://www.gov.uk/government/uploads/system/uploads/attachment_data/file/264677/ACMD_ketamine_report_dec13.pdf (accessed January 23, 2014).

Baker, SC, J Stahlschmidt, J Oxley, J Hinley, I Eardley, F Marsh, D Gillatt, S Fulford, and J Southgate. 2013. "Nerve hyperplasia: A unique feature of ketamine cystitis." *Acta Neuropathol Commun* 1: 64.

Braden, GL, MH O'Shea, and JG Mulhern. 2005. "Tubulointerstitial diseases." *Am J Kidney Dis* 46: 560–72.

Chan, YC. 2012. "Acute and chronic toxicity pattern in ketamine abusers in Hong Kong." *J Med Toxicol* 8: 267–70.

Chen, CH, MH Lee, YC Chen, and Lin MF. 2011. "Ketamine-snoring associated cystitis." *J Formos Med Assoc* 110: 787–91.

Chen, WJ, TC Fu, TT Ting, WL Huang, GM Tang, CK Hsiao, and CY Chen. 2009. "Use of ecstasy and other psychoactive substances among school-attending adolescents in Taiwan: National surveys 2004–2006." *BMC Public Health* 9: 27.

Cheung, RY, SS Chan, JH Lee, AW Pang, KW Choy, and TK Chung. 2011. "Urinary symptoms and impaired quality of life in female ketamine users: Persistence after cessation of use." *Hong Kong Med J* 17 (4): 267–73.

Chiew, YW, and CS Yang. 2009. "The case: Disabling frequent urination in a young adult." *Kidney Int* 76 (1): 123–4.

Chu, PSK, SC Kwok, KM Lam, TY Chu, SWC Chan, CM Man, WK Ma, KL Chui, MK Yiu, YC Chan, ML Tse, and FL Lau. 2007. "'Street ketamine'-associated bladder dysfunction: A report of ten cases." *Hong Kong Med J* 13 (4): 311–3.

Chu, PSK, WK Ma, KC To, MK Yiu, and CW Man. 2011. "Research on urological sequelae of ketamine abuse." Project Ref. No.: BDF080050, Beat Drugs Fund Association, the Government of the Hong Kong Special Administrative Region. Available at http://www.nd.gov.hk/pdf/20110307_beat_drug_fund_report.pdf (accessed January 29, 2014).

Chu, PSK, WK Ma, SCW Wong, RWH Chu, CH Cheng, S Wong, JML Tse, FL Lau, MK Yiu, and CW Man. 2008. "The destruction of the lower urinary tract by ketamine abuse: A new syndrome?" *BJU Int* 102: 1616–22.

Chung, SD, HC Chang, B Chiu, PH Chan, and PF Liu. 2007. "Ketamine-related urinary bladder ulceration." *Incont Pelvic Floor Dysfunct* 1 (4): 153.

Colebunders, B, and P Van Erps. 2008. "Cystitis due to the use of ketamine as a recreation drug: A case report." *J Med Case Reports* 26 (2): 219.

Domino, EF, P Chodoff, and G Carssen. 1965. "Pharmacologic effects of Ci-581, a new dissociative anesthetic in man." *Clin Pharmacol Ther* 6: 279–91.

Flecknell, P. 1998. *Laboratory Animal Anaesthesia*. London: Academic Press.

Garcia-Larrosa, A, C Castillo, M Ventura, JA Lorente, O Bielsa, and O Arango. 2012. "Cystitis and ketamine associated bladder dysfunction." *Actas Urol Esp* 36 (1): 60–4.

Ho, CCK, H Pezhman, S Praveen, EH Goh, and B Lee. 2010. "Ketamine-associated ulcerative cystitis: A case report and literature review." *Malaysia J Med Sci* 17 (2): 61–5.

Hopcroft, SA, AM Cottrell, K Mason, P Abrams, and JD Oxley. 2011. "Ureteric intestinal metaplasia in association with chronic recreational ketamine abuse." *J Clin Pathol* 64 (6): 551–2.

Huang, YC, CM Jeng, and TC Cheng. 2008. "Ketamine-associated ulcerative cystitis." *Tzu Chi Med J* 20 (2): 144–6.

Jansen, KL. 2000. "A review of the nonmedical use of ketamine: Use, users and consequences." *J Psychoactive Drugs* 32: 419–33.

Johnston, LD, PM O'Malley, JG Bachman, and JE Schulenberg. 2011. "Monitoring the future—National results on adolescent drug use overview of key findings." Available at http://www.monitoringthefuture.org/pubs/monographs/mtfoverview2011.pdf (accessed February 2, 2014).

Kalsi, SS, DM Wood, and PI Dargan. 2011. "The epidemiology and patterns of acute and chronic toxicity associated with recreational ketamine use." *Emerg Health Threats J* 4: 7107.

Lee, CL, YH Jiang, and HC Huo. 2013. "Increased apoptosis and subepithelial inflammation in patients with ketamine-related cystitis: A comparison with non-ulcerative cystitis and controls." *BJU Int* 112: 1156–62.

Ma, WK, PSK Chu, S Wong, KT Loo, SWH Chan, MK Yiu, and CW Man. 2008. "Street ketamine-associated bladder dysfunction and cystitis: Pathological review." *Indian J Urol* 24 (Suppl 2): 210.

Mak, SK, MTY Chan, WF Bower, SKH Yip, SSM Hou, BBB WU, and CY Man. 2011. "Lower urinary tract changes in young adults using ketamine." *J Urol* 186 (2): 601–4.

Malchow, RJ, and IH Black. 2008. "The evolution of pain management in the critically ill trauma patient: Emerging concepts from the global was on terror-
ism." *Crit Care Med* 36: S346–57.

Malinovsky, JM, F Servin, A Cozia, JY Lepage, and M Pinaud. 1996. "Ketamine and
norketamine plasma concentrations after i.v., nasal and rectal administration
in children." *Br J Anaesth* 77: 203–7.

McTavish, JD, M Jinzaki, KH Zou, RD Nawfel, and SG Silverman. 2002. "Multi-
detector row CT urography: Comparison of strategies for depicting the normal urinary collecting system." *Radiology* 225: 783–90.

Meng, E, HY Chang, SY Chang, GH Sun, DS Yu, and Cha TL. 2011. "Involvement
of purinergic neurotransmission in ketamine induced bladder dysfunction."
J Urol 186: 1134–41.

Middela, S, and I Pearce. 2011. "Ketamine-induced vesiculopathy: A literature
review." *Int J Clin Pract* 65 (1): 27–30.

Mui, WH, KC Lee, SC Chiu, CY Pang, PSK Chu, CW Man, CS Wong, WK Sze, and
Y Tung. 2014. "Primary yolk sac tumor of the urinary bladder: A case report
and review of the literature." *Oncol Lett* 7: 199–202.

Narcotics Division, Security Bureau, the Government of Hong Kong Special
Administrative Region of the People's Republic of China. 2014. "Central registry of drug abuse." Accessed February 18, 2014. Available at http://www
.nd.gov.hk/en/drugstatistics.htm.

Nemitz, DG, C Westmoreland, E George, J Kapwijk, and N Watts. 2002. "Toxic
effects of ketamine hydrochloride (Vetalar TMV) on cultures of marmoset
bladder epithelial cells and whole rat bladders in vitro." *Proceedings of the
4th World Congress on Alternatives and Animal Use in the Life Sciences*, New
Orleans.

Ng, CF, PKF Chiu, ML Li, CW Man, SSM Hou, ESY Chan, and PSK Chu. 2013.
"Clinical outcomes of augmentation cystoplasty in patients suffering from
ketamine-related bladder contractures." *Int Urol Nephrol* 45 (5): 1245–51.

Ng, CM, WK Ma, KC To, and MK Yiu. 2012. "The Chinese version of the pelvic
pain and urgency/frequency symptom scale: A useful assessment tool for
street-ketamine abusers with lower urinary tract symptoms." *Hong Kong
Med J* 18: 123–30.

Ng, SH, ML Tse, HW Ng, and FL Lau. 2010. "Emergency department presentation
of ketamine abusers in Hong Kong: A review of 233 cases." *Hong Kong Med J*
16: 6–11.

Oxley, JD, AM Cottrell, S Adams, and D Gillatt. 2009. "Ketamine cystitis as a mimic
of carcinoma in situ." *Histopathology* 55: 705–8.

Parkin, MC, SC Turfus, NW Smith, Halket JM, RA Braithwaite, and SP Elliott. 2008.
"Detection of ketmaine and its metabolites in urine by ultra-high pressure
liquid chromatography-tandem mass spectrometry." *J Chromatogr B Analyt
Technol Biomed Life Sci* 876: 137–42.

Parsons, CL, EJ Stanford, BS Kahn, and PK Sand. 2002. "Tools for diagnosis and
treatment." *Female Patient* 27(Suppl): 12–7.

Reich, DL, and G Silvay. 1989. "Ketamine: An update on the first twenty-five years
of clinical experience." *Can J Anaesth* 36: 186–97.

Selby, NM, J Anderson, P Bungay, LJ Chesterton, and NV Kolhe. 2008. "Obstructive
nephropathy and kidney injury associated with ketamine abuse." *NDT Plus*
5: 310–2.

Shahani, R, C Streutker, B Dickson, and Stewart RJ. 2007. "Ketamine-associated ulcerative cystitis: A new clinical entity." *Urology* 69: 810–2.

Shorn, TOF, and JG Whitwam. 1980. "Are there long-term effects of ketamine on the central nervous system?" *Br J Anaseth* 52: 967–8.

Siegel, PK. 1978. "Phencyclidine and ketamine intoxication: A study of four populations of recreational users." *NIDA Res Monogr* 21: 119–47.

Sinner, B, and BM Graf. 2008. "Ketamine." *Handb Exp Pharmacol* 182: 313–33.

Tam, YH, CF Ng, KKY Pang, Yee CH, WCW Chu, VYF Leung, GLH Wong, VWS Wong, HLY Chan, and BPS Lai. 2014. "One-stop clinic for ketamine-associated uropathy: Report on service delivery model, patients' characteristics and non-invasive investigations at baseline by a cross-sectional sutdy in a prospective cohort of 318 teenages and young adults." *BJU Int* 114: 754–60.

Tan, S, WM Chan, MSM Wai, LKK Hui, VWK Hui, AE James, LY Yeung, and DT Yew. 2011. "Ketamine effects on the urogenital system—Changes in the urinary bladder and sperm motility." *Microsc Res Tech* 74: 1192–8.

Tsai, TH, TL Cha, CM Lin, CW Tsao, SH Tang, FP Chuang, ST Wu, GH Sun, DS Yu, and SY Chang. 2009. "Ketamine-associated bladder dysfunction." *Int J Urol* 16 (10): 826–9.

Winstock, AR, L Mitcheson, DA Gillatt, and AM Cottrell. 2012. "The prevalence and the natural history of urinary symptoms among recreational ketamine users." *BJU Int* 110: 1762–66.

Wood, D, A Cottrell, SC Baker, J Southgate, M Harris, S Fulford, C Woodhouse, and D Gillat. 2011. "Recreational ketamine: From pleasure to pain." *BJU Int* 107: 1881–4.

World Health Organization. 2012. "Ketamine—Expert peer review on critical review report." 35th Expert Committee on Drug Dependence, Hammamet, Tunisia. Available at http://www.who.int/medicines/areas /quality_safety/4.2Ketamine_criticalreview.pdf?ua=1 (accessed January 31, 2014).

Yeung, JTH, JKF Ma, KKM Kwok, AWT Yung, RLF Cheng, KC To, and MK Yiu. 2012. "Ketamine-related urological complications: Radiological features." *Hong Kong J Radiol* 15: 4–9.

Yeung, LY, JA Rudd, WP Lam, YT Mak, and DT Yew. 2009. "Mice are prone to kidney pathology after prolonged ketamine addiction." *Toxicol Lett* 191: 275–8.

Yew, DT. 2010. "Long-term ketamine abuse and apoptosis in cynomologus monkeys and mice." Project Ref. No.: BDF080048, Beat Drugs Fund Association, the Government of the Hong Kong Special Administrative Region. Available at http://www.nd.gov.hk/pdf/long_term_ketamine_abuse_apoptosis_in _cynomologus_monkeys_and_mice/final_report_with_all_attachments.pdf (accessed December 12, 2013).

chapter ten

Ketamine and the lower urinary tract

Summary of pathophysiological evidence in humans and animal models

Hong Chai Tang, Phoebe Y.H. Lam, and Willmann Liang

Contents

10.1 Background and literature search

The harmful consequences of ketamine abuse in the lower urinary tract were first reported in a 2007 case study of nine patients in Canada (Shahani et al. 2007). At about the same time, similar observations were made in the United Kingdom and Hong Kong (Chu et al. 2007; Selby et al. 2008). These case studies spurred a series of reports centering on lower urinary tract changes in humans. As of May 2014, a search of the PubMed database using the keywords "ketamine" and "bladder" or "urinary tract" yielded more than 300 publications. After screening, only those concerning bladder and lower urinary tract (patho)physiology were included in the discussion here. In total, 36 relevant studies involving humans and 6 involving laboratory animals were identified.

10.2 Human studies

Table 10.1 lists the relevant reports involving human subjects or human tissue samples. Summarized in this table are diagnostic methodologies and major pathophysiological findings among the different studies, together with the number of subjects and the average duration of ketamine use.

10.2.1 General characteristics of research
             methodology and clinical findings

Most of the studies documented the number of subjects, their age, and gender information. The majority of ketamine abusers were in their early 20s to 30s, with men being more prevalent. Duration of ketamine usage was the most common determinant of the severity of drug use, with only two of the studies reporting daily dosage (Chang et al. 2012; Tsai et al. 2009). Interestingly, both of them (Chang et al. 2012; Tsai et al. 2009) make up only a handful of studies listing data from individual subjects. (The other examples being Huang et al. 2014, Mason et al. 2010, Misra et al. 2014, and Oxley et al. 2009.) Where appropriate, correlation analyses were applied for these studies. (Refer to Section 10.2.3.)

In the urine of virtually all ketamine users tested, no bacterial culture was found. Another feature shared among a predominant number of studies (and subjects) listed in Table 10.1 was the presence of lower urinary tract symptoms (LUTS), which was often the very reason the subjects were recruited or admitted in the first place. Ketamine-associated LUTS may include increased frequency, urgency, urge incontinence, dysuria, bladder pain, and hematuria (Bokor and Anderson 2014; Skeldon and Goldenberg 2014). In many of the studies concerned, whether the individual subjects possessed one or more of the LUTS was not considered in

Table 10.1 Published Studies of Lower Urinary Tract Pathology Associated with Ketamine Usage in Humans (from Year 2007 Onward)

Publication (first author listed, arranged by accepted date)	No. of patients (male:female)	Average duration of ketamine use (years) [range]	LUTS	Urine culture	Imaging methods CT	IVU	US	Cystos-copy	Urody-namics	Biopsy	Bladder capacity (mL)	Bladder over-activity	Bladder inflammation	Bladder wall thickening	Urothelial denudation	Bladder ulceration	Other points to note
Shahani	9 (NK)	NK	Y	-ve	Y	—	—	Y	—	Y	Small	—	Y	Y	Y	Y	Fibrosis, metaplasia
Chu	10 (7:3)	1.8 [1–4]	Y	-ve	—	—	—	Y	Y	Y	30–100	Y	Y	—	—	—	MRI
Selby	1 M	2	Y	—	—	—	—	Y	—	Y	150	—	Y	—	—	—	—
Chu	59 (38:21)	3.5 [0.5–10]	Y	-ve	—	—	—	Y	Y	Y	14–600	Y	Y	Y	Y	—	↓Compliance
Colebunders	1 M	NK	Y	-ve	—	—	—	Y	—	Y	Small	—	Y	Y	—	—	
Cottrell	1 M	NK	Y	-ve	—	—	—	Y	—	Y	Small	—	Y	—	—	Y	
Chen	4 F	NK	Y	-ve	—	—	—	—	—	—	—	—	Y	—	—	—	
Tsai	11 (7:4)	1–4	Y	-ve	—	—	—	Y	Y	Y	</>150 ind.	Y	Y	—	Y	—	Cor.
Chen	4 (1:3)	1	Y	-ve	—	—	—	Y	Y	Y	26–68 ind.	—	Y	—	—	Y	Atypia, CM, Cor.
Oxley	17 (12:5)	NK	—	—	—	—	—	—	—	Y	—	—	—	—	Y	—	
Robson	1 F	1	Y	-ve	—	—	—	Y	—	Y	—	—	Y	—	—	—	
Ho	1 M	Est. 0.75	Y	-ve	—	—	Y	Y	—	Y	Small	—	Y	Y	—	Y	
Shahzad	1 M	2	Y	-ve	—	—	Y	Y	—	Y	—	—	Y	Y	—	Y	MRI
Storr	3 (1:2)	NK	Y	-ve	—	—	—	Y	—	Y	Small	—	Y	—	Y	—	Legitimate use, carcinoma
Chiew	1 M	4	Y	-ve	Y	Y	—	Y	—	Y	Small	—	Y	—	—	Y	Fibrosis
Ng	233 (159:74)	NK	Y	—	—	—	—	—	—	—	—	—	—	—	—	—	A&E

(Continued)

Table 10.1 (Continued) Published Studies of Lower Urinary Tract Pathology Associated with Ketamine Usage in Humans (from Year 2007 Onward)

Publication (first author listed, arranged by accepted date)	No. of patients (male:female)	Average duration of ketamine use (years) [range]	LUTS	Urine culture	Imaging methods			Cystos-copy	Urody-namics	Biopsy	Bladder capacity (mL)	Bladder over-activity	Bladder inflammation	Bladder wall thickening	Urothelial denudation	Bladder ulceration	Other points to note
					CT	IVU	US										
Mason	23 (16:7)	Md.: 2 [1–15]	Y	Mostly –ve	Y	Y	Y	–	–	Y	Small	–	Y	Y	–	Y	Atypia, CM
Chao	1 F	3	Y	–ve	–	–	Y	–	–	–	Small	–	Y	Y	–	–	
Jenyon	1 M	6	Y	–ve	Y	–	Y	Y	–	Y	150	–	Y	–	Y	–	Bladder stone, calcification
Chang	20 (13:7)	NK	Y	–	Y	–	Y	Y	Y	Y	11–163	–	Y	–	Y	–	Cor.
Nomiya	1 M	30	Y	–	Y	–	Y	Y	–	–	250	Y	Y	–	–	–	
Wei	1 M	3	Y	–	–	–	–	–	–	–	–	–	–	–	–	–	
Venyo	1 M	NK	Y	–ve	–	–	–	Y	–	Y	–	–	Y	–	Y	–	
Lai	6 (4:2)	3 [1.5–4.5]	Y	–ve	–	Y	–	Y	Y	Y	25–144 ind.	Y	Y	–	–	–	Contracted bladder
Wei	1 M	2	Y	–	Y	–	–	Y	–	–	Small	–	Y	Y	–	–	Bladder diverticulum
Ng	50 (20:30)	4.7	Y	–	–	–	–	Y	Y	–	–	Y	Y	–	–	–	↓ Compliance, PUF

Study	N (M:F)	Duration									Bladder capacity							Findings	Bladder trabeculation
Peng	1 M	3	Y	Y	-ve	Y	—	—	—	—	—	Small	—	—	—	—	—	—	—
Chen	1 M	3	Y	Y	-ve	Y	—	—	—	—	Y	Small	Y	Y	—	—	—	—	—
Srirangam	1 M	5	Y	Y	—	—	—	—	—	—	—	Small	—	—	—	Y	—	Metaplasia	Y
Jalil	4 (2:2)	NK	Y	Y	-ve	Y	IVU	—	—	—	—	Small	—	—	—	Y	—	—	Y
Lee	16 (9:7)	Md.: 3.5	Y	Y	—	Y	—	Y	Y	Y	Y	Avg: 110	Y	Y	Y	—	—	↓Compliance, apoptosis, E-cad.	—
Baker	NK	NK	—	Y	-ve	—	—	—	—	—	Y	—	Y	—	—	Y	—	NFP+, NGFR+	—
Mui	1 F	7; 12	—	Y	—	Y	—	Y	—	Y	Y	—	—	Y	—	Y	—	Yolk sac tumor	—
Wei	1 M	>3	—	Y	—	—	Y	—	—	—	—	Small	—	—	Y	—	—	—	—
Huang	27 (11:16)	3.4 [0.5–12]	Y	—	—	Y	—	Y	Y	Y	—	13–187	Y	Y	Y	Y	—	Cor.	—
Misra	25 (19:6)	2.6 [1–10]; NK	Y	Y	Mostly -ve	Y	Y	—	Y	Y	Y	60–350	—	Y	Y	Y	Y	Atypia, rigid bladder, CM	Y

Note: —, not observed or not reported; </>, categorized in the study as less than or greater than; –ve, negative bacterial culture; A&E, only involved cases admitted to accident and emergency departments of hospitals; Avg., average; CM, carcinoma markers measured; Cor., correlation analysis performed (refer to Tables 10.2 through 10.5); CT, computed tomography; E-cad., E-cadherin; Est., estimated; ind., individual data available (for studies with more than 1 subject); IVU, intravenous urography; LUTS, lower urinary tract symptoms; Md., median; MRI, magnetic resonance imaging; NFP+, positive neurofilament protein staining; NGFR+, positive nerve growth factor receptor staining; NK, not known; PUF, primarily pelvic pain and urgency/frequency symptom scale results in this study; US, ultrasound; Y, technique used or finding explicitly reported in the study.

conjunction with the specific LUT pathophysiological finding. Such information would have enabled the examination of any correlation between presenting LUTS and underlying pathophysiological mechanisms. A large-scale retrospective study documented the percentage of patients admitted to hospital accident and emergency departments with each of the different LUTS, although no LUT pathological data were recorded (Ng et al. 2010). Knowing how LUTS of individual subjects relate to LUT pathology would add much value to our understanding of ketamine-associated toxicity.

All studies employed one or more diagnostic techniques in determining the pathophysiological characteristics of the LUT. Morphological aberration of the bladder such as bladder trabeculation and diverticulum was visible as well (Chen et al. 2012; Wei et al. 2012). The most popular imaging method used was computed tomography (CT) (in 11 studies), followed by ultrasound (US) (in 9 studies), intravenous urography (IVU) (in 5 studies), and magnetic resonance imaging (MRI) (in 2 studies). The rationale behind the choice of imaging modality, or if imaging was performed at all, was not explicitly mentioned by the investigators. Neither was it explained why subsets of patients in the same study were subjected to different imaging methods, such examples as in Chang et al. (2012), Mason et al. (2010), and Misra et al. (2014). On the other hand, it was not uncommon in single case reports where the patient was diagnosed with more than one imaging method (e.g., in Chiew and Yang 2009, Jenyon and Sole 2013, and Shahzad et al. 2009). As revealed in the report of Mason et al. (2010), the use of one imaging method might not be sufficient in identifying abnormality in the LUT. In addition, imaging was best combined with other techniques in order to reach the correct diagnosis, as demonstrated by Misra et al. (2014) where inflamed bladder could sometimes be missed if only CT or US was performed without cystoscopy.

Cystoscopy was used in more occasions than imaging in identifying LUT pathology as shown in Table 10.1. Bladder inflammation and ulceration were common among ketamine users. While imaging could reveal a smaller than normal bladder size, urodynamics study provided a more quantitative measurement of bladder capacity. Among studies that reported, bladder capacity of ketamine users could be as low as 14 mL as compared to the normal volume of approximately 500 mL. A number of studies (Chiew and Yang 2009; Chu et al. 2008; Lee et al. 2013; Misra et al. 2014; Shahani et al. 2007) explicitly described fibrosis of the bladder wall, decreased bladder compliance, or bladder rigidity, which could contribute to small bladder capacity. Monitoring the urodynamics also enabled bladder overactivity to be seen, as demonstrated in nearly all of the studies concerned. The final LUT characteristic commonly observed among ketamine users was bladder wall thickening, best demonstrated by staining of biopsy samples. This phenomenon could be a result of bladder inflammation and proliferation, the latter being of notable interest since metaplasia and atypia were reported by a few (Jalil and Gupta 2012; Mason et al. 2010; Misra et al. 2014; Oxley et al. 2009; Shahani et al. 2007).

10.2.2 Is bladder carcinoma a manifestation of chronic ketamine use?

While squamous metaplasia in the bladder was observed in a small number of ketamine users (of unknown duration of usage) (Jalil and Gupta 2012; Shahani et al. 2007), which might be suggestive of malignant development, only one confirmed case of yolk sac tumor that metastasized to the para-aortic lymph node was reported thus far (Mui et al. 2014). This case involved a patient presented twice to a clinic; the first time 7 years after she started taking ketamine, and the second time a further 5 years later (Mui et al. 2014). On the first visit, urothelial metaplasia was revealed by bladder biopsy, whereas a yolk sac tumor was identified on the second visit (Mui et al. 2014). The only other published case of ketamine-related bladder carcinoma was reported by Storr (2009), in which a lung cancer patient was put on ketamine for its analgesic effect. Because of the existing medical condition of this latter patient, and since the carcinoma in this case occurred only 7 months into ketamine treatment, it remained to be determined if a causal link exists between prolonged ketamine use and bladder carcinoma.

In other reports that recruited long-time ketamine users, that is, those with 10 or more years of use as listed in Table 10.1, most did not describe the appearance of carcinoma or carcinoma-like phenomenon. Urothelial atypia was observed in some studies that recruited short- to long-time ketamine users (Mason et al. 2010; Misra et al. 2014). It was not clear whether these atypia cases came from the long-time users since individual data were not revealed (Mason et al. 2010; Misra et al. 2014). Nevertheless, the lack of cytokeratin 20 (CK20) was found to be distinct from what was expected in true urothelial dysplasia and carcinoma (Mason et al. 2010; Misra et al. 2014). In particular, elevated CK20 expression was considered a signature marker of bladder (urothelial) carcinoma, although high levels of p53 and Ki67 were also suggestive of potential malignancy (Lopez-Beltran et al. 2014). Accordingly, carcinoma was ruled out for samples with no CK20 expression (Mason et al. 2010; Misra et al. 2014; Oxley et al. 2009). In a study where only cases of urothelial atypia (but with an unknown ketamine usage duration) were examined, the lacked CK20 expression was observed in all cases, while that of Ki67 and p53 was variable (Oxley et al. 2009). Therefore, existing information fails to establish a relationship between ketamine use and carcinoma just yet. Possibly, closer monitoring of the atypia cases over an extended period is necessary if any malignant progression is to be spotted. Meanwhile, other markers may be employed to better distinguish nonmalignant atypia from carcinoma, for example, with CD44 staining, which, unlike CK20, was absent in carcinoma (Lopez-Beltran et al. 2014).

10.2.3 Correlation analyses concerning ketamine-associated pathological outcomes in humans

Based on available data of individual subjects in selected studies, analyses were performed in order to reveal any correlation that was present between various pathological parameters.

10.2.3.1 Atypia markers, inflammatory infiltration, and bladder ulceration

In several studies mentioned earlier, Ki67 and p53 expression was measured in assessing cases of urothelial atypia. In one report (Oxley et al. 2009), in addition to Ki67 and p53, inflammatory cell (eosinophil) infiltration and bladder ulceration were also documented for each subject. Correlation analysis was performed and results are shown in Table 10.2. A strong and significant positive correlation ($r = 0.870$; $P < 0.0001$) was present, as expected, between eosinophil presence and ulceration. Although Ki67 and p53 were present in all subjects, no significant correlation was found between the two, nor was any correlation identified between ulceration or eosinophil presence and the atypia markers. Studies with a larger sample size and with normal subjects inclusive will aid in the determination of any correlational relationship between ulceration/inflammation and atypia.

Table 10.2 Correlation Analysis of Data from 10 Ketamine Abusers: Relationships among the Levels of Ki67 and p53 Expression and Presence of Eosinophils and Bladder Ulceration

	Ulceration	Eosinophil	Ki67	p53
Ulceration		0.870[a] (0.00001)	0.102 (0.779)	−0.167 (0.645)
Eosinophil			0.102 (0.779)	−0.167 (0.645)
Ki67				0.408 (0.242)

Note: Nominal values (1 = high, −1 = low to medium) were used to represent the expression levels of Ki67 and p53 in the analysis. On the other hand, nominal values (1 = yes, −1 = no) were used to indicate whether eosinophils and ulceration were present. Values shown are Spearman correlation coefficients; *P* values are in parentheses.

[a] Significant correlation (i.e., $P < 0.05$).

10.2.3.2 Gender, duration of ketamine usage, and bladder pathology

As shown in Table 10.1, the three prominent aspects of bladder pathology, namely, inflammation, wall thickening, and small capacity, were reported by many groups. The correlation between inflammation and ulceration, potentially also with atypia markers, was shown after analyzing the data by Oxley et al. (2009). In another study (Huang et al. 2014), where pathological data of the individual subjects were available, correlation analysis was performed among five parameters: gender, duration of ketamine usage, bladder inflammation, wall thickening, and small capacity, and the results are summarized in Table 10.3. No gender–duration of use correlation was present. Appearance of bladder inflammation and bladder wall thickening was also not related to gender or duration of ketamine use. While bladder inflammation did not necessarily yield a small bladder, incidence of the latter was positively correlated with bladder wall thickening ($r = 0.500$; $P = 0.008$). That a larger volume occupied by the thickened (and more rigid) bladder wall would render the intravesical volume smaller than normal was intuitive. The interesting finding here was that bladder inflammation and wall thickening did not correlate. It is possible that a more quantitative comparison, for example, using inflammatory cell count and numerical measurement of wall thickness, may give a better picture of the relationship between these parameters.

Table 10.3 Correlation Analysis of Data from 27 Ketamine Abusers: Relationships among Gender, Duration of Ketamine Usage, and Bladder Pathology (Including Inflammation, Wall Thickening, and Small Capacity)

	Gender	Duration	Inflammation	Wall thickening	Small capacity
Gender		−0.128 (0.524)	−0.135 (0.502)	0.053 (0.792)	−0.213 (0.286)
Duration			0.102 (0.611)	0.092 (0.646)	0.185 (0.356)
Inflammation				0.316 (0.108)	0.158 (0.431)
Wall thickening					0.500[a] (0.008)

Note: Nominal values (1 = male, −1 = female) were used to represent the two genders. Presence of the particular bladder pathology was represented by nominal values of 1 = yes and −1 = no. Values shown are Spearman correlation coefficients; *P* values are in parentheses.

[a] Significant correlation (i.e., $P < 0.05$).

10.2.3.3 Bladder and upper urinary tract pathology

In the same study (Huang et al. 2014), upper urinary tract (UUT) pathological parameters such as ureter wall thickening and hydronephrosis were also examined. These abnormal UUT features may correlate with the development of bladder pathology. In Table 10.4, results from correlation analysis among bladder wall thickening, ureter wall thickening, and hydronephrosis are shown. There was no correlation at all ($r = 0.000$; $P = 1.000$) between thickening of bladder and ureter wall, though not surprisingly, a strong correlation was found between ureter wall thickening and hydronephrosis ($r = 0.791$; $P < 0.0001$), probably attributed to the proximity of these UUT structures. The authors of this study (Huang et al. 2014) did not specify whether different portions of the ureter wall were distinguished when determining the presence of thickening. It will be interesting to examine if there is correlation between the thickening of bladder and distal ureter wall, and likewise between hydronephrosis and proximal ureter wall thickening. If the onset of LUTS is taken into consideration besides duration of ketamine usage, a clearer understanding on the progression of ketamine-associated urinary tract pathology will be plausible.

10.2.3.4 Ketamine dosage, onset of LUTS, and hydronephrosis

Whereas most published reports documented ketamine usage in terms of duration, two studies considered ketamine dosage instead (Chang et al. 2012; Tsai et al. 2009). In both of these studies, the onset of LUTS and

Table 10.4 Correlation Analysis of Data from the Same 27 Ketamine Abusers Referred to in Table 10.3: Relationship between Bladder and UUT Pathology

	Gender	Duration	Bladder wall thickening	Ureter wall thickening	Hydronephrosis
Gender		−0.128 (0.524)	0.053 (0.792)	−0.107 (0.592)	0.017 (0.933)
Duration			0.092 (0.646)	−0.200 (0.316)	−0.297 (0.132)
Bladder wall thickening				0.000 (1.000)	0.079 (0.695)
Ureter wall thickening					0.791[a] (0.000001)

Note: Bladder wall thickening was used to compare the presence (represented by nominal values of 1 = yes and −1 = no) of UUT pathology, including ureter wall thickening and hydronephrosis. Values shown are Spearman correlation coefficients; *P* values are in parentheses.

[a] Significant correlation (i.e., $P < 0.05$).

Table 10.5 Correlation Analysis of Data from 31 Ketamine Abusers in Two Studies: Relationships among Gender, Ketamine Dosage, Onset of LUTS, and Hydronephrosis

	Gender	Dosage	LUTS onset	Hydronephrosis
Gender		−0.00380 (0.984)	0.167 (0.368)	−0.367 (0.05)
Dosage			0.081 (0.665)	0.117 (0.545)
LUTS onset				0.345 (0.067)

Note: Nominal values (1 = male, −1 = female; and 1 = with hydronephrosis, −1 = without hydronephrosis) were used to represent the nonnumerical parameters. Values shown are Spearman correlation coefficients; *P* values are in parentheses.

hydronephrosis were reported, which might provide some insight into the progression of ketamine-associated urinary tract pathology. Results of the correlation analysis are shown in Table 10.5. Overall, no significant correlation was found among the different parameters, but two interesting observations can be made. First, the data suggested a borderline correlation between females and hydronephrosis ($r = -0.367$; $P = 0.05$). However, when data from another study (Huang et al. 2014) documenting hydronephrosis were pooled together, the correlation coefficient was closer to zero ($r = -0.172$) with a *P* value of 0.204, indicative of no meaningful relationship between gender and hydronephrosis. The second observation was regarding the onset of LUTS, which showed a tendency of negative correlation with hydronephrosis ($r = -0.345$; $P = 0.067$). If data from a larger patient group yield a similar and statistically significant result, it may suggest that kidney damage occurs before LUTS becomes observable. Assuming LUTS are the outcomes of underlying bladder pathophysiological problems, it is possible that a negative correlation also exists between bladder pathology (e.g., inflamed bladder, thickened bladder wall) and onset of LUTS. A similar methodology employed by Chang et al. (2012) and Tsai et al. (2009) extends to identifying bladder pathological features that may reveal their relationships with the onset of LUTS.

10.3 *Laboratory evidence complementary to clinical findings with new implications*

Ever since the observation of LUTS in humans, numerous laboratory studies have been conducted aiming to, first, establish a valid animal model of ketamine abuse and, second, examine features and underlying mechanisms of ketamine-associated LUT toxicity. Both rats and mice having undergone chronic ketamine treatment have been used as tools to mimic pathological changes observed in human bladders. A consensus on the dose and duration of ketamine treatment is yet to be reached, with ranges (in human equivalent doses) between 1 and 8 mg/kg over a treatment period from 0.5 to 6 months having been reported (Chuang et al. 2013; Gu et al. 2013; Kekesi et al. 2011; Meng et al. 2011; Tan et al. 2011; Tang et al. 2014). Nevertheless, notable findings concerning histological, functional, and molecular characteristics of the animal bladders provide vital information that may assist in clinical diagnosis and management.

In the bladders of human ketamine users, it is common to observe thinning or shedding of the urothelium and inflammatory cell infiltration. These distinct features were also reproduced in both rat (Chuang et al. 2013) and mouse (Tan et al. 2011) models. Interstitial fibrosis and a hardened, rigid bladder have been reported explicitly in humans (Chiew and Yang 2009; Misra et al. 2014; Shahani et al. 2007). Similar observations were made in the animal studies as well, as evidenced by increased collagen deposits within the bladder wall (Tan et al. 2011). Apoptosis was apparent in the ketamine-treated mouse bladder (Tan et al. 2011), a finding echoed by Lee et al. (2013), which showed that urothelial apoptosis was more prominent in the bladders of ketamine users than in those of other cystitis patients. Furthermore, there was negative correlation between apoptosis and bladder capacity (Lee et al. 2013), in line with the observation of small bladder volume also found in animals (Chuang et al. 2013; Kekesi et al. 2011; Meng et al. 2011).

Functionally, the decrease in bladder capacity is likely attributable to poor bladder compliance, as observed in humans and in animals. A trend of decreasing bladder compliance was demonstrated as ketamine treatment prolonged in rats (Chuang et al. 2013) and in mice (Meng et al. 2011). Nonvoiding contractions also became more frequently (Chuang et al. 2013), which corresponded with overactive bladder symptoms in humans (Table 10.1). Characteristics of bladder contractions were examined further in the mouse models (Meng et al. 2011; Tang et al. 2014). While the amplitudes of K^+, carbachol, and electrically stimulated contractions did not differ between untreated and ketamine-treated bladders (Meng et al. 2011; Tang et al. 2014), response times were slower in the latter tissues (Tang et al. 2014). Recovery times after induction of electrically stimulated contractions were also slower in ketamine-treated bladders (Tang et al. 2014). In a study of ketamine-treated rat bladders, the expression of phosphorylated transgelin was increased (Gu et al. 2013). Since transgelin is involved in contraction development, increased levels of the phosphorylated form may render less of the active protein to be involved in initiation of a contractile response, as demonstrated by the slower response times in mice (Tang et al. 2014). Changes in the expression levels of other proteins were also reported in ketamine-treated rat bladders although their implications are yet to be determined (Gu et al. 2013). Studying the proteomic characteristics in human ketamine users may reveal key information about the underlying mechanisms of impaired bladder function.

Although bladder contraction occurs primarily under the influence of cholinergic neurotransmission, the role of noncholinergic neurotransmitters, namely, ATP, may be increased in ketamine-associated bladder dysfunction. Using ATP as a stimulus, contraction was more pronounced in the ketamine-treated bladder (Meng et al. 2011), and a similar effect was observed when the atropine-sensitive component of electrical stimulation was abolished (Meng et al. 2011). The change in contractility may be explained by the increased expression of P2X1 receptor after ketamine treatment (Meng et al. 2011). Expression of other purinoceptor subtypes, for example, P2X3 (Cockayne et al. 2000), may also be increased, contributing to increased bladder pain experienced by human ketamine users. In another study, ketamine-treated rat bladders showed higher expression of cyclo-oxygenase 2 and endothelial and inducible nitric oxide synthase, the activity of which results in the production of inflammatory and pain mediators (Chuang et al. 2013). It is worth mentioning that in a number of patients, bladder pain was so severe that cystoscopy could not be performed (Chen et al. 2011; Tsai et al. 2009). It was also revealed in humans that ketamine use resulted in decreased E-cadherin expression in the bladder (Lee et al. 2013), suggesting disintegration of the urothelium– urine barrier, increased exposure to urine by nerve endings, and, thus, increased pain sensation. The findings of increased expression of pain- and inflammation-related proteins in the bladders of ketamine-treated animals are therefore worth validating in humans, which may provide a new direction in the development of better clinical management strategies.

10.4 Concluding remarks

Since 2007, many useful findings have been reported from both human and animal studies concerning LUT damages induced by ketamine. However, current intervention measures remain less than optimal, both in the alleviation of LUTS and in the preservation of bladder function. For example, it is not uncommon to resort to cystectomy and enterocystoplasty as a last treatment option for some patients (Chung et al. 2013). It is of great clinical interest to the concerned patients if further research can lead to better treatment or management strategies. Investigation into relationships between onsets of LUTS, bladder pathology, and UUT pathology may yield important findings concerning progression of ketamine-associated urinary tract damage, noting that vesicoureteric reflux as reported in some of the studies (e.g., Ho et al. 2010; Lee et al. 2013; Ng et al. 2012; Selby et al. 2008) arises from small bladder capacity (Middela and Pearce 2011). Accumulation of ketamine metabolites may elicit their respective toxic effects (Middela and Pearce 2011), and this postulation needs to be examined further in laboratory studies. Animal models of ketamine also provide an ideal platform for monitoring duration- and dose-dependent progression of tissue damages and for screening of potential drug candidates. It is envisaged that with more clinical and animal studies conducted, significant progress can be made with regard to management strategies of LUTS and the associated tissue damages.

References

Baker SC, Stahlschmidt J, Oxley J et al. 2013. Nerve hyperplasia: A unique feature of ketamine cystitis. *Acta Neuropathol Commun* 1:64.
Bokor G, Anderson PD. 2014. Ketamine: An update on its abuse. *J Pharm Pract* 27:582–586.
Chang T, Lin C-C, Lin AT-L et al. 2012. Ketamine-induced uropathy: A new clinical entity causing lower urinary tract symptoms. *LUTS* 4:19–24.
Chao J-Y, Shai H-A. 2010. Duloxetine treatment of long-term ketamine abuse-related lower urinary tract symptoms: A case report. *Gen Hosp Psychiatry* 32:647.e5–647.e6.
Chen C-H, Lee M-H, Chen Y-C et al. 2011. Ketamine-snorting associated cystitis. *J Formosa Med Assoc* 110:787–791.
Chen KT, Foo NP, Lin HJ. 2008. Frequent visits with urinary symptoms: Subtle signs of ketamine abuse. *Am J Emerg Med* 26:1061–1062.
Chen YC, Chen YL, Huang GS et al. 2012. Ketamine-associated vesicopathy. *Q J Med* 105:1023–1024.
Chiew Y-W, Yang C-S. 2009. Disabling frequent urination in a young adult. *Kidney Int* 76:123–124.
Chu PSK, Kwok SC, Lam KM et al. 2007. "Street ketamine"-associated bladder dysfunction: A report of ten cases. *Hong Kong Med J* 13(4):311–313.
Chu PSK, Ma WK, Wong SCW et al. 2008. The destruction of the lower urinary tract by ketamine abuse: A new syndrome? *BJU Int* 102:1616–1622.
Chuang S-M, Liu K-M, Li Y-L et al. 2013. Dual involvements of cyclooxygenase and nitric oxide synthase expressions in ketamine-induced ulcerative cystitis in rat bladder. *Neurourol Urodyn* 32:1137–1143.
Chung S-D, Wang C-C, Kuo H-C. 2013. Augmentation enterocystoplasty is effective in relieving refractory ketamine-related bladder pain. *Neurourol Urodyn* 33:1207–1211.
Cockayne DA, Hamilton SG, Zhu QM et al. 2000. Urinary bladder hyporeflexia and reduced pain-related behaviour in P2X3-deficient mice. *Nature* 407:1011–1015.
Colebunders B, Van Erps P. 2008. Cystitis due to the use of ketamine as a recreational drug: A case report. *J Med Case Rep* 2:219.
Cottrell AM, Gillatt DA. 2008. Ketamine-associated urinary tract pathology: The tip of the iceberg for urologists. *Br J Med Surg Urol* 1:136–138.
Gu D, Huang J, Shan Z et al. 2013. Effects of long-term ketamine administration on rat bladder protein levels: A proteomic investigation using two-dimensional difference gel electrophoresis system. *Int J Urol* 20:1024–1031.
Ho CCK, Pezhman H, Praveen S et al. 2010. Ketamine-associated ulcerative cystitis: A case report and literature review. *Malaysian J Med Sci* 17(2):61–65.

Huang L-K, Wang J-H, Shen S-H et al. 2014. Evaluation of the extent of ketamine-induced uropathy: The role of CT urography. *Postgrad Med J* 90:185–190.

Jalil R, Gupta S. 2012. Illicit ketamine and its bladder consequences: Is it reversible? *BMJ Case Rep* doi:10.1136/bcr-2012-007244.

Jenyon T, Sole G. 2013. Bladder calcification secondary to ketamine. *Urol J* 10(2):912–914.

Kekesi O, Tuboly G, Szucs M et al. 2011. Long-lasting, distinct changes in central opioid receptor and urinary bladder functions in models of schizophrenia in rats. *Eur J Pharmacol* 661:35–41.

Lai Y, Wu S, Ni L et al. 2012. Ketamine-associated urinary tract dysfunction: An underrecognized clinical entity. *Urol Int* 89:93–96.

Lee C-L, Jiang Y-H, Kuo H-C. 2013. Increased apoptosis and suburothelial inflammation in patients with ketamine-related cystitis: A comparison with non-ulcerative interstitial cystitis and controls. *BJU Int* 112:1156–1162.

Lopez-Beltran A, Montironi R, Vidal A et al. 2014. Urothelial dysplasia of the bladder. *Anal Quant Cytopathol Histopathol* 35:121–129.

Mason K, Cottrell AM, Corrigan AG et al. 2010. Ketamine-associated lower urinary tract destruction: A new radiological challenge. *Clin Radiol* 65:795–800.

Meng E, Chang H-Y, Chang S-Y et al. 2011. Involvement of purinergic neurotransmission in ketamine induced bladder dysfunction. *J Urol* 186:1134–1141.

Middela S, Pearce I. 2011. Ketamine-induced vesicopathy: A literature review. *Int J Clin Pract* 65(1):27–30.

Misra S, Chetwood A, Coker C et al. 2014. Ketamine cystitis: Practical considerations in management. *Scand J Urol* 48(5):482–488.

Mui WH, Lee KC, Chiu SC et al. 2014. Primary yolk sac tumour of the urinary bladder: A case report and review of the literature. *Oncol Lett* 7:199–202.

Ng CM, Ma WK, To KC et al. 2012. The Chinese version of the pelvic pain and urgency frequency symptom scale: A useful assessment tool for street-ketamine abusers with lower urinary tract symptoms. *Hong Kong Med J* 18(2):123–130.

Ng SH, Tse ML, Ng HW et al. 2010. Emergency department presentation of ketamine abusers in Hong Kong: A review of 233 cases. *Hong Kong Med J* 16(1):6–11.

Nomiya A, Nishimatsu H, Homma Y. 2011. Interstitial cystitis symptoms associated with ketamine abuse: The first Japanese case. *Int J Urol* 18:735.

Oxley JD, Cottrell AM, Adams S et al. 2009. Ketamine cystitis as a mimic of carcinoma in situ. *Histopathology* 55:705–708.

Peng T-R, Lee M-C, Wu T-W et al. 2014. Suspected ketamine-associated lower urinary tract symptoms. *Urol J* 11(2):1508–1510.

Robson N, Vicknasingam B, Narayanan S. 2010. Illicit ketamine induced frequency of micturition in a young Malay woman. *Drug Alcohol Rev* 29:334–336.

Selby NM, Anderson J, Bungay P et al. 2008. Obstructive nephropathy and kidney injury associated with ketamine abuse. *NDT Plus* 5:310–312.

Shahani R, Streutker C, Dickson B et al. 2007. Ketamine-associated ulcerative cystitis: A new clinical entity. *Urology* 69:810–812.

Shahzad S, Antil S, Ferrie B et al. 2009. Dystrophic calcification of urinary bladder associated with ketamine use. *Urology* 74(Suppl 4A):S317.

Skeldon SC, Goldenberg SL. 2014. Urological complications of illicit drug use. *Nat Rev Urol* 11:169–177.

Srirangam S, Mercer J. 2012. Ketamine bladder syndrome: An important differential diagnosis when assessing a patient with persistent lower urinary tract symptoms. *BMJ Case Rep* doi:10.1136/bcr-2012-006447.

Storr TM. 2009. Can ketamine prescribed for pain cause damage to the urinary tract? *Palliat Med* 23:670–672.

Tan S, Chan WM, Wai MSM et al. 2011. Ketamine effects on the urogenital system—changes in the urinary bladder and sperm motility. *Microsc Res Tech* 74:1192–1198.

Tang HC, Lam WP, Zhang X et al. 2014. Chronic ketamine treatment-induced changes in contractility characteristics of the mouse detrusor. *Int Urol Nephrol* 46(8):1563–1571.

Tsai T-H, Cha T-L, Lin C-M et al. 2009. Ketamine-associated bladder dysfunction. *Int J Urol* 16:826–829.

Venyo AKG, Benatar B. 2012. Severe lower urinary tract symptoms associated with bilateral hydronephrosis and renal impairment: A case report of ketamine abuse uropathy with a review of the literature. *WebmedCentral Urol* 3(2):WMC002969.

Wei Y, Yang J, Song W et al. 2012. Gourd-shaped bladder associated with ketamine abuse. *Urol Int* 89:123–124.

Wei YB, Yang JR. 2011. Moxifloxacin relieves the persistent symptoms of lower urinary tract after cessation of ketamine abuse. *Hong Kong Med J* 17(6):515.

Wei YB, Yang JR. 2013. Dependence and urinary symptoms among recreational ketamine users. *BJU Int* 111(3):E21–E22.

chapter eleven

Postmortem toxicology of ketamine

James Watterson

Contents

11.1   Introduction to postmortem toxicology

Postmortem toxicology involves application of analytical toxicology measurements to postmortem tissues or fluids for purposes of identifying and quantifying any drugs, drug metabolites, or toxins present and interpreting the extent to which those compounds may have been contributory to the death of the individual. Postmortem toxicology is complex, involving significant challenges that are both analytical and interpretive in nature.[1]

Analytically, the challenge is primarily associated with the accurate determination of the drugs and metabolites that may be relevant to the death investigation. Postmortem blood and tissues are often significantly degraded, which leads to substantial matrix effects in chromatography and mass spectrometry. Accordingly, stringent sample preparation approaches based on multistep solvent extraction or solid-phase extraction (SPE) are required.

Interpretation of analytical results requires consideration of a variety of factors.[1,2]

1. The nature of the samples assayed is critical in terms of assessment of toxicity (i.e., urine drug concentrations correlate poorly with toxicity, while blood drug concentrations are a much better indicator of toxicity).

2. The site of collection of specimens (e.g., blood or liver) can have significant consequences on the measured concentrations of drugs and their metabolites, as a result of postmortem redistribution (PMR) processes that result in temporal and site-dependent differences in blood or liver drug concentration. For example, peripheral blood samples (e.g., femoral blood) and samples of liver collected from the deep segments of the right lobe are preferred in order to minimize this effect.[3]

3. Pharmacological factors such as the potential for development of tolerance to drug effects and specific toxic symptoms associated with particular physiological systems (e.g., central nervous system or cardiovascular system) must be considered.

4. The potential influence of drug interactions on the overall toxicity observed must also be considered.

5. The stability of drugs and drug metabolites in the postmortem environment must be considered, where the impact of postmortem interval (prior to autopsy) and sample storage and handling, are relevant parameters.

11.2   Analysis of ketamine and metabolites in biological samples

Identification of ketamine and its metabolites in biological tissues has been established.[4–10] Ketamine ($pK_a \sim 7.5$) and its metabolites norketamine and dehydronorketamine are weakly basic and therefore may be isolated in a standard basic drug screen, which isolates such compounds for analysis by LC/MS/MS or GC/MS. Extraction of ketamine and metabolites has been achieved using liquid–liquid extraction[4,8–10] and SPE.[7] A variety of tissues have been used in these applications, including blood,[4,8–10] urine,[5] liver,[4,8,9] kidney,[4,8,9] hair,[6] bone marrow, and bone.[7] One study has examined the stability of ketamine and metabolites in blood and serum.[11] The authors showed that ketamine and its metabolites were generally stable to decomposition at −80°C, −20°C, 4°C, and ambient temperatures, but dehydronorketamine may display a tendency to diffuse into blood cells.

11.3   Relevance of ketamine to postmortem toxicology

In addition to its various therapeutic applications, which include anesthesia,[12] and its more recent candidacy as a potential drug of interest in treating depression,[13] ketamine has established itself clearly as a drug of abuse. From the perspective of the recreational user, the effects of interest include dissociation, depersonalization, and hallucinations described as feelings of lightness and out-of-body experiences (also referred to as the so-called K-hole[14]). Tolerance, dependence, and indicia of withdrawal, including flashbacks, may occur with long-term users.[9] Distortions of perceptions and schizotypal symptoms and perceptual distortions may persist after cessation of ketamine use.[15] Ketamine overdose can result in a variety of toxic sequelae, including both excessive cardio-vascular stimulation (elevated heart rate and blood pressure) and dose-dependent respiratory depression.[4] This latter property becomes critical in postmortem investigations involving drugs of abuse, since it is not uncommon to observe polypharmacy in those instances. Accordingly, central nervous system depressants, including ethanol, opioids, and benzodiazepines may contribute in an additive or synergistic fashion to enhance respiratory depression and cause death where individual drug concentrations are lower than those associated with fatality when observed alone.

In postmortem toxicological analyses, drugs of abuse are commonly detected and found to be contributory to death, although ketamine has not been widely associated with fatal intoxications. Reports of postmor-tem toxicological analysis where ketamine was detected are not com-mon, and those found in the peer-reviewed literature are summarized in Table 11.1.[4] From these studies, it appears that deaths caused exclusively by ketamine overdose are quite rare. Lalonde and Wallage[10] reported two postmortem cases where ketamine was detected in both central (heart) and peripheral (femoral) blood. In one of those cases, death was attributed to ketamine intoxication and ketamine concentrations in heart blood and femoral blood were 6.9 and 1.8 mg/L, respectively. In this case, as has been reported elsewhere,[16] it is clear that differences in measured blood ketamine concentrations may be observed depend-ing on the anatomic site of selection, suggestive of PMR. The propensity of the ketamine metabolites (norketamine and dehydronorketamine) toward PMR is not clear at this time. Accordingly, as is the practice in many forensic toxicology laboratories, sampling of peripheral blood for analysis is advised.

Another important factor in the assessment of the role of ketamine in postmortem toxicology is the potential for incurring injury or physi-cal harm while under the influence of ketamine. The reduction in awareness and perceptual distortions that accompany the dissociative effects of ketamine have been suggested to put users at risk of danger-ous falls, impaired driving, or drowning.[17] The impairing effects may be significantly enhanced when ketamine is coadministered with other psychoactive drugs, including, but not limited to, ethanol, THC (can-nabis), cocaine, or amphetamine-related drugs (e.g., methamphetamine, MDMA).

Table 11.1 Summary of Ketamine-Related Fatalities

Report	Ketamine concentration	Notes
Moore et al.[4]	1.8 mg/L (B)	Mixed-drug fatality
	2.0 mg/L (U)	EtOH—170 mg/dL
	4.3 mg/kg (Br)	
	6.1 mg/kg (Spl)	
	4.9 mg/L (L)	
	3.6 mg/kg (K)	
Licata et al.[8]	27.4 mg/L (B)	Single fatality
	8.5 mg/L (U)	Norketamine detected in all
	15.2 mg/kg (Bile)	specimens but not quantified
	3.2 mg/kg (Br)	
	6.6 mg/kg (L)	
	3.4 mg/kg (K)	
Peyton et al.[9]	Case 1:	Case 1: Fatal intravenous
	7 mg/L (B)	ketamine overdose
	6.3 mg/kg (L)	Case 2: Ketamine was
	3.2 mg/kg (K)	administered as an anesthetic
	1.6 mg/L (Lung)	in treatment of a gunshot
	2.4 mg/L (H)	wound
	Case 2:	
	3.0 mg/L (B)	
	0.8 mg/L (L)	
	0.6 mg/L (K)	
	0.8 mg/kg (Spl)	
	4.0 mg/kg (Br)	
	2.2 mg/kg (Lung)	
	3.5 mg/kg (H)	
Lalonde and	Case 1:	Case 1: Ketamine-related
Wallage[10]	1.8 mg/L (FB)	fatality
	6.9 mg/L (HB)	Case 2: Ketamine was
	Case 2:	incidental to death
	0.6mg/L (FB)	
	1.6 mg/L (HB)	

Note: B, blood; Br, brain; FB, femoral blood; H, heart; HB, heart blood; K, kidney; L, liver;
Spl, spleen; U, urine.

11.4   Conclusions and summary

Cases of fatal ketamine toxicity are relatively rare. Nonetheless, fatal ketamine overdoses have been reported in the literature, with blood ketamine
concentrations of 1.8 mg/L or greater being associated with fatal toxicity.
It is important to note that ketamine may be susceptible to PMR, implying
that peripheral postmortem blood samples are the preferred matrix for
analysis where possible. Additionally, ketamine toxicity may be subject

to tolerance, leading to some overlap between blood ketamine concentrations associated with fatalities and those observed where ketamine use was incidental to the death. Finally, since ketamine causes respiratory depression at high doses, coadministration of other central nervous system depressants (e.g., ethanol, opiates, or benzodiazepines) as is often observed in cases involving drugs of abuse, may lead to enhanced respiratory depression and death at lower blood ketamine concentrations than may be observed in cases where ketamine alone was administered.

References

1. Drummer, O. 2007. Post-mortem Toxicology. *Forensic Sci. Int.* 165:199–203.
2. Skopp, G. 2004. Preanalytic Aspects in Postmortem Toxicology. *Forensic Sci. Int.* 142(2–3):75–100.
3. Flanagan, R.J., Connally, G. and Evans, J.M. 2005. Analytical Toxicology: Guidelines for Sample Collection Postmortem. *Toxicol. Rev.* 24:63–71.
4. Moore, K.A., Kilbane, E.M., Jones, R., Kunsman, G.W., Levine, B. and Smith, M. 1997. Tissue Distribution of Ketamine in a Mixed-Drug Fatality. *J. Forensic Sci.* 2(6):1183–1185.
5. Lin, H.R., Lin, H.L., Lee, S.F., Liu, C. and Lua, A.C. 2010. A Fast Screening Procedure for Ketamine and Metabolites in Urine Samples with Tandem Mass Spectrometry. *J. Anal Toxicol.* 34:149–154.
6. Wu, Y.H., Lin, K.L., Chen, S.C. and Chang, Y.Z. 2008. Simultaneous Quantitative Determination of Amphetamines, Ketamine, Opiates and Metabolites in Human Hair by Gas Chromatography/Mass Spectrometry. *Rapid Commun. Mass Spectrom.* 22(6):887–897.
7. Cornthwaite, H.C. and Watterson, J.H. 2014. Microwave-assisted Extraction of Ketamine and Metabolites from Skeletal Tissues. *Anal. Meth.* 6:1142–1148.
8. Licata, M., Pierini, G. and Popoli, G. 1994. A Fatal Ketamine Poisoning. *J. Forensic Sci.* 39(5):1314–1320.
9. Peyton, S.H., Couch, A.T. and Bost, R.O. 1988. Tissue Distribution of Ketamine: Two Case Reports. *J. Anal. Toxicol.* 12(5):268–269.
10. Lalonde, B.R. and Wallage, H.R. 2004. Postmortem Blood Ketamine Distribution in Two Fatalities. *J. Anal. Toxicol.* 28:71–74.
11. Hijazi, Y., Bolon, M. and Boulieu, R. 2001. Stability of Ketamine and Its Metabolites Norketamine and Dehydronorketamine in Human Biological Samples. *Clin. Chem.* 47(9):1713–1715.
12. Kalsi, S.S., Wood, D.M. and Dargan, P.I. 2011. The Epidemiology and Patterns of Acute and Chronic Toxicity Associated with Recreational Ketamine Use. *Emerg. Health Threats J.* 4:7107–7117.
13. Aan Het Rot, M., Zarate, C.A., Charney, D.S. and Mathew, S.J. 2012. Ketamine for Depression: Where Do We Go From Here? *Biol. Psychiatry* 72(7):537–547.
14. Schifano, F., Corkery, J., Oyefeso, A., Tonia, T. and Ghodse, T. 2008. Trapped in the "K-hole": Overview of Deaths Associated with Ketamine Misuse in the UK (1993–2006). *J. Clin. Psychopharmacol.* 28(1):114–116.
15. Curran, H.V. and Morgan, C. 2000. Cognitive, Dissociative and Psychotogenic Effects of Ketamine in Recreational Users on the Night of Drug Use and 3 Days Later. *Addiction* 95(4):575–590.

16. Dalpe-Scott, M., Degouffe, M., Garbutt, D. and Drost, M. 1995. A Comparison of Drug Concentrations in Postmortem Cardiac and Peripheral Blood in 320 Cases. *Can. Soc. Forensic Sci. J.* 28(2):113–121.
17. Mozayani, A. 2002. Ketamine: Effects of Ketamine on Human Performance and Behavior. *Forensic Sci. Rev.* 14:123–131.

chapter twelve

The antidepressant effects of ketamine and the underlying mechanisms

Nan Wang and Jian-Jun Yang

Contents

12.1   Introduction

Depression is a worldwide and heterogeneous disorder with severe health and socioeconomic consequences (Kessler et al. 2003). Despite years of research and efforts to develop effective treatments, available antidepressant medications have serious limitations, including a low rate of treatment response (i.e., just one in three respond to the first medication, and only up to two in three respond to multiple medications) (Trivedi et al. 2006). Moreover, there is an average time lag of 3–4 weeks before a therapeutic effect is observed, which is a serious problem given the high rate of suicide in patients with depression. Thus, there is a critical and still unmet need to identify and test novel drug targets for mood disorders in order to develop more effective medications.

Ketamine is a voltage-dependent, high-affinity, noncompetitive N-methyl-D-aspartate (NMDA) receptor antagonist that inhibits the flux of cations (such as Ca^{2+} and Na^+) in the presence of the coagonists glutamate and glycine. The racemic mixture (i.e., equal amounts of the S(+)- and R(−) isomers) of ketamine has been approved for the induction and maintenance of anesthesia via intramuscular or intravenous administration and has been marketed worldwide since the 1970s. The World Health Organization has listed ketamine as a core medicine (i.e., a minimum medical requirement for a basic health care system) for both adults and children. As ketamine has a short systemic half-life of just 180 min and is extensively metabolized in the liver, which results in a high first-pass effect, this drug is not suitable for oral delivery.

The package insert for ketamine describes the drug as an approved anesthetic for diagnostic and surgical procedures and that an intravenous bolus injection of 1–2 mg/kg within 1 min can rapidly induce anesthesia that lasts for 5–10 min. This prescribed application has been linked to several dissociative reactions, which can occur on emerging from anesthesia. However, ketamine still remains a desirable anesthetic because of its short half-life and the lack of associated respiratory depression (Clemens et al. 1982). Over the past 40 years, ketamine has been administered as an anesthetic worldwide to several million people and has a good safety profile (Lahti et al. 2001; White et al. 1982).

12.2   Ketamine exerts a rapid and robust antidepressant effect in depressed patients

Converging lines of evidence suggest that major depression is associated with abnormalities in glutamatergic synaptic transmission (Sanacora et al. 2008) resulting in a loss of synaptic plasticity in the mood and emotion circuits (Kavalali and Monteggia 2012; Lee et al. 2012). As a glutamatergic NMDA receptor (NMDAR) antagonist, ketamine has been widely studied, showing rapid and robust antidepressant responses, which occur within hours and that are possibly mediated by changes in glutamate transmission (Berman et al. 2000; Maeng et al. 2008; Zarate et al. 2006b). Moreover, the antidepressant properties of ketamine have been observed in patients who are resistant to two or more of the other typical antidepressants (i.e., they are considered to have treatment-resistant depression [TRD]). The identification of a rapid-acting, efficacious antidepressant with a completely different mechanism of action is a significant development for the treatment of depression. This therefore represents an exciting prospect for improving the outcome of depressed patients and thus leads to a completely new approach with regard to the way that mental health services are delivered.

12.3   Efficacy of ketamine in TRD and bipolar depression

Many small clinical studies and case reports have found that slow subanesthetic infusions of ketamine (at 0.5 mg/kg over a period of 40 min) provide effective antidepressant properties, even in patients who have previously responded poorly to conventional antidepressant drugs (Diazgranados et al. 2010a; Ibrahim et al. 2011; Zarate et al. 2006a). In selected patients, the antidepressant effects of ketamine have been shown to last for 4–7 days after a single dose infusion, and they can be maintained for several weeks with repeated infusions (Murrough et al. 2013b). Interestingly, the antidepressant effects of a single dose persist significantly longer (i.e., for approximately 5 days) than the half-life of ketamine (i.e., approximately 3 h). Ketamine has a rapid onset of efficacy of within a few hours and its effect is both robust and clinically meaningful in patients with TRD or bipolar depression. Nearly two-thirds of all patients respond and one-third achieve remission within just 1 day.

The adverse effects of ketamine, including dissociation and hallucinations, were exposed in early trials that include unblinding of both the test subjects and the clinicians involved. However, in a recent study, a single dose of ketamine was compared with the short-acting benzodiazepine, midazolam, which was used as an active placebo. The ketamine treatment group exhibited a greater improvement in the Montgomery–Asberg Depression Rating Scale (MADRS) score than the midazolam group within 24 h after treatment (Murrough et al. 2013a).

An open-label study was also used to test whether a repeated dose of ketamine could sustain the single-dose response. The study started with just 10 subjects (aan het Rot et al. 2010) and then added 14 more (Murrough et al. 2013b), to give a total of 24 subjects with TRD. Subjects initially went through a washout period when they took no antidepressant medication; they were then administered with up to six intravenous infusions of ketamine at 0.5 mg/kg over 40 min, and three times a week over a 12-day period. At the end of the study, the overall response rate was 71%. Among responders, the median time to relapse after the last ketamine infusion was 18 days. Patients responding to the first ketamine administration ($n = 17$) all responded to the consecutive treatments, whereas those who did not respond to the first administration and who completed the six intravenous infusions study ($n = 4$) consistently failed to respond. This multiple-dose study suggested that the initial antidepressant response can be maintained with multiple doses. However, there was significant variability in the duration of response, confirming previous evidence observed in the single-dose studies (Ibrahim et al. 2012; Mathew et al. 2010; Zarate et al. 2006a). The data suggested that there seems to be a subset of subjects who are able to sustain the response for several weeks to months after the last dose of ketamine, which might indicate that individualized ketamine treatment frequencies might be required to maintain the clinical benefit over longer periods. On the whole, studies whereby ketamine was administered open-label showed the highest interindividual variability in the duration of response (Ibrahim et al. 2012; Mathew et al. 2010; Murrough et al. 2013b). In this situation, longer-term placebo-controlled studies are needed to investigate this approach further.

In studies on bipolar disorder, patients were initially treated with lithium or valproate for 4 weeks before dosing with ketamine (Diazgranados et al. 2010a). The relapse rates were similar in subjects on either lithium or valproate and were also comparable to those reported during the TRD study with ketamine monotherapy.

12.4   *Effects of ketamine on suicidal ideation*

Interest regarding the effect of ketamine on suicidal ideation has also been generated. Although such patients were excluded from initial randomized controlled trials (RCTs), ketamine was reported to quickly reduce suicidal ideation scores on the depression rating scales (Berman et al. 2000; Zarate et al. 2006a). This has been replicated in two open-label trials of patients with major depressive disorder (MDD) and in one RCT of bipolar patients (Diazgranados et al. 2010a; Price et al. 2009; Zarate et al. 2012), demonstrating the robust and rapid antisuicidal effect of ketamine. Diazgranados et al. (2010b) gave 33 TRD subjects a single infusion of ketamine (at 0.5 mg/ kg). Patients were rated at baseline, and then at 40, 80, 120, and 230 min postinfusion using the scale for suicide ideation (Beck et al. 1979). Scores significantly decreased within 40 min of ketamine infusion and remained low for up to 4 h postinfusion. In another study, early antisuicidal effects (i.e., within 1 day) were found with ketamine, and these effects remained significant for several weeks (aan het Rot et al. 2010). Repeated ketamine infusions also reduced the suicidality ratings but only as long as the duration of the 12-day repeated infusion trial (Price et al. 2009). In addition, a study investigating the efficacy of intravenous ketamine in the emergency room showed beneficial results, but this study was not without its flaws, most notably the lack of a control group (Larkin and Beautrais 2011). However, these original studies have led the way for further work to explore the effect of ketamine on suicidal ideation in more detail.

More recent studies have provided us with clinical evidence to suggest that ketamine exerts a rapid antidepressant effect in affective disorders and it appears to reduce suicidal ideation. The studies to date have shown that the effect of ketamine peaks at 24 h postinfusion and in general lasts for 1–2 weeks. Clearly, maintaining the transient antidepressant response of ketamine is required. Additionally, the psychotomimetic side effects of ketamine and the concerns regarding abuse potential of this drug make the long-term clinical use of ketamine unfeasible. Mechanistic insights are thus urgently required in order to produce a similarly acting agent without the associated safety issues.

12.5 Ketamine's safety profile

Ketamine has been used as an anesthetic since 1963, and it is considered to have a very good medical safety profile (Morgan et al. 2012; Strayer and Nelson 2008). However, limited data regarding its chronic or long-term use are available. Ketamine has a wide therapeutic range and patients have recovered fully after receiving even 10 times the normal dose (Perry et al. 2007).

Ketamine appears to have a similar safety profile in patients with mood disorders as it does in healthy volunteers: the most common adverse events reported from depression studies include perceptual disturbances, drowsiness, dizziness, confusion, and elevations in blood pressure and pulse; however, these are usually resolved shortly after the end of the infusion period (Diazgranados et al. 2010a; Ibrahim et al. 2011; Murrough et al. 2013a).

12.5.1   Adverse events associated with short-term use of ketamine at subanesthetic doses

In a series of studies in 469 healthy volunteers, ketamine was reported to be associated with short-lasting, dose-dependent psychosis/dissociative effects after a single administration of a subanesthetic dose. These side effects were shown to be reversible upon the cessation of drug administration (Perry et al. 2007). However, the adverse effects of ketamine sedation/dissociation require careful monitoring and observation of subjects during the period of drug administration and up to 1 h after the completion of exposure. No evidence of ketamine abuse or psychiatric problems related to single-dose ketamine exposure has been found in follow-up (i.e., up to 6 months) data collected.

12.5.2   Adverse events associated with chronic use of ketamine

Much of the literature regarding chronic ketamine use is derived from ketamine abusers rather than from clinically treated patients. Frequent ketamine users (i.e., those who take the drug more than 5 times per week) exhibit both short- and long-term memory impairments. This might be attributed to the dosages used by street users being at a much higher level than those prescribed for TRD treatment (Morgan and Curran 2006). Fortunately, the memory impairments seem to be reversible, as they were not found in 30 ex-ketamine users who were abstinent for at least a year. The infrequent or recreational use of ketamine does not appear to be associated with long-term cognitive impairments.

12.5.3   Ketamine combined with electroconvulsive therapy

Electroconvulsive therapy (ECT) is generally considered to have a more rapid onset of action than standard antidepressant agents, and it is a highly effective treatment of unipolar and bipolar depression (Weiner 2001). The data regarding the onset time of ECT antidepressant action suggest that a range of five to seven treatments within approximately 3 weeks are required to significantly ameliorate the severity of symptoms (Nobler et al. 1997). This is considerably longer than the onset of the antidepressant effect of ketamine. One of the most promising aspects of ketamine is that even patients resistant to ECT may benefit from ketamine administration (Ibrahim et al. 2011; Zarate et al. 2006a).

Considering that ketamine is a licensed anesthetic agent, it is reasonable to question whether ketamine coadministration can expedite the antidepressant action of ECT. Interestingly, ketamine anesthesia has been used with ECT for more than 40 years (Brewer et al. 1972; Green 1973), and more recent preliminary evidence suggests that ketamine anesthesia may improve seizure duration relative to other common anesthetic agents used in ECT (Krystal et al. 2003; McDaniel et al. 2006). However, the direct antidepressant effects of ketamine have not been considered as a potential benefit to the use of ketamine in these studies.

Two uncontrolled studies have shown that depression scores decrease faster during ECT in patients given ketamine than those given propofol or thiopentone (Kranaster et al. 2011; Okamoto et al. 2010). However, an RCT with either thiopental alone or thiopental plus ketamine (at 0.5 mg/kg) for anesthesia did not appear to either enhance or expedite the antidepressant effect of ECT (Abdallah et al. 2012).

12.5.4 Antidepressant-like behavioral effects of ketamine in rodents

Since 2002, the antidepressant effect of ketamine has been widely investigated in a number of preclinical studies with animal models. For example, it was shown that acute administration of a single dose of ketamine produced antidepressant-like behavior in rodents. The protracted period used for these experiments, which lasted from a few days to several weeks, was considered to resemble the time course required for the clinical effects of ketamine to take place (Maeng et al. 2008; Yilmaz et al. 2002).

12.5.5 Forced swimming test

The forced swimming test (FST) is the most frequently used behavioral test for assessing depressive-like behaviors in rodents. Mice and rats placed in cylinders containing water rapidly become immobile, as demonstrated by either floating passively or making only the few movements necessary to remain afloat. On the basis of an immobility response induced by inescapable exposure to stress, the FST has strong predictive validity because short-term administration of antidepressant compounds such as tricyclic compounds, monoamine oxidase inhibitors, selective serotonin reuptake inhibitors (SSRIs), and atypical antidepressants has been shown to reduce the immobility response time during the FST (Cryan et al. 2005b).

A single administration of ketamine may produce an acute reduction in immobility during the FST, which occurs shortly after injection. Although the majority of these studies used a 10-mg/kg dose of ketamine, subanesthetic doses of 10–50 mg/kg ketamine have also been reported to exert antidepressant-like effects in the FST (Garcia et al. 2008b; Iijima et al. 2012; Li et al. 2011). A few studies, on the other hand, failed to detect any acute antidepressant-like effects after administration of ketamine in mice or rats when using the FST (Bechtholt-Gompf et al. 2011; Popik et al. 2008).

One feature of the pharmacology of ketamine that is distinct from other, commonly used antidepressants is that it produces protracted antidepressant-like behavioral actions, which persist for at least several days after administration. These protracted effects have been reported to persist for various lengths of time ranging between 8 days and 2 weeks (Garcia et al. 2008a; Ma et al. 2013; Maeng et al. 2008; Yilmaz et al. 2002). Interestingly, the antidepressant-like effects of ketamine can still be observed in the FST, 2 months after the cessation of a 15-day ketamine treatment in adolescent rats (Parise et al. 2013). Consistent with this result, other studies have used a 10- or 12-day dosing regimen to establish longer-lasting antidepressant-like effects of chronic ketamine in the FST (Akinfiresoye and Tizabi 2013; Tizabi et al. 2012). Only one study evaluating these protracted effects of ketamine failed to observe this effect (Lindholm et al. 2012).

Chronic stress has been shown to facilitate detection of the antidepressant-like effects of ketamine in the FST (Koike et al. 2013). There are also some differences in the strain of rats used, with regard to their sensitivity to ketamine. For example, Wistar rats are insensitive to the antidepressant-like properties of ketamine at low doses (i.e., 2.5 to 5 mg/kg) after chronic treatment. In contrast, Wistar–Kyoto (WKY) rats have been shown to be extremely sensitive to the ketamine-induced reduction in FST immobility (Tizabi et al. 2012). WKY rats have a high baseline immobility level in the FST, allowing for a greater sensitivity to compounds. Moreover, these rats are a genetic model of pathological depression and anxiety (Solberg et al. 2004; Will et al. 2003), which might provide them with an enhanced sensitivity to the antidepressant-like effects of ketamine. Finally, WKY rats are insensitive to SSRIs (López-Rubalcava and Lucki 2000; Will et al. 2003), indicating that ketamine is active under conditions where conventional antidepressants are ineffective. This feature makes the WKY rat a useful strain to assess novel compounds resembling ketamine.

12.5.6 Tail suspension test

The tail suspension test (TST) is a less stressful test of behavioral despair than the FST and involves mice that are suspended from their tail (Cryan et al. 2005a; Steru et al. 1985). Using the TST, ketamine was shown to reduce behavioral despair in mice acutely, with studies reporting a reduction in immobility time of as little as 30 min (Cruz et al. 2009; Koike et al. 2011a; Mantovani et al. 2003; Rosa et al. 2003) and up to 24 h (Koike et al. 2011b) after a single administration.

The most effective dose of ketamine in the TST is approximately 30 mg/kg. ICR mice were shown to be sensitive to ketamine as they continued to exhibit reduced immobility at 72 h after treatment (Koike et al. 2011b). Furthermore, a lower dose of ketamine (10 mg/kg) was also shown to be effective in reducing TST immobility, which was increased by chronic mild stress (CMS) 48 h after ketamine administration (Ma et al. 2013). In contrast, two other studies have indicated that the acute reduction in immobility by higher doses of ketamine (i.e., 50 and 160 mg/kg) is not maintained 1 week later in mice (Bechtholt-Gompf et al. 2011; Popik et al. 2008). These data suggest that the TST is most valuable in measuring the more immediate antidepressant actions of ketamine. However, exposure to stress might increase the sensitivity to ketamine in the TST.

12.5.7 Sucrose preference test

Sucrose consumption is widely used to assess anhedonia in rodents and it is sensitive to chronic stress and antidepressant treatment. Repeated administration of ketamine for 7 days was shown to reverse the decrease in sucrose consumption in CMS rats. However, it should be noted that this dosing regimen also increased sweet food consumption in both stressed and nonstressed rats (Garcia et al. 2009). Moreover, administration of 0.5 mg/kg for 10 days significantly increased sucrose consumption in WKY rats (Akinfiresoye and Tizabi 2013). A dose of 10 mg/kg ketamine also significantly increased sucrose consumption at 1, 3, 5, and 7 days after administration in rats (Li et al. 2011) and at 4, 6, and 8 days after administration in the CMS mouse model (Ma et al. 2013), suggesting the possibility of protracted effects of ketamine on this behavior. These data support the use of the sucrose preference test (SPT) as a sensitive screening test for rapid-acting antidepressant-like drugs including ketamine. However, in another study, the consumption of sugar pellets in rats exposed to CMS was not altered by ketamine treatment (Rezin et al. 2009).

12.5.8 Novelty-suppressed feeding

Exposure to a novel environment produces an anxiety-like phenotype in rodents known as hyponeophagia. In the novelty-suppressed feeding (NSF) and novelty-induced hypophagia tests, the latency to feed is increased and the amount of food consumed is reduced in a novel environment. NSF requires acute food deprivation for 24 h before testing, whereas novelty-induced hypophagia utilizes an 8- to 10-day training period without deprivation. Hyponeophagia is an anxiety-related test that is reliably attenuated after chronic, but not acute, administration of antidepressant agents (Dulawa and Hen 2005).

Ketamine was shown to reduce the latency to eat within hours after administration. The effective dose of ketamine in this task varied across studies, with 30 min and 24 h reported after 5–10 mg/kg (Carrier and Kabbaj 2013; Li et al. 2010) and 30 mg/kg (Iijima et al. 2012), respectively, but all tests led to a reduced latency to feed in the novel environment. Moreover, 10 mg/kg ketamine successfully reduced the latency to eat in the novelty-induced hypophagia just 1 h after administration (Burgdorf et al. 2013). More protracted effects of ketamine (3 mg/kg) were observed 48 h after administration in mice exposed to CMS, although it did not reduce feeding latency in nonstressed mice (Autry et al. 2011).

Together, these data suggest that hyponeophagia is highly sensitive to a single dose of ketamine. The fact that ketamine also rapidly produces anxiolytic effects whereas conventional antidepressants require chronic treatment for weeks agrees with a more rapid onset of clinical effects. As patients with TRD exhibit more comorbid anxiety than those with treatment-responsive MDD, the usefulness of measuring ketamine in anxiety tests should be noted.

12.5.9   Locomotor activity

The antidepressant-like effects of ketamine are frequently measured in conjunction with spontaneous activity, as increased locomotor activity may produce false-positive effects in the aforementioned behavioral tests. Ketamine (at 5–15 mg/kg) has been reported to produce hyperactivity within 10 min after intraperitoneal injection (da Silva et al. 2010). In addition, repeated ketamine (50 mg/kg) administration has been suggested to sensitize rats to its hyperactive effects (Popik et al. 2008). In contrast, the majority of studies did not find hyperactivity after ketamine administration. A reduction of open-field behavior was observed in rats within 30 min after administration of 50 mg/kg ketamine (Engin et al. 2009) and within 1 h after 10 and 25 mg/kg ketamine treatment (Gigliucci et al. 2013). In addition, a single dose of ketamine did not alter locomotor activity beyond 30 min postinjection in either rats (Akinfiresoye and Tizabi 2013; Réus et al. 2011; Tizabi et al. 2012; Yang et al. 2012) or mice (Lindholm et al. 2012). Moreover, chronic use of low-dose ketamine did not alter the locomotor activity in adult rats (Garcia et al. 2008b; Ma et al. 2013). However, hyperactivity was displayed in adolescent but not adult rats after chronic ketamine use (Parise et al. 2013).

Overall, these data suggest that it is important to assess changes in locomotor activity to identify and eliminate the involvement of any potential locomotor effect in the behavioral responses to ketamine administration.

12.5.10   Chronic mild stress

Exposure to CMS (i.e., a series of stressors in an unpredictable sequence over a prolonged period) induces depressive behavior in rodents. The CMS model satisfies most of the criteria of validity for an animal model of depression and it produces a number of behavioral changes in rodents thought to resemble features of depressed patients (Willner 1997, 2005). The behavioral and molecular alterations induced by CMS are reversed by conventional antidepressants after treatment for several weeks. In contrast, acute administration of ketamine has been shown to reverse the behavioral and physiological changes induced by CMS in rats within hours, and this was maintained after chronic ketamine administration (Garcia et al. 2009; Parise et al. 2013). For example, the reversal of CMS-induced depressive-like phenotypes (evaluated by the FST, NSF, and SPT) has been reported by use of ketamine in stress-naïve mice (Autry et al. 2011). Consistent with these findings, two similar studies have indicated an increased sensitivity of CMS-exposed mice to ketamine (Li et al. 2011; Ma et al. 2013). Taken together, the CMS model is the most consistent and possibly the most valid method of examining the antidepressant-like effects of ketamine in preclinical studies.

12.5.11   Learned helplessness

The learned helplessness (LH) model of depression produces escape deficits in rodents exposed to unpredictable and uncontrollable stress (Seligman et al. 1980). This is a popular model of depression as it possesses good validity and induces a number of endophenotypes that can be assessed in other behavioral tests, including the FST and NSF. Repeated routine treatment with antidepressants was reported to reverse the coping behavior deficits in rats and mice (Caldarone et al. 2000; Shanks and Anisman 1988). In contrast, an acute single dose of ketamine (10 mg/kg) was reported to reverse the coping behavior deficits within 30–60 min (Beurel et al. 2011; Koike et al. 2011b) and 24 h after treatment (Li et al. 2010; Maeng et al. 2008). Furthermore, ketamine at a lower dose (i.e., 3 mg/kg) was reported to exert antidepressant-like effects in the LH model, in mice exposed to CMS (Autry et al. 2011).

12.6   Preclinical and clinical trials to sustain the antidepressant effects of ketamine

Strategies to sustain the rapid antidepressant effects of ketamine primarily focus on repeated infusions, usually administered on alternate days over an extended period. The ketamine repeated infusion protocol was evaluated preclinically by Parise et al. (2013) in adolescent male rats (postnatal day 35) receiving two ketamine injections (at doses of 0, 5, 10, or 20 mg/kg), 4 h apart, 1 day after the FST. Separate groups exposed to CMS were assessed to confirm the findings from the FST. After these experiments, adolescent naive rats were exposed either to just 1 day or to 15 consecutive days of 20 mg/kg ketamine twice daily. The effects of ketamine on the behavioral responses to rewarding (sucrose preference) and aversive (elevated plus-maze) circumstances were determined 2 months later. Ketamine was shown to reverse the CMS-induced depression-like behavior normally associated with the FST. In addition, repeated ketamine use led to anxiolytic- and antidepressant-like responses regardless of the age of rats on exposure. On the other hand, none of the ketamine doses elicited drug-seeking behaviors, measured by place preference conditioning (Parise et al. 2013). These preliminary findings indicate that repeated ketamine exposure serves as a potentially useful antidepressant for use in adolescents.

The clinical efficacy of repeated ketamine infusions has a less positive result. The reactivity of repeated ketamine infusions in 10 patients with TRD was investigated and the MADRS score was shown to be altered from the baseline (aan het Rot et al. 2010). Nine of the patients presented a more than 50% reduction in their MADRS score on day 2 and then they received five additional infusions on days 3, 5, 8, 10, and 12. Among these patients, eight of the nine relapsed within 19 days after the sixth infusion of ketamine. Murrough et al. (2013b) showed a similar result in 24 patients with TRD where the antidepressant effects lasted on average for 18 days after the last infusion. Segmiller et al. (2013) also infused S-ketamine into six participants with TRD six times over a 4-week period and observed an improvement in the 21-item Hamilton depression rating scale scores both before and 120 min after the sixth infusion.

Riluzole is a glutamate-modulating agent with neuroprotective and synaptic plasticity-enhancing effects, which is used in the treatment of amyotrophic lateral sclerosis (Du et al. 2007; Mizuta et al. 2001). Clinical studies have attempted to sustain the antidepressant actions of ketamine with riluzole. Thus, in one study, riluzole was administered orally for 4 weeks to lasting ketamine responders (Mathew et al. 2010), whereas in another study, all participants were treated with the drug within 4–6 h after ketamine infusion (Ibrahim et al. 2012). Both of these studies failed to show any benefit over the placebo in maintaining the response to ketamine infusion.

Memantine is a low-to-moderate affinity and noncompetitive NMDA antagonist. One case study from Kollmar et al. (2008) found positive results after oral administration of memantine. Unfortunately, however, a placebo-controlled study could not show any antidepressant effects of memantine (Zarate et al. 2006c). To date, efforts to maintain the antidepressant effect of ketamine are disappointing.

12.7   Mechanism of the antidepressant effects of ketamine: Preclinical studies

12.7.1   Subanesthetic ketamine acutely increases glutamate release

The clinical observations that indicate that ketamine has a very different mechanism of action from other antidepressants have fueled tremendous effort in understanding the mechanism in preclinical models. Recent studies have generally tried to link intracellular phenomena observed to behavioral alterations that reflect antidepressant efficacy. Acute subanesthetic doses of ketamine trigger a complex intracellular cascade that ultimately induces synaptogenesis and dendritic spine formation, which both seem to be essential for the antidepressant effects to happen (Duman and Aghajanian 2012).

In one of the first reports investigating the effects of subanesthetic doses of ketamine on the glutamatergic system, Moghaddam et al. (1997) demonstrated an increased release of extracellular glutamate in the prefrontal cortex of conscious animals; which was observed with low doses of ketamine (i.e., 10, 20, and 30 mg/kg) but not with intermediate (50 mg/kg) or anesthetic doses (200 mg/kg). Interestingly, the ketamine-induced increase in glutamate and dopamine release and impairment in working memory were blocked by an α-amino-3-hydroxy-5-methyl-4-isoxazolepropionic acid (AMPA) receptor antagonist, suggesting that ketamine-induced increases in glutamate levels reflect heightened glutamatergic neurotransmission and not simple metabolic status changes (Moghaddam et al. 1997). The effects of subanesthetic doses of ketamine were also recently investigated in vivo via the use of ^{1}H-magnetic resonance spectroscopy (MRS). Kim et al. (2011) reported an increase in the prefrontal glutamate level with a corresponding decrease in glutamine/glutamate ratio, after subchronic (i.e., 7-day) administration of 30 mg/kg ketamine in anesthetized rats. In some ways, however, it is hard to interpret the findings of this report because the ketamine dose used in the experiments was higher than the one used to induce antidepressant effects (i.e., 10 mg/kg); also, no measurement was performed after a single dose of ketamine (Kim et al. 2011). These issues, as well as the difficulty of understanding how there is an increase in total glutamate without an equivalent increase in glutamine, make the relevance of these findings for depression rather uncertain. Interestingly, a ketamine-triggered acute increase in glutamatergic transmission seems to be a relatively transient phenomenon, only lasting up to 2 h, which is well below the average 4- to 7-day duration of antidepressant action in humans after a single dose of ketamine (Zarate et al. 2010).

Recently, Chowdhury et al. (2011) used ^{1}H-MRS to study the effects of subanesthetic (30 mg/kg) and anesthetic (80 mg/kg) doses of ketamine on glutamate release in the prefrontal cortex in rats waking from anesthesia. An acute increase in glutamate, glutamine, and γ-aminobutyric acid (GABA) labeling was observed in the medial prefrontal cortex of rats treated with subanesthetic but not anesthetic doses of ketamine, suggesting that the drug increases the glutamate/glutamine cycle as well as the glutamine/GABA cycle. In addition, the effects observed seemed to be specific to the medial prefrontal cortex, as the hippocampus was not affected. A subanesthetic dose of ketamine was not associated with any changes in the total concentration of glutamate, glutamine, or GABA, consistent with some negative findings reported in humans using ^{1}H-MRS. These findings replicate and extend the original findings demonstrating ketamine-induced glutamate release (Moghaddam et al. 2012).

The biological determinants of increased glutamate release are not entirely clear, but one of the most plausible hypotheses is that ketamine inhibits the NMDARs on tonic GABAergic interneurons, which then disinhibits glutamatergic pyramidal cells in the prefrontal cortex (Homayoun and Moghaddam 2007). From the published literature, it appears that there is a clear distinction between the effect of subanesthetic and anesthetic doses of ketamine on cortical disinhibition, with only subanesthetic doses producing an acute increase in glutamate release. Among the subanesthetic doses, there is uncertainty whether the ketamine-induced increase in glutamate release only provokes the psychotic-like symptoms or if it also plays a role in the antidepressant effects. Future studies are needed to investigate the effect of the lower subanesthetic doses of ketamine (i.e., 10 mg/kg or lower) on the glutamate/glutamine cycle to determine whether an increase in glutamate release is relevant to the antidepressant effects of ketamine.

12.7.2 AMPA receptor activation is necessary for the antidepressant effects of ketamine

Recent studies have shown that AMPA receptor (AMPAR) activation is essential for the antidepressant effects of ketamine and that these effects might be abolished by pretreatment with the AMPAR antagonist, NBQX (Maeng et al. 2008). This represents one of the possible underlying mechanisms of the antidepressant effects mediated by ketamine (Autry et al. 2011; Koike et al. 2011b; Li et al. 2010). AMPAR blockade can also inhibit the ketamine-induced activation of the mammalian target of rapamycin (mTOR) pathway, as well as increased synaptogenesis and synaptic protein increases that are believed to be essential for its antidepressant actions (Li et al. 2010).

12.7.3 Ketamine promotes synaptic plasticity

Subanesthetic doses of ketamine (i.e., 3 mg/kg) also promote a rapid induction in the translation of brain-derived neurotrophic factor (BDNF) through the desensitization of eukaryotic elongation factor 2 kinase, as well as the dephosphorylation of eukaryotic elongation factor 2, both of which are induced via NMDAR inhibition. Increased levels of BDNF protein were shown in the hippocampus within 30 min after the administration of ketamine but not after 24 h, while no change was apparent in the levels of BDNF mRNA at either time point (Autry et al. 2011). In addition, the ketamine-induced antidepressant effects and induction of BDNF were both abolished by administration of the protein synthesis inhibitor anisomycin during the FST, providing further evidence that increased protein synthesis is necessary for the effects of ketamine on behavior (Autry et al. 2011). It is also possible that anisomycin pretreatment decreases the basal levels of BDNF protein synthesis in dendrites and spines, which might reduce the activity-dependent release of BDNF. The inhibition of protein translation during neural activity did not abolish the behavioral effects of ketamine, which suggests that the antidepressant effects are triggered by NMDA blockade specifically during resting spontaneous glutamatergic signaling (Autry et al. 2011). Increased levels of BDNF protein were found in the hippocampus but not in the nucleus accumbens, which might indicate a region-specific effect (Autry et al. 2011). In addition, behavioral antidepressant effects and increased synaptogenesis induced by ketamine were abolished in mice with the BDNF Val66Met Met/Met genotype, which had previously been shown to be associated with lower BDNF release (Liu et al. 2011). This provides further evidence that BDNF might be a key player in the mechanism of action responsible for the antidepressant properties of ketamine.

Autry and colleagues tested the hypothesis that stimulation of the mTOR pathway is necessary for the antidepressant effects of ketamine. However, they did not observe increased mTOR pathway activation after ketamine administration (Autry et al. 2011). This finding is in contrast to what had been reported previously (Li et al. 2010), which demonstrated activation of the mTOR complex and key components of the initiation of translation (i.e., p70 ribosomal S6 kinase and eukaryotic initiation factor 4E binding), after administration of 10 mg/kg ketamine. Activation of the mTOR pathway was both rapid (being evident just 30 and 60 min after administration) and transient, with the mTOR phosphoproteins returning to the baseline (nonstimulated) levels within 2 h. Further studies demonstrated an inverted U-dose response for ketamine, with low doses (i.e., 5 and 10 mg/kg) stimulating mTOR signaling and higher, anesthetic doses (such as 80 mg/kg) having no effect.

Ketamine was also shown to induce an mTOR-dependent increase in synaptic proteins such as PSD95, GluR1, and the presynaptic protein synapsin I. In addition, the elevation of synaptic proteins was delayed after mTOR activation, starting 2 h after ketamine administration and (somewhat surprisingly) persisting up to 72 h after ketamine treatment. In addition to the elevation of synaptic proteins, ketamine also promoted the formation of spines on the apical dendrites of layer V pyramidal neurons, which resulted in an increase in neurotransmitter-induced excitatory postsynaptic currents. This effect was investigated at only 24 h after ketamine treatment and so it is not known what happens at earlier time points or indeed how long the effect persists after a single dose of sub-anesthetic ketamine.

Additionally, the preclinical data suggest similarities in the mechanisms of action of lithium and ketamine via glycogen synthase kinase-3β (GSK-3β) inhibition (Beurel et al. 2011). Ketamine was reported to increase phosphorylation of GSK-3, and the antidepressant effects of the drug were shown to be absent in GSK-3 knock-in mice, suggesting that inhibition of this pathway might be a key downstream effect of NMDAR blockade, which underlies the antidepressant properties of ketamine (Beurel et al. 2011). However, the fact that administration of lithium, a potent GSK-3 inhibitor, does not induce a rapid antidepressant response might also indicate that ketamine-induced GSK-3 inhibition is a simple epiphenomenon and not particularly relevant for the antidepressant action of ketamine. Further studies are needed to clarify the role of GSK-3 inhibition in the antidepressant effects of ketamine and the underlying mechanisms involved.

From the existing evidence, it is apparent that ketamine triggers a complex series of intracellular and extracellular biological processes that ultimately result in an increase in synaptic plasticity. In the future, it will be extremely important for us to better understand the temporal dynamics of the individual cellular and molecular events that are triggered by ketamine and to investigate their regional specificity, as well as to provide a more thorough characterization of the dose-dependent effects and their link to behavioral changes.

12.8   Mechanism of action of ketamine: Clinical studies

On account of its unique rapid and efficacious antidepressant effects even in the more refractory patient populations, ketamine represents an extraordinary "tool" to investigate the biology of the antidepressant response as well as the underlying mechanism. This knowledge is key for the development of novel antidepressant compounds, which might act more quickly than the existing ones and thus provide clinical improvements for targeting TRD. Given the very complex temporal dynamics of the downstream effects of ketamine, it is clear that serial measurements of its pharmacodynamic effects are essential to better understand the mechanism of action as well as gain a more accurate interpretation of the findings.

Although ketamine has been studied as a challenge agent to model psychotic-like symptoms in healthy volunteers for more than two decades (Moghaddam and Krystal 2012), it is unclear whether the findings in healthy volunteers provide meaningful information for its antidepressant properties rather than simply reflect its dissociative and psychotic-like effects. Typically, ketamine is used as a challenge agent for psychosis as a bolus infusion (i.e., normally 0.23–0.26 mg/kg, given over 1–2 min), followed by a lower maintenance dose with constant infusion (usually 0.4–0.6 mg/kg provided over 1 h) (Krystal et al. 2005). This infusion schema is different from the way that ketamine has been administered in depressed patients, typically with a constant infusion rate of 0.5 mg/kg over 40 min. In the absence of a direct comparison of the pharmacokinetic and pharmacodynamic effects of these two infusion protocols, it is unclear whether the challenge protocol is at all informative for depression. Nevertheless, as psychotic-like effects usually dissipate within 1 h from the beginning of ketamine administration, next-day findings in healthy volunteers might still hold some value and shed some light on the antidepressant properties of ketamine.

By reason of these important caveats, understanding the biology of the antidepressant action of ketamine needs to rely on studies performed in depressed patients, where an assessment of the clinical response can also be performed.

12.8.1   Studies in depressed patients

A few studies investigated whether ketamine increased the level of peripheral BDNF in depressed patients and whether changed levels in BDNF correlated with the antidepressant improvement of ketamine. Previous preclinical experiments have implicated an increase in BDNF as an early effect triggered by subanesthetic doses of ketamine, while studies in ketamine abusers also show an increase in BDNF over normative values, which is associated with repeated ketamine use (Ricci et al. 2011). On the other hand, an early report in 23 patients with TRD, whereby several time points up to 230 min postinfusion of ketamine were investigated, failed to show an increase in BDNF (Machado-Vieira et al. 2009). A subsequent study, representing an extension of the previous study, showed a modest increase in the level of BDNF at 230 min after ketamine infusion although no correlation between BDNF changes and a clinical response was observed (Duncan et al. 2013). However, none of these studies factored Val66Met single nucleotide polymorphism (SNP) into the analysis; hence, it is not known whether there was any interaction between genotype and response status on BDNF levels. Future studies are warranted to investigate the effects of repeated ketamine dosing, as well as measuring time points beyond 230 min.

To our knowledge, there is only a single study that investigated changes in amino acid neurotransmitter levels after ketamine administration in patients with depression (Valentine et al. 2011). Unfortunately, this study did not detect any changes in glutamate, glutamine, or GABA in the occipital cortex at 3 and 48 h postinfusion of 0.5 mg/kg ketamine. However, the interpretation of these negative results is somewhat challenging as other brain regions, which might be more relevant for depression pathophysiology and for the mechanism of action of ketamine, were not investigated in this study. Changes in glutamine or glutamate/glutamine might be very sensitive measures for detecting the effect of acute drug action, but their accurate quantification remains highly challenging with MRS at medium field strengths. Future studies at ultrahigh field strengths or using [^{13}C]-MRS, which allows more accurate quantification of the glutamate/glutamine cycle, will be very informative.

A recent study used a highly innovative method for investigating brain plasticity in depressed patients who received subanesthetic doses of ketamine (Cornwell et al. 2012). Previous studies showed that both motor and somatosensory evoked potentials are amenable to changes after interventions that promote or decrease brain plasticity, such as repetitive transcranial magnetic stimulation or arm immobilization, respectively (Esser et al. 2006; Huber et al. 2006). As ketamine is thought to exert its antidepressant effects via increasing brain plasticity, Cornwell et al. (2012) tested the hypothesis that the neuroplastic changes induced by ketamine could also be detectable using evoked potentials. They measured somatosensory evoked fields (SEFs) using magnetoencephalography (MEG) recordings and showed that ketamine responders displayed increased SEFs at 6–7 h after ketamine administration, while this effect was not detected in nonresponders. However, SEF measurements were not acquired at later time points in a sizable group of patients; thus, it remains unknown whether maintenance of the response is dependent on sustaining the cortical plasticity enhancement (Cornwell et al. 2012).

These findings are consistent with an early study showing that ketamine potentiated motor evoked potentials and decreased the resting motor threshold in healthy volunteers (Di Lazzaro et al. 2003). It was subsequently suggested that these phenomena reflected AMPA transmission potentiation, one of the strongest candidates to explain the mechanism of action of ketamine in depression (Murrough et al. 2013b).

Future studies will also be needed to investigate the effects of ketamine on activity/metabolism in the anterior cingulate cortex (ACC), amygdala, and other regions of the brain that are part of the visceromotor network and involved in mood regulation (Price and Drevets 2012), as well as to establish a time course of effects induced by ketamine in depressed patients. Preliminary results have recently been presented at international conferences but are not yet published. Finally, the resting-state functional connectivity, which is measured via magnetic resonance imaging (MRI), is also a technique of great interest for detecting the pharmacodynamic effects of ketamine that are related to its antidepressant activity. Recently, studies showed abnormal functional connectivity of the dorsomedial prefrontal cortex in depressed patients (Sheline et al. 2010), which might be reversed by traditional antidepressants (Di Simplicio et al. 2012).

12.8.2   Studies in healthy subjects

The glutamatergic metabolite changes detected with ^{1}H-MRS in healthy subjects after ketamine administration is conflicting; however, there is also significant methodological heterogeneity across studies. Studies in healthy subjects showed a very transient effect of ketamine on amino acid neurotransmitters, which appeared to return to baseline levels within 1 h after drug administration (Rowland et al. 2005; Stone et al. 2012). Rowland et al. (2005) used a loading dose and low-dose maintenance protocol and showed ACC glutamine elevation during the loading phase but not at the beginning of the maintenance period. These results are consistent with a more recent study showing an increase in ACC glutamate within 35 min after the start of ketamine infusion in 13 healthy subjects (Stone et al. 2012); however, glutamine yielded acceptable quality results only in a minority of subjects (i.e., $n = 3$ and $n = 4$ before and after ketamine infusion, respectively). This study used a slightly different infusion protocol (i.e., the Clements 250 model), which is supposed to yield more stable ketamine levels during the maintenance infusion (Absalom et al. 2007). The temporal dynamic observed is consistent with evidence from preclinical studies, which shows that the glutamate surge is rapid and returns to baseline within 2 h (Moghaddam et al. 1997). However, questions remain whether ^{1}H-MRS at 3T or 4T is an appropriate method to investigate the acute changes induced by ketamine. This is because the glutamate MRS signal reflects neurotransmission only minimally, while the vast majority of the signal is represented by the intracellular neuronal glutamate content, which also supports basal metabolic activity. In the only study in healthy subjects that investigated an infusion protocol consistent with how ketamine is administered in depressed patients (i.e., 0.5 mg/kg over 40 min), no changes in glutamine, glutamic acid, or glutamate were detected either during or after ketamine infusion in the ACC (Taylor et al. 2012).

A recent study used resting-state functional connectivity to investigate the effect of ketamine on large-scale networks at 24 h after a single intravenous ketamine administration (Scheidegger et al. 2012). This time point was chosen because previous clinical studies had demonstrated a maximal antidepressant response after 1 day (Zarate et al. 2006a). Ketamine was shown to decrease functional connectivity between the posterior cingulate cortex and the "dorsal nexus" (Sheline et al. 2010), as well as the pregenual ACC (pgACC) and the medial prefrontal cortex, which are all areas previously implicated in the pathophysiology of depression (Price and Drevets 2012). As the default mode network is associated with self-referential processes and ruminations (Raichle 2010) and shows increased functional connectivity in patients with depression, a decrease in functional connectivity after ketamine treatment might reflect the restoration of a normative pattern, which might be associated with an antidepressant response.

12.9   Prediction of response to ketamine use

After the replication study conducted at the National Institute of Mental Health to determine the antidepressant action of ketamine (Zarate et al. 2006a), several studies investigated whether the pretreatment baseline measures (i.e., genetic polymorphisms, brain activity and metabolites, and sleep electroencephalography measures) in depressed patients might help identify those who would respond favorably to a single intravenous ketamine infusion.

Several preliminary studies have started to investigate the genetic predictors of the antidepressant response to ketamine. The Val66Met SNP has been linked to psychiatric disorders and impaired trafficking/regulation of BDNF. An obvious candidate tested by Laje et al. (2012) is the BDNF Val66Met SNP rs6265, which had previously been shown to influence some downstream effects of ketamine, as well as its antidepressant-like activity (Liu et al. 2011). Patients with MDD carrying the Met allele showed a lower percentage of improvement (24%) after ketamine treatment when compared with Val/Val homozygous subjects (41%).

In addition to studying genetics, MEG recordings were obtained in 11 unmedicated patients with TRD while they viewed emotionally charged faces, to investigate whether baseline ACC activity might predict a response to ketamine (Salvadore et al. 2009), similar to that demonstrated for other conventional antidepressants (Mayberg 2009). Patients with depression showed an abnormal habituation pattern when compared with healthy controls. Activity in the pgACC in response to viewing fearful faces was positively correlated with the magnitude of the antidepressant response 230 min after ketamine treatment, while an inverse correlation was found for activity in the right amygdala (Salvadore et al. 2009). Overall, responders to ketamine showed an activity pattern that more closely resembled normative responses in healthy subjects, when compared with nonresponders. In a subsequent study, whether pgACC activity predicted a response to ketamine during a nonemotional task such as the spatial working memory n-back task was also investigated (Salvadore et al. 2010). Previous studies had demonstrated that the pgACC displayed either increased or decreased activity in relation to the emotional demands of the tasks (Drevets and Raichle 1998) and that patients with MDD exhibited pgACC hyperactivity during working memory tasks, when compared with control subjects (Rose et al. 2006). In addition, patients who are more likely to respond to ketamine displayed decreased pgACC activity and lower pgACC connectivity when compared with non-responders. Thus, these findings might indicate that ketamine responders show functional integrity of mood-regulating circuitry, while nonresponders show abnormal activation of the mood-regulating circuitry even in the absence of emotionally arousing stimuli.

It was recently suggested that ^{1}H-MRS might also be a valuable technique to investigate promising predictors of the response to ketamine. Patients who showed the greatest clinical improvement 230 min after ketamine administration displayed a low Glx (glutamine and glutamate)/glutamate ratio in the dorsoanterolateral and dorsomedial areas of the prefrontal cortex, when compared with patients who did not show clinical improvement (Salvadore et al. 2012). As the Glx peak is largely constituted by glutamate and glutamine, the Glx/glutamate ratio was used as a proxy for glutamine in the study, given the low spectral resolution of glutamine with 3T MRI fields. Findings in the ventromedial prefrontal cortex did not reach statistical significance but showed a similar direction. Pretreatment levels of GABA did not correlate with the magnitude of antidepressant improvement in either region. An important caveat that needs to be considered is that for technical feasibility, amino acid neurotransmitters were measured in relatively large voxels, thus yielding suboptimal spatial resolution, which did not allow for making any inference with regard to the role of specific anatomical regions in predicting an antidepressant response to ketamine.

A prominent feature of depression is a decreased slow wave sleep. Recent evidence indicates that pretreatment sleep electroencephalography might be able to differentiate between ketamine responders and nonresponders. The baseline delta sleep ratio in particular was inversely correlated with the magnitude of clinical improvement at just 1 day after ketamine administration. This evidence suggests that subjects with a more disrupted pattern of slow wave activity are the ones who display better clinical improvement after ketamine treatment. According to the synaptic homeostasis hypothesis, slow wave activity might reflect brain plasticity (Tononi and Cirelli 2006). This suggests that responders to ketamine are characterized by lower brain plasticity, which might then be increased by ketamine (Cornwell et al. 2012).

12.10   Future directions for research on the use of ketamine as an antidepressant

12.10.1   Stereoisomers of ketamine

Ketamine is a chiral compound and its *R*- and *S*-stereoisomers have different binding affinities. *S*-Ketamine has a greater affinity for the phencyclidine site of the NMDAR than *R*-ketamine (Kohrs and Durieux 1998). It has been reported that R-ketamine, when given in similar analgesic doses as S-ketamine, has a higher incidence of psychedelic side effects (Raeder and Stenseth 2000). A case study of four patients who were administered oral S-ketamine (at 1.25 mg/kg) for 14 days alongside traditional anti-depressants in an effort to shorten the time to the onset of action showed that there was a rapid and sustained improvement in mood within the first week of treatment in two patients (50%). However, the limited number of patients who took part in the study did not allow any conclusions to be made about efficacy of the treatment (Paslakis et al. 2010). Oral S-ketamine has been reported to be well tolerated. Paul et al. (2009) described the effect of racemic ketamine and S-ketamine infusion therapy in two patients with TRD. Both patients experienced psychotomimetic side effects during keta-mine infusion that were absent during treatment with S-ketamine. They concluded that S-ketamine might exert similar antidepressant effects as ketamine in patients with TRD but that it might be better tolerated by the patients. Further studies of S-ketamine are required.

12.10.2   Route of administration of ketamine

There are a limited number of reports describing the effect of orally administered ketamine in depression as this would be the preferable route of administration for patients. However, it is known that oral ketamine is used off-label in palliative care and treatment of therapy-resistant pain. A small case study with just two patients in a hospice setting who were given a single dose of oral ketamine (at 0.5 mg/kg) for depression dem-onstrated a rapid and modestly sustained relief of depression and anxiety (Irwin and Iglewicz 2010). This was followed up more recently by a larger, open-label study (Irwin et al. 2013) where 14 subjects with depression or depression mixed with anxiety that warranted psychopharmacological intervention received daily oral ketamine hydrochloride (0.5 mg/kg) over a 28-day period. The investigators found that over the 28-day trial, there was significant improvement in both the depressive and anxious symp-toms for the eight subjects who completed the trial. There were few side effects, with the most common being diarrhea, trouble sleeping, and trou-ble sitting still. The response rate for depression in this study was simi-lar to that after intravenously administered ketamine; however, the time to response was more protracted. The findings of the potential efficacy of oral ketamine for depression and the response of anxiety symptoms are novel. Further research, with randomized, controlled clinical trials, is necessary to establish the efficacy and safety of oral ketamine for the treatment of depression and anxiety in patients receiving hospice care or other subject populations.

Another recent study where ketamine was administered sublin-gually in 26 patients with refractory unipolar and bipolar depression demonstrated clear and sustained antidepressant effects of the drug on mood, sleep, and cognition in 20 patients (77%), with only mild and transient lightheadedness as a common side effect (i.e., no psychotic, euphoria, or dissociative symptoms were observed) (Lara et al. 2013). There are other studies now ongoing to assess the optimal oral route of administration. Intranasally administered ketamine is also being evaluated as it has previously been shown to benefit analgesic–refractory chronic pain patients at a dose comparable to that used in most intravenous ketamine studies (Carr et al. 2004).

12.10.3   Caution when using ketamine in cancer patients

Concerns have been raised with regard to the up-regulation of mTOR by ketamine (Yang et al. 2011) in patients with cancer. Studies have shown that up-regulated mTOR may cause the acceleration of tumor growth (Shor et al. 2009). Thus, additional more specific studies that address patient prognosis are required to investigate whether ketamine is suitable for the treatment of refractory depression in patients with cancer.

12.10.4   Biomarkers of ketamine's antidepressant effects

The general consensus of opinion is that ketamine may never reach mainstream use as a treatment for TRD or bipolar depression. However, it is generally believed that the research conducted on ketamine is a means for learning more about the pathophysiology of depression. This will hopefully lead to the development of novel therapeutic agents that act in a similar fashion to ketamine, but that have a longer duration of action and fewer side effects, or that act on the effectors of ketamine known to be involved in its mechanism of action. Translational research is currently exploring the identification of biomarkers that might be involved in the prevention, diagnosis, treatment responses, and severity as well as the prognosis of depression. Zarate et al. (2013) have identified a range of potential biomarkers from genetic, functional neuroimaging, sleep, and clinical studies that are implicated in the antidepressant actions of ketamine. While these biomarker studies have limited practical implications at present, they offer hope that we will eventually be able to successfully apply these techniques to clinical populations. Focusing research in this direction offers opportunities that might ultimately facilitate the development of personalized treatment and thus increase the probability of success.

Acknowledgment

This work was supported by grant nos. 30872424 and 81271216 from the National Natural Science Foundation of China.

References

aan het Rot, M., Collins, K.A., Murrough, J.W., Perez, A.M., Reich, D.L., Charney, D.S., and Mathew, S.J. 2010. Safety and efficacy of repeated-dose intravenous ketamine for treatment-resistant depression. *Biol. Psychiatry* 67:139–145.

Abdallah, C.G., Fasula, M., Kelmendi, B., Sanacora, G., and Ostroff, R. 2012. Rapid antidepressant effect of ketamine in the electroconvulsive therapy setting. *J. ECT* 28:157–161.

Absalom, A.R., Lee, M., Menon, D.K., Sharar, S.R., De Smet, T., Halliday, J., Ogden, M., Corlett, P., Honey, G.D., and Fletcher, P.C. 2007. Predictive performance of the Domino, Hijazi, and Clements models during low-dose target-controlled ketamine infusions in healthy volunteers. *Br. J. Anaesth.* 98:615–623.

Akinfiresoye, L., and Tizabi, Y. 2013. Antidepressant effects of AMPA and ketamine combination: Role of hippocampal BDNF, synapsin, and mTOR. *Psychopharmacology (Berl.)* 230:291–298.

Autry, A.E., Adachi, M., Nosyreva, E., Na, E.S., Los, M.F., Cheng, P., Kavalali, E.T., and Monteggia, L.M. 2011. NMDA receptor blockade at rest triggers rapid behavioural antidepressant responses. *Nature* 475:91–95.

Bechtholt-Gompf, A.J., Smith, K.L., John, C.S., Kang, H.H., Carlezon, W.A. Jr., Cohen, B.M., and Ongür, D. 2011. CD-1 and Balb/cJ mice do not show enduring antidepressant-like effects of ketamine in tests of acute antidepressant efficacy. *Psychopharmacology (Berl.)* 215:689–695.

Beck, A.T., Kovacs, M., and Weissman, A. 1979. Assessment of suicidal intention: The scale for suicide ideation. *J. Consult. Clin. Psychol.* 47:343–352.

Berman, R., Cappiello, A., Anand, A., Oren, D.A., Heninger, G.R., Charney, D.S., and Krystal, J.H. 2000. Antidepressant effects of ketamine in depressed patients. *Biol. Psychiatry* 47:351–354.

Beurel, E., Song, L., and Jope, R.S. 2011. Inhibition of glycogen synthase kinase-3 is necessary for the rapid antidepressant effect of ketamine in mice. *Mol. Psychiatry* 16:1068–1070.

Brewer, C.L., Davidson, J.R., and Hereward, S. 1972. Ketamine ("Ketalar"): A safer anaesthetic for ECT. *Br. J. Psychiatry* 120:679–680.

Burgdorf, J., Zhang, X.L., Nicholson, K.L., Balster, R.L., Leander, J.D., Stanton, P.K., Gross, A.L., Kroes, R.A., and Moskal, J.R. 2013. GLYX-13, a NMDA receptor glycine-site functional partial agonist, induces antidepressant-like effects without ketamine-like side effects. *Neuropsychopharmacology* 38:729–742.

Caldarone, B.J., George, T.P., Zachariou, V., and Picciotto, M.R. 2000. Gender differences in learned helplessness behavior are influenced by genetic background. *Pharmacol. Biochem. Behav.* 66:811–817.

Carr, D.B., Goudas, L.C., Denman, W.T., Brookoff, D., Staats, P.S., Brennan, L., Green, G., Albin, R., Hamilton, D., Rogers, M.C., Firestone, L., Lavin, P.T., and Mermelstein, F. 2004. Safety and efficacy of intranasal ketamine for the treatment of breakthrough pain in patients with chronic pain; a randomized, double-blind, placebo-controlled, cross-over study. *Pain* 108:17–27.

Carrier, N., and Kabbaj, M. 2013. Sex differences in the antidepressant-like effects of ketamine. *Neuropharmacology* 70:27–34.

Chowdhury, G.M., Behar, K.L., Cho, W., Thomas, M., Rothman, D.L., and Sanacora, G. 2011. ^{1}H-[^{13}C]-NMR spectroscopy measures of ketamine's effect on amino acid neurotransmitter metabolism. *Biol. Psychiatry* 71:1022–1025.

Clemens, J., Nimmo, W.S., and Grant, I.S. 1982. Bioavailability, pharmacokinetics, and analgesic activity of ketamine in humans. *J. Pharm. Sci.* 71:539–542.

Cornwell, B.R., Salvadore, G., Furey, M., Marquardt, C.A., Brutsche, N.E., Grillon, C., and Zarate, C.A. Jr. 2012. Synaptic potentiation is critical for rapid antidepressant response to ketamine in treatment-resistant major depression. *Biol. Psychiatry* 72:555–561.

Cruz, S.L., Soberanes-Chávez, P., Páez-Martinez, N., and López-Rubalcava, C. 2009. Toluene has antidepressant-like actions in two animal models used for the screening of antidepressant drugs. *Psychopharmacology (Berl.)* 204:279–286.

Cryan, J.F., Mombereau, C., and Vassout, A. 2005a. The tail suspension test as a model for assessing antidepressant activity: Review of pharmacological and genetic studies in mice. *Neurosci. Biobehav. Rev.* 29:571–625.

Cryan, J.F., Valentino, R.J., and Lucki, I. 2005b. Assessing substrates underlying the behavioral effects of antidepressants using the modified rat forced swimming test. *Neurosci. Biobehav. Rev.* 29:547–569.

da Silva, F.C., do Carmo de Oliveira Cito, M., da Silva, M.I., Moura, B.A., de Aquino Neto, M.R., Feitosa, M.L., de Castro Chaves, R., Macedo, D.S., de Vasconcelos, S.M., de França Fonteles, M.M., and de Sousa, F.C. 2010. Behavioral alterations and pro-oxidant effect of a single ketamine administration to mice. *Brain Res. Bull.* 83:9–15.

Di Lazzaro, V., Oliviero, A., Profice, P., Pennisi, M.A., Pilato, F., Zito, G., Dileone, M., Nicoletti, R., Pasqualetti, P., and Tonali, P.A. 2003. Ketamine increases human motor cortex excitability to transcranial magnetic stimulation. *J. Physiol.* 547:485–496.

Di Simplicio, M., Norbury, R., and Harmer, C.J. 2012. Short-term antidepressant administration reduces negative self-referential processing in the medial prefrontal cortex in subjects at risk for depression. *Mol. Psychiatry* 17:503–510.

Diazgranados, N., Ibrahim, L., Brutsche, N.E., Newberg, A., Kronstein, P., Khalife, S., Kammerer, W.A., Quezado, Z., Luckenbaugh, D.A., Salvadore, G., Machado-Vieira, R., Manji, H.K., and Zarate, C.A. Jr. 2010a. A randomized add-on trial of an N-methyl-D-aspartate antagonist in treatment-resistant bipolar depression. *Arch. Gen. Psychiatry* 67:793–802.

Diazgranados, N., Ibrahim, L.A., Brutsche, N.E., Ameli, R., Henter, I.D., Luckenbaugh, D.A., Machado-Vieira, R., and Zarate, C.A. Jr. 2010b. Rapid resolution of suicidal ideation after a single infusion of an N-methyl-D-aspartate antagonist in patients with treatment-resistant major depressive disorder. *J. Clin. Psychiatry* 71:1605–1611.

Drevets, W.C., and Raichle, M.E. 1998. Reciprocal suppression of regional cerebral blood flow during emotional versus higher cognitive processes: Implications for interactions between emotion and cognition. *Cogn. Emot.* 12:353–385.

Du, J., Suzuki, K., Wei, Y., Wang, Y., Blumenthal, R., Chen, Z., Falke, C., Zarate, C.A. Jr., and Manji, H.K. 2007. The anticonvulsants lamotrigine, riluzole, and valproate differentially regulate AMPA receptor membrane localization: Relationship to clinical effects in mood disorders. *Neuropsychopharmacology* 32:793–802.

Dulawa, S.C., and Hen, R. 2005. Recent advances in animal models of chronic antidepressant effects: The novelty-induced hypophagia test. *Neurosci. Biobehav. Rev.* 29:771–783.

Duman, R.S., and Aghajanian, G.K. 2012. Synaptic dysfunction in depression: Potential therapeutic targets. *Science* 338:68–72.

Duncan, W.C., Sarasso, S., Ferrarelli, F., Selter, J., Riedner, B.A., Hejazi, N.S., Yuan, P., Brutsche, N., Manji, H.K., Tononi, G., and Zarate, C.A. 2013. Concomitant BDNF and sleep slow wave changes indicate ketamine-induced plasticity in major depressive disorder. *Int. J. Neuropsychopharmacol.* 16:301–311.

Engin, E., Treit, D., and Dickson, C.T. 2009. Anxiolytic- and antidepressant-like properties of ketamine in behavioral and neurophysiological animal models. *Neuroscience* 161:359–369.

Esser, S.K., Huber, R., Massimini, M., Peterson, M.J., Ferrarelli, F., and Tononi, G. 2006. A direct demonstration of cortical LTP in humans: A combined TMS/EEG study. *Brain Res. Bull.* 69:86–94.

Garcia, L.S., Comim, C.M., Valvassori, S.S., Réus, G.Z., Andreazza, A.C., Stertz, L., Fires, G.R., Gavioli, E.C., Kapczinski, F., and Quevedo, J. 2008a. Chronic administration of ketamine elicits antidepressant-like effects in rats without affecting hippocampal brain-derived neurotrophic factor protein levels. *Basic Clin. Pharmacol. Toxicol.* 103:502–506.

Garcia, L.S., Comim, C.M., Valvassori, S.S., Reus, G.Z., Barbosa, L.M., Andreazza, A.C., Stertz, L., Fries, G.R., Gavioli, E.C., Kapczinski, F., and Quevedo, J. 2008b. Acute administration of ketamine induces antidepressant-like effects in the forced swimming test and increases BDNF levels in the rat hippocampus. *Prog. Neuropsychopharmacol. Biol. Psychiatry* 32:140–144.

Garcia, L.S., Comim, C.M., Valvassori, S.S., Réus, G.Z., Stertz, L., Kapczinski, F., Gavioli, E.C., and Quevedo, J. 2009. Ketamine treatment reverses behavioral and physiological alterations induced by chronic mild stress in rats. *Prog. Neuropsychopharmacol. Biol. Psychiatry* 33:450–455.

Gigliucci, V., O'Dowd, G., Casey, S., Egan, D., Gibney, S., and Harkin, A. 2013. Ketamine elicits sustained antidepressant-like activity via a serotonin-dependent mechanism. *Psychopharmacology (Berl.)* 228:157–166.

Green, C.D. 1973. Ketamine as an anaesthetic for ECT. *Br. J. Psychiatry* 122:123–124.

Homayoun, H., and Moghaddam, B. 2007. NMDA receptor hypofunction produces opposite effects on prefrontal cortex interneurons and pyramidal neurons. *J. Neurosci.* 27:11496–11500.

Huber, R., Ghilardi, M.F., Massimini, M., Ferrarelli, F., Riedner, B.A., Peterson, M.J., and Tononi, G. 2006. Arm immobilization causes cortical plastic changes and locally decreases sleep slow wave activity. *Nat. Neurosci.* 9:1169–1176.

Ibrahim, L., Diazgranados, N., Franco-Chaves, J., Brutsche, N., Henter, I.D., Kronstein, P., Moaddel, R., Wainer, I., Luckenbaugh, D.A., Manji, H.K., and Zarate, C.A. Jr. 2012. Course of improvement in depressive symptoms to a single intravenous infusion of ketamine vs add-on riluzole: Results from a 4-week, double-blind, placebo-controlled study. *Neuropsychopharmacology* 37:1526–1533.

Ibrahim, L., Diazgranados, N., Luckenbaugh, D.A., Machado-Vieira, R., Baumann, J., Mallinger, A.G., and Zarate, C.A. Jr. 2011. Rapid decrease in depressive symptoms with an N-methyl-D-aspartate antagonist in ECT-resistant major depression. *Prog. Neuropsychopharmacol. Biol. Psychiatry* 35:1155–1159.

Iijima, M., Fukumoto, K., and Chaki, S. 2012. Acute and sustained effects of a metabotropic glutamate 5 receptor antagonist in the novelty-suppressed feeding test. *Behav. Brain Res.* 235:287–292.

Irwin, S.A., and Iglewicz, A. 2010. Oral ketamine for the rapid treatment of depression and anxiety in patients receiving hospice care. *J. Palliat. Med.* 13:903–908.

Irwin, S.A., Iglewicz, A., Nelesen, R.A., Lo, J.Y., Carr, C.H., Romero, S.D., and Lloyd, L.S. 2013. Daily oral ketamine for treatment of depression and anxiety in patients receiving hospice care: A 28 day open-label proof-of-concept trial. *J. Palliat. Med.* 16:958–965.

Kavalali, E.T., and Monteggia, L.M. 2012. Synaptic mechanisms underlying rapid antidepressant action of ketamine. *Am. J. Psychiatry* 169:1150–1156.

Kessler, R., Berglund, P., Demler, O., Jin, R., Koretz, D., Merikangas, K.R., Rush, A.J., Walters, E.E., Wang, P.S., and National Comorbidity Survey Replication. 2003. The epidemiology of major depressive disorder: Results for the National Comorbidity Survey Replication (NCS-R). *JAMA* 289:3095–3105.

Kim, S.Y., Lee, H., Kim, H.J., Bang, E., Lee, S.H., Lee, D.W., Woo, D.C., Choi, C.B., Hong, K.S., Lee, C., and Choe, B.Y. 2011. In vivo and ex vivo evidence for ketamine-induced hyperglutamatergic activity in the cerebral cortex of the rat: Potential relevance to schizophrenia. *NMR Biomed.* 24:1235–1242.

Kohrs, R., and Durieux, M.E. 1998. Ketamine: Teaching an old drug new tricks. *Anesth. Analg.* 87:1186–1193.

Koike, H., Iijima, M., and Chaki, S. 2011a. Involvement of the mammalian target of rapamycin signaling in the antidepressant-like effect of group II metabotropic glutamate receptor antagonists. *Neuropharmacology* 61:1419–1423.

Koike, H., Iijima, M., and Chaki, S. 2011b. Involvement of AMPA receptor in both the rapid and sustained antidepressant-like effects of ketamine in animal models of depression. *Behav. Brain Res.* 224:107–111.

Koike, H., Iijima, M., and Chaki, S. 2013. Effects of ketamine and LY341495 on the depressive-like behavior of repeated corticosterone-injected rats. *Pharmacol. Biochem. Behav.* 107:20–23.

Kollmar, R., Markovic, K., Thürauf, N., Schmitt, H., and Kornhuber, J. 2008. Ketamine followed by memantine for the treatment of major depression. *Aust. N. Z. J. Psychiatry* 42:170. doi: 10.1080/00048670701787628.

Kranaster, L., Kammerer-Ciernioch, J., Hoyer, C., and Sartorius, A. 2011. Clinically favourable effects of ketamine as an anaesthetic for electroconvulsive therapy: A retrospective study. *Eur. Arch. Psychiatry Clin. Neurosci.* 261:575–582.

Krystal, A.D., Weiner, R.D., Dean, M.D., Lindahl, V.H., Tramontozzi, L.A. 3rd, Falcone, G., and Coffey, C.E. 2003. Comparison of seizure duration, ictal EEG, and cognitive effects of ketamine and methohexital anaesthesia with ECT. *J. Neuropsychiatry Clin. Neurosci.* 15:27–34.

Krystal, J.H., Abi-Saab, W., Perry, E., D'Souza, D.C., Liu, N., Gueorguieva, R., McDougall, L., Hunsberger, T., Belger, A., Levine, L., and Breier, A. 2005. Preliminary evidence of attenuation of the disruptive effects of the NMDA glutamate receptor antagonist, ketamine, on working memory by pretreatment with the group II metabotropic glutamate receptor agonist, LY354740, in healthy human subjects. *Psychopharmacology (Berl.)* 179:303–309.

Lahti, A.C., Warfel, D., Michaelidis, T., Weiler, M.A., Frey, K., and Tamminga, C.A. 2001. Long-term outcome of patients who receive ketamine during research. *Biol. Psychiatry* 49:869–875.

Laje, G., Lally, N., Mathews, D., Brutsche, N., Chemerinski, A., Akula, N., Kelmendi, B., Simen, A., McMahon, F.J., Sanacora, G., and Zarate, C. Jr. 2012. Brain-derived neurotrophic factor Val66Met polymorphism and antidepressant efficacy of ketamine in depressed patients. *Biol. Psychiatry* 72:e27–e28.

Lara, D.R., Bisol, L.W., and Munari, L.R. 2013. Antidepressant, mood stabilising and precognitive effects of very lose dose sublingual ketamine in refractory unipolar and bipolar depression. *Int. J. Neuropsychopharmacol.* 16:2111–2117.

Larkin, G.L., and Beautrais, A.L. 2011. A preliminary naturalistic study of low-dose ketamine for depression and suicide ideation in the emergency department. *Int. J. Neuropsychopharmacol.* 14:1127–1131.

Lee, P.H., Perlis, R.H., Jung, J.Y., Byrne, E.M., Rueckert, E., Siburian, R., Haddad, S., Maverfeld, C.E., Heath, A.C., Pergadia, M.L., Madden, P.A., Boomsma, D.I., Penninx, B.W., Sklar, P., Martin, N.G., Wray, N.R., Purcell, S.M., and Smoller, J.W. 2012. Multi-locus genome-wide association analysis supports the role of glutamatergic synaptic transmission in the etiology of major depressive disorder. *Transl. Psychiatry* 2:e184. doi: 10.1038/tp.2012.95.

Li, N., Lee, B., Liu, R.J., Banasr, M., Dwyer, J.M., Iwata, M., Li, X.Y., Aghajanian, G., and Duman, R.S. 2010. mTOR-dependent synapse formation underlies the rapid antidepressant effects of NMDA antagonists. *Science* 329:959–964.

Li, N., Liu, R.J., Dwyer, J.M., Banasr, M., Lee, B., Son, H., Li, X.Y., Aghajanian, G., and Duman, R.S. 2011. Glutamate N-methyl-D-aspartate receptor antagonists rapidly reverse behavioral and synaptic deficits caused by chronic stress exposure. *Biol. Psychiatry* 69:754–761.

Lindholm, J.S., Autio, H., Vesa, L., Antila, H., Lindemann, L., Hoener, M.C., Skolnick, P., Rantamäki, T., and Castrén, E. 2012. The antidepressant-like effects of glutamatergic drugs ketamine and AMPA receptor potentiator LY 451646 are preserved in bdnf$^{+/-}$ heterozygous null mice. *Neuropharmacology* 62:391–397.

Liu, R.J., Lee, F.S., Li, X.Y., Bambica, F., Duman, R.S., and Aghajanian, G.K. 2011. Brain-derived neurotrophic factor Val66Met allele impairs basal and ketamine-stimulated synaptogenesis in prefrontal cortex. *Biol. Psychiatry* 71:996–1005.

López-Rubalcava, C., and Lucki, I. 2000. Strain differences in the behavioral effects of antidepressant drugs in the rat forced swimming test. *Neuropsychopharmacology* 22:191–199.

Ma, X.C., Dang, Y.H., Jia, M., Ma, R., Wang, F., Wu, J., Gao, C.G., and Hashimoto, K. 2013. Long-lasting antidepressant action of ketamine, but not glycogen synthase kinase-3 inhibitor SB216763, in the chronic mild stress model of mice. *PLoS One* 8:e56053. doi: 10.1371/journal.pone.0056053.

Machado-Vieira, R., Manji, H.K., and Zarate, C.A. 2009. The role of the tripartite glutamatergic synapse in the pathophysiology and therapeutics of mood disorders. *Neuroscientist* 15:525–539.

Maeng, S., Zarate, C.A. Jr., Du, J., Schloesser, R.J., McCammon, J., Chen, G., and Manji, H.K. 2008. Cellular mechanisms underlying the antidepressant effects of ketamine: Role of α-amino-3-hydroxy-5-methylisoxazole-4-propionic acid receptors. *Biol. Psychiatry* 63:349–352.

Mantovani, M., Pértile, R., Calixto, J.B., Santos, A.R., and Rodrigues, A.L. 2003. Melatonin exerts an antidepressant-like effect in the tail suspension test in mice: Evidence for involvement of N-methyl-D-aspartate receptors and the L-arginine-nitric oxide pathway. *Neurosci. Lett.* 343:1–4.

Mathew, S.J., Murrough, J.W., aan het Rot, M., Collins, K.A., Reich, D.L., and Charney, D.S. 2010. Riluzole for relapse prevention following intravenous ketamine in treatment-resistant depression: A pilot randomized, placebo-controlled continuation trial. *Int. J. Neuropsychopharmacol.* 13:71–82.

Mayberg, H.S. 2009. Targeted electrode-based modulation of neural circuits for depression. *J. Clin. Invest.* 119:717–725.

McDaniel, W.W., Sahota, A.K., Vyas, B.V., Laguerta, N., Hategan, L., and Oswald, J. 2006. Ketamine appears associated with better word recall than etomidate after a course of 6 electroconvulsive therapies. *J. ECT* 22:103–106.

Mizuta, I., Ohta, M., Ohta, K., Nishimura, M., Mizuta, E., and Kuno, S. 2001. Riluzole stimulates nerve growth factor, brain-derived neurotrophic factor and glial cell line-derived neurotrophic factor synthesis in cultured mouse astrocytes. *Neurosci. Lett.* 310:117–120.

Moghaddam, B., Adams, B., Verma, A., and Daly, D. 1997. Activation of glutamatergic neurotransmission by ketamine: A novel step in the pathway from NMDA receptor blockade to dopaminergic and cognitive disruptions associated with the prefrontal cortex. *J. Neurosci.* 17:2921–2927.

Moghaddam, B., and Krystal, J.H. 2012. Capturing the angel in "angel dust": Twenty years of translational neuroscience studies of NMDA receptor antagonists in animals and humans. *Schizophr. Bull.* 38:942–949.

Morgan, C.J., and Curran, H.V. 2006. Acute and chronic effects of ketamine upon human memory: A review. *Psychopharmacology (Berl.)* 188:408–424.

Morgan, C.J., Curran, H.V., and Independent Scientific Committee on Drugs. 2012. Ketamine use: A review. *Addiction* 107:27–38.

Murrough, J.W., Iosifescu, D.V., Chang, L.C., Al Jurdi, R.K., Green, C.E., Perez, A.M., Igbal, S., Pillemer, S., Foulkes, A., Shah, A., Charney, D.S., and Mathew, S.J. 2013a. Antidepressant efficacy of ketamine in treatment-resistant major depression: A two-site randomized controlled trial. *Am. J. Psychiatry.* 170:1134–1142.

Murrough, J.W., Perez, A.M., Pillemer, S., Stern, J., Paridea, M.K., aan het Rot, M., Collins, K.A., Mathew, S.J., Charney, D.S., and Iosifescu, D.V. 2013b. Rapid and longer-term antidepressant effects of repeated ketamine infusions in treatment-resistant depression. *Biol. Psychiatry* 74:250–256.

Nobler, M.S., Sackeim, H.A., Moeller, J.R., Prudic, J., Petkova, E., and Waternaux, C. 1997. Quantifying the speed of symptomatic improvement with electroconvulsive therapy: Comparison of alternative statistical methods. *Convuls. Ther.* 13:208–221.

Okamoto, N., Nakai, T., Sakamoto, K., Nagafusa, Y., Higuchi, T., and Nishikawa, T. 2010. Rapid antidepressant effect of ketamine anaesthesia during electroconvulsive therapy of treatment-resistant depression: Comparing ketamine and propofol anaesthesia. *J. ECT* 26:223–227.

Parise, E.M., Alcantara, L.F., Warren, B.L., Wright, K.N., Hadad, R., Sial, O.K., Kroeck, K.G., Iñiguez, S.D., and Bolaños-Guzmán, C.A. 2013. Repeated ketamine exposure induces an enduring resilient phenotype in adolescent and adult rats. *Biol. Psychiatry* 74:750–759.

Paslakis, G., Gilles, M., Meyer-Lindenberg, A., and Deuschle, M. 2010. Oral administration of the NMDA receptor antagonist S-ketamine as add-on therapy of depression: A case series. *Psychopharmacology* 43:33–35.

Paul, R., Schaaff, N., Padberg, F., Möller, H.J., and Frodl, T. 2009. Comparison of racemic ketamine and S-ketamine in treatment-resistant major depression: Report of two cases. *World J. Biol. Psychiatry* 10:241–244.

Perry, E.B. Jr., Cramer, J.A., Cho, H.S., Petrakis, I.L., Karper, L.P., Genovese, A., O'Donnell, E., Krystal, J.H., D'Souza, D.C., and Yale Ketamine Study Group. 2007. Psychiatric safety of ketamine in psychopharmacology research. *Psychopharmacology (Berl.)* 192:253–260.

Popik, P., Kos, T., Sowa-Kućma, M., and Nowak, G. 2008. Lack of persistent effects of ketamine in rodent models of depression. *Psychopharmacology (Berl.)* 198:421–430.

Price, J.L., and Drevets, W.C. 2012. Neural circuits underlying the pathophysiology of mood disorders. *Trends Cogn. Sci.* 16:61–71.

Price, R.B., Nock, M.K., Charney, D.S., and Mathew, S.J. 2009. Effects of intravenous ketamine on explicit and implicit measures of suicidality in treatment-resistant depression. *Biol. Psychiatry* 66:522–526.

Raeder, J.C., and Stenseth, L.B. 2000. Ketamine: A new look at an old drug. *Curr. Opin. Anaesthesiol.* 13:463–468.

Raichle, M.E. 2010. Two views of brain function. *Trends Cogn. Sci.* 14:180–190.

Réus, G.Z., Stringari, R.B., Ribeiro, K.F., Ferraro, A.K., Vitto, M.F., Cesconetto, P., Souza, C.T., and Quevedo, J. 2011. Ketamine plus imipramine treatment induces antidepressant-like behavior and increases CREB and BDNF protein levels and PKA and PKC phosphorylation in rat brain. *Behav. Brain Res.* 221:166–171.

Rezin, G.T., Gonçalves, C.L., Daufenbach, J.F., Fraga, D.B., Santos, P.M., Ferreira, G.K., Hermani, F.V., Comom, C.M., Quevedo, J., and Streck, E.L. 2009. Acute administration of ketamine reverses the inhibition of mitochondrial respiratory chain induced by chronic mild stress. *Brain Res. Bull.* 79:418–421.

Ricci, V., Martinotti, G., Gelfo, F., Tonioni, F., Caltagirone, C., Bria, P., and Angelucci, F. 2011. Chronic ketamine use increases serum levels of brain-derived neurotrophic factor. *Psychopharmacology (Berl.)* 215:143–148.

Rosa, A.O., Lin, J., Calixto, J.B., Santos, A.R., and Rodrigues, A.L. 2003. Involvement of NMDA receptors and L-arginine-nitric oxide pathway in the antidepressant-like effects of zinc in mice. *Behav. Brain Res.* 144:87–93.

Rose, E.J., Simonotto, E., and Ebmeier, K.P. 2006. Limbic over-activity in depression during preserved performance on the nback task. *NeuroImage* 29:203–215.

Rowland, L.M., Bustillo, J.R., Mullins, P.G., Jung, R.E., Lenroot, R., Landgraf, E., Barrow, R., Yeo, R., Lauriello, J., and Brooks, W.M. 2005. Effects of ketamine on anterior cingulate glutamate metabolism in healthy humans: A 4-T proton MRS study. *Am. J. Psychiatry* 162:394–396.

Salvadore, G., Cornwell, B.R., Colon-Rosario, V., Coppola, R., Grillon, C., Zarate, C.A. Jr., and Manji, H.K. 2009. Increased anterior cingulate cortical activity in response to fearful faces: A neurophysiological biomarker that predicts rapid antidepressant response to ketamine. *Biol. Psychiatry* 65:289–295.

Salvadore, G., Cornwell, B.R., Sambataro, F., Latov, D., Colon-Rosario, V., Carver, F., Holroyd, T., Diazgranados, N., Machado-Vieira, R., Grillon, C., Drevets, W.C., and Zarate, C.A. Jr. 2010. Anterior cingulate desynchronization and functional connectivity with the amygdala during a working memory task predict rapid antidepressant response to ketamine. *Neuropsychopharmacology* 35:1415–1422.

Salvadore, G., van der Veen, J.W., Zhang, Y., Marenco, S., Machado-Vieira, R., Baumann, J., Ibrahim, L.A., Luckenbaugh, D.A., Shen, J., Drevets, W.C., and Zarate, C.A. Jr. 2012. An investigation of amino-acid neurotransmitters as potential predictors of clinical improvement to ketamine in depression. *Int. J. Neuropsychopharmacol.* 15:1063–1072.

Sanacora, G., Zarate, C.A., Krystal, J.H., and Manji, H.K. 2008. Targeting the glutamatergic system to develop novel, improved therapeutics for mood disorders. *Nat. Rev. Drug Discov.* 7:426–437.

Scheidegger, M., Walter, M., Lehmann, M., Metzger, C., Grimm, S., Boeker, H., Boesiger, P., Henning, A., and Seifritz, E. 2012. Ketamine decreases resting state functional network connectivity in healthy subjects: Implications for antidepressant drug action. *PLoS One* 7:e44799. doi: 10.1371/journal.pone.0044799.

Segmiller, F., Rüther, T., Linhardt, A., Padberg, F., Berger, M., Pogarell, O., Möller, H.J., Kohler, C., and Schüle, C. 2013. Repeated S-ketamine infusions in therapy resistant depression: A case series. *J. Clin. Pharmacol.* 53:996–998.

Seligman, M.E., Weiss, J., Weinraub, M., and Schulman, A. 1980. Coping behavior: Learned helplessness, physiological change and learned inactivity. *Behav. Res. Ther.* 18:459–512.

Shanks, N., and Anisman, H. 1988. Stressor-provoked behavioral changes in six strains of mice. *Behav. Neurosci.* 102:894–905.

Sheline, Y.I., Price, J.L., Yan, Z., and Mintun, M.A. 2010. Resting-state functional MRI in depression unmasks increased connectivity between networks via the dorsal nexus. *Proc. Natl. Acad. Sci. U.S.A.* 107:11020–11025.

Shor, B., Gibbons, J.J., Abraham, R.T., and Yu, K. 2009. Targeting mTOR globally in cancer: Thinking beyond rapamycin. *Cell Cycle* 8:3831–3837.

Solberg, L.C., Baum, A.E., Ahmadiyeh, N., Shimomura, K., Li, R., Turek, F.W., Churchill, G.A., Takahashi, J.S., and Redei, E.E. 2004. Sex- and lineage-specific inheritance of depression-like behavior in the rat. *Mamm. Genome* 15:648–662.

Steru, L., Chermat, R., Thierry, B., and Simon, P. 1985. The tail suspension test: A new method for screening antidepressants in mice. *Psychopharmacology (Berl.)* 85:367–370.

Stone, J.M., Dietrich, C., Edden, R., Mehta, M.A., De Simoni, S., Reed, L.J., Krystal, J.H., Nutt, D., and Barker, G.J. 2012. Ketamine effects on brain GABA and glutamate levels with ^{1}H-MRS: Relationship to ketamine-induced psychopathology. *Mol. Psychiatry* 17:664–665.

Strayer, R.J., and Nelson, L.S. 2008. Adverse events associated with ketamine for procedural sedation in adults. *Am. J. Emerg. Med.* 26:985–1028.

Taylor, M.J., Tiangga, E.R., Mhuircheartaigh, R.N., and Cowen, P.J. 2012. Lack of effect of ketamine on cortical glutamate and glutamine in healthy volunteers: A proton magnetic resonance spectroscopy study. *J. Psychopharmacol.* 26:733–737.

Tizabi, Y., Bhatti, B.H., Manaye, K.F., Das, J.R., and Akinfiresove, L. 2012. Antidepressant-like effects of low ketamine dose is associated with increased hippocampal AMPA/NMDA receptor density ratio in female Wistar-Kyoto rats. *Neuroscience* 213:72–80.

Tononi, G., and Cirelli, C. 2006. Sleep function and synaptic homeostasis. *Sleep Med. Rev.* 10:49–62.

Trivedi, M., Rush, A.J., Wisniewski, S.R., Nierenberg, A.A., Warden, D., Ritz, L., Norquist, G., Howland, R.H., Lebowitz, B.D., McGrath, P.J., Shores-Wilson, K., Biggs, M.M., Balasubramani, G.K., Fava, M., and STAR*D Study Team. 2006. Evaluation of outcomes with citalopram for depression using measurement-based care in STAR*D: Implications for clinical practice. *Am. J. Psychiatry* 163:28–40.

Valentine, G.W., Mason, G.F., Gomez, R., Fasula, M., Watzl, J., Pittman, B., Krystal, J.H., and Sanacora, G. 2011. The antidepressant effect of ketamine is not associated with changes in occipital amino acid neurotransmitter content as measured by [(1)H]-MRS. *Psychiatry Res.* 191:122–127.

Weiner, R.D. 2001. *The Practice of Electroconvulsive Therapy: Recommendations for Treatment, Training and Privileging: A Task Force Report of the American Psychiatric Association,* second ed. American Psychiatric Publishing, Inc., Arlington, VA.

White, P.F., Way, W.L., and Trevor, A.J. 1982. Ketamine—Its pharmacology and therapeutic uses. *Anaesthesiology* 56:119–136.

Will, C.C., Aird, F., and Redei, E.E. 2003. Selectively bred Wistar-Kyoto rats: An animal model of depression and hyper-responsiveness to antidepressants. *Mol. Psychiatry* 8:925–932.

Willner, P. 1997. Validity, reliability and utility of the chronic mild stress model of depression: A 10-year review and evaluation. *Psychopharmacology (Berl.)* 134:319–329.

Willner, P. 2005. Chronic mild stress (CMS) revisited: Consistency and behavioural-neurobiological concordance in the effects of CMS. *Neuropsychobiology* 52:90–110.

Yang, C., Li, X., Wang, N., Xu, S., Yang, J., and Zhou, Z. 2012. Tramadol reinforces antidepressant effects of ketamine with increased levels of brain-derived neurotrophic factor and tropomyosin-related kinase B in rat hippocampus. *Front. Med.* 6:411–415.

Yang, C., Zhou, Z.Q., and Yang, J.J. 2011. Be prudent of ketamine in treating resistant depression in patients with cancer. *J. Palliat. Med.* 14:537. doi: 10.1089 /jpm.2010.0525.

Yilmaz, A., Schulz, D., Aksoy, A., and Canbeyli, R. 2002. Prolonged effect of an anesthetic dose of ketamine on behavioral despair. *Pharmacol. Biochem. Behav.* 71:341–344.

Zarate, C.A. Jr., Singh, J.B., Carlson, P.J., Brutsche, N.E., Ameli, R., Luckenbaugh, D.A., Charney, D.S., and Manji, H.K. 2006a. A randomized trial of an N-methyl-D-aspartate antagonist in treatment resistant major depression. *Arch. Gen. Psychiatry* 63:856–864.

Zarate, C.A. Jr., Singh, J., and Manji, H.K. 2006b. Cellular plasticity cascades: Targets for the development of novel therapeutics for bipolar disorder. *Biol. Psychiatry* 59:1006–1020.

Zarate, C.A. Jr., Singh, J.B., Quiroz, J.A., De Jesus, G., Denicoff, K.K., Luckenbaugh, D.A., Manji, H.K., and Charney, D.S. 2006c. A double-blind, placebo-controlled study of memantine in the treatment of major depression. *Am. J. Psychiatry* 163:153–155.

Zarate, C. Jr., Machado-Vieira, R., Henter, I., Ibrahim, L., Diazgranados, N., and Salvadore, G. 2010. Glutamatergic modulators: The future of treating mood disorders? *Harv. Rev. Psychiatry* 18:293–303.

Zarate, C.A. Jr., Brutsche, N.E., Ibrahim, L., Franco-Chaves, J., Diazgranados, N., Cravchik, A., Selter, J., Marquardt, C.A., Liberty, V., and Luckenbaugh, D.A. 2012. Replication of ketamine's antidepressant efficacy in bipolar depression: A randomized controlled add-on trial. *Biol. Psychiatry* 71:939–946.

Zarate, C.A. Jr., Mathews, D.C., and Furey, M.L. 2013. Human biomarkers of rapid antidepressant effects. *Biol. Psychiatry* 73:1142–1155.

chapter thirteen

Social correlates of ketamine and other psychoactive drug abuse in Hong Kong

Yuet-Wah Cheung

Contents

13.1 Introduction

The rapid rise in the use of cannabis and cough medicine among young people in Hong Kong in the mid-1990s caught the attention of the public and the government to the emerging youth drug abuse problem in Hong Kong (Cheung and Ch'ien 1996). Chris Patten, then Governor of Hong Kong, held two governor summits on illicit drugs in 1995 and 1996 to tackle the growing seriousness of young people's drug use (Cheung 1998). This was the beginning of the "New Drug Era" (Cheung and Zhong 2014), characterized by the soaring popularity of psychoactive drugs among young people. The high tide of psychoactive drug abuse arrived in 2000, when ecstasy and ketamine made their initial appearance on the Hong Kong drug scene, and these immediately became the drugs of choice among young drug abusers. Since the early 2000s, among the various psychoactive drugs available, ketamine has become the dominant drug for young drug users. As the prevalence of use of ketamine and other psychoactive drugs by young people increased, an interdepartmental Task Force on Youth Drug Abuse, led by the Secretary for Justice, was established in 2007 to tackle the youth problem of psychotropic drug abuse (Task Force on Youth Drug Abuse 2008). A trial scheme on drug testing in schools was implemented in secondary schools in the Tai Po District between 2009 and 2010, and then it was extended from 2010 to 2011 (Narcotics Division 2010, 2011). In the past decade, an extensive number of resources have been devoted to drug prevention, treatment, and research via the Beat Drugs Fund (BDF), which was established by the government in 1996.

When ketamine first landed on the Hong Kong drug scene in the early 2000s, little was known about its characteristics and physical effects. Since the mid-2000s, there have been a flurry of studies performed with this drug, but most of them are clinical studies to investigate the physical harm or cognitive impairment (Liang et al. 2013, 2014; Poon et al. 2010; Tang et al. 2013; Wang et al. 2013; Wei et al. 2013; Wong et al. 2014). Very few studies have systematically examined and compared the influence of social, demographic, and psychological factors on young people's ketamine abuse. To bridge this research gap, a study entitled "A Longitudinal Survey of Psychoactive Drug Abusers in Hong Kong," funded by the BDF, was conducted by this author (Cheung 2012). This 3-year study, which was carried out between early 2009 and the end of 2011, was the first longitudinal survey of psychoactive drug use and its social and psychological correlates in Hong Kong. The objective of this chapter is to present some of the major findings of the survey pertaining to the sociodemographic and psychosocial factors affecting ketamine and other psychoactive drug use among a sample of identified drug abusers.

13.2   Study and methods

The longitudinal survey of psychoactive drug abusers in Hong Kong recruited a sample of psychoactive drug abusers from outreach youth agencies as well as various treatment and rehabilitation programs and followed up on them for a total of six interviews, spaced out at 6-month intervals. As many as 36 youth outreach agencies and drug treatment and rehabilitation programs participated in the survey. Subjects were selected if they "had ever used a psychoactive drug." The definition of "a psychoactive drug" used in the survey was adopted from the classification of drugs used by the Narcotics Division, Security Bureau. Substances of abuse are broadly divided into two types, namely, "narcotic analgesics" and "psychotropic substances." The former type includes heroin, opium, morphine, and physeptone/methadone, whereas the latter includes hallucinogens, depressants, stimulants, tranquilizers, and other substances such as ketamine, cough medicine, and organic solvents (Narcotics Division 2012). The term *psychoactive drug* in this chapter is the same as the term *psychotropic drug* that is used by the Narcotics Division.

The sample size of the baseline survey at time point 1 (T1) was 754, whereas those of the subsequent time points were as follows: T2 = 600, T3 = 434, T4 = 376, T5 = 347, and T6 = 288. The retention rate between two time points was 79.6% from T1 to T2, 72.3% from T2 to T3, 86.6% from T3 to T4, 92.3% from T4 to T5, and 86.0% from T5 to T6. The overall retention rate from T1 to T6 was 38.2%. Given that this sample was not a stable one (e.g., during the 3-year study, some subjects left the agency because of program completion or premature departure), these retention rates were quite satisfactory.

Data collection began in April 2009, and finished in December 2011, after six series of interviews were completed. The social worker staff of most of the participating agencies served as interviewers. They had a good rapport with their clients and so had the advantage of gaining the subjects' trust in the interviews, thereby yielding more reliable responses to the questionnaire. In rare cases where the social worker staff were not available for conducting interviews, subjects were interviewed by the staff of a professional research organization. Each subject was given HK$100 as remuneration for every successful interview.

13.2.1   Independent variables

13.2.1.1   Sociodemographic variables
A number of commonly used sociodemographic variables were used in the study. These included sex, age, number of siblings, marital status (never married, married), student status (whether still a student or not and whether regularly attending school), employment (for nonstudents and those 16 years or older), religion (yes or no), and housing type (rented or owned private housing; public housing, or other).

13.2.1.2   Psychosocial variables
13.2.1.2.1   Permissiveness to use drugs.   This variable refers to the extent to which drug use is acceptable to the individual. It was formed by asking four questions: "Abusing drugs is not a proper behavior (reverse scoring)," "Using drugs occasionally is acceptable," "Abusing drugs frequently is acceptable," and "Abusing drugs when having fun with friends on recreational occasions is okay." Each question was measured by a four-point scoring system (1 = strongly disagree, 2 = disagree, 3 = agree, 4 =

strongly agree). The scale had a score range of 4–16. The α value of the scale was 0.704 at the baseline survey, indicating a high degree of reliability. The higher the degree of permissiveness, the more positive the attitude toward drug use.

13.2.1.2.2 Educational strain. This is a socially induced strain experienced by the individual because of the discrepancy between aspired educational level and perceived educational level that will be achieved. This concept was constructed in the sociological tradition of strain theory, which posits that the discrepancy between the aspired goal of success in society and the availability of legitimate means to achieve the goal is a source of strain in the individual (Agnew 1992; Merton 1957). If the strain is not handled properly, then anger and other negative emotions might develop, resulting in various kinds of deviant behavior, including drug abuse. Two questions were asked in the questionnaire, one on the aspired education level and the other on how greatly the subject would be disappointed if he or she could not achieve the desired education level in future. The latter question, with a four-point scoring system from 1 (no big deal) to 4 (very disappointed), was used to measure this variable.

13.2.1.2.3 Discrimination by others. Public discrimination experienced by active drug addicts may pose obstacles in their recovery process. Discrimination can produce a "labeling" effect on drug abusers, confining them to continuous drug use, or for addicts who have gone through treatment, driving them to relapse (Becker 1963; Cheung and Cheung 2003). In this study, the subject was asked whether or not he or she had ever experienced any discrimination from other people (1 = never, 2 = some, 3 = a lot).

13.2.1.2.4 Concerning views of life. Several variables were asked about whether the subjects agreed with the questions about a few aspects of life, including "already found goal in life," "satisfied with life," "sense of uncertainty about future," "doing extreme things such as drug use shows vitality of young people," "wish to have own family, job, and normal life," "think that their parents don't know how to teach their children," and "think that their schools don't know how to teach their students." Each of these items used a five-point scoring system (1 = strongly disagree, 2 = disagree, 3 = no opinion, 4 = agree, and 5 = strongly agree).

13.2.1.2.5 Self-esteem. The Rosenberg's self-esteem scale (Rosenberg 1965) was used to measure the self-esteem of subjects. The scale consists of 10 questions that the subject is asked to evaluate himself or herself with respect to self-image and self-confidence. A five-point scoring system was used for each of the 10 questionnaire items, and the scores of the scale ranged from 5 to 50. The scale had an α value of 0.826 in the baseline survey, showing a very high degree of reliability for this sample.

13.2.1.2.6 *Hopelessness.* Shek (2005) developed and validated a Chinese version of the Hopelessness Scale in Hong Kong, which was originally designed by Beck et al. (1974). A simplified version using 4 of the 16 questions in Shek's Chinese scale was used. These questions assessed how pessimistic the subject was about things he or she wished to obtain now or in the future. A five-point scoring system was used for each questionnaire item. Scores of the scale ranged from 4 to 20. The α value of this scale in the baseline survey was 0.704, representing a high degree of reliability for this sample.

13.2.1.2.7 *Subjective weathering.* This is the feeling of being more "weathered" by a tough life when compared with other people in the same age group. Subjects were asked to report whether (i) they feel that they are more mature; (ii) they had begun to take up household and other duties at a younger age; and (iii) they feel older at heart, compared with people in their similar age cohorts. Each of these three questions had a score range of "1" (more weathered than people in their same age group) to "5" (much less weathered). The α value of this scale in the baseline survey was 0.563, which is of acceptable reliability.

13.2.1.2.8 *Depression.* A simplified version of the Beck Depression Scale (Beck et al. 1961) was used to measure the depression level of subjects in our study. The simplified version asked the subject if he or she feels sad, feels discouraged about the future, has a hard time enjoying things, feels guilty, is disappointed in himself or herself, blames himself or herself for his or her faults and weaknesses, thinks about killing himself or herself, has no interest in people, puts off making decisions, worries about his or her looks, has to push himself or herself to do things, feels tired, and has no appetite. The Chinese version of the simplified scale was previously validated in a study of secondary school students in Hong Kong (Mok et al. 2008). The α value of the Depression Scale was 0.850 in the baseline survey sample, indicating a high level of reliability.

13.2.1.2.9 *Stricken by traumatic events.* The subjects were asked if they had been stricken by any distressing or traumatic events in the past 6 months, such as the divorce of parents or the loss of a family member. This is one of Agnew's (1992) types of strain, generated by "exposure to negative stimuli."

13.2.1.3 *Drug use history variable*
The number of years of drug use by a subject was also included as an independent variable because it shows the history of drug use.

13.2.2   Dependent variable

The dependent variable was "drug use in the last 30 days." Subjects were asked to indicate whether or not they had used illicit drugs in the period of 30 days before the day of the interview. The answers were "yes" or "no." In the questionnaire, the frequency of drug use was also asked, but not the amount. Thus, it was not possible to construct a drug use variable, which would better establish the level of use. In view of this, and noting that any use is not desirable, the "yes" and "no" dichotomy would serve the purpose in the present analysis.

13.2.3   Selection of subjects

Samples were recruited from outreach and residential treatment and rehabilitation agencies. For subjects in a residential treatment program, if they had been in the program for more than 30 days, then they would not have used drugs in the last 30 days in the residential treatment setting. Thus, not having used drugs in the last 30 days could not reflect the true ability to remain drug free, as they were kept away from drugs in the residential program. Thus, these subjects were not included in the analysis. On the other hand, if a subject was in a residential program during the interview but reported drug use in the last 30 days, then he or she was eligible to enter the program, as his or her drug use might have occurred before entering the program. These subjects were therefore included in the analysis. In summary, subjects selected for the present analysis were those who either were not in a residential treatment and rehabilitation program or were in a residential program during the interviewing period but had reported drug use in the last 30 days.

13.2.4   Method of data analysis

Data analysis was divided into two parts. First, for each of the sociodemographic, psychosocial, and drug use history variables, the bivariate association with the dependent variable (i.e., drug use in the last 30 days) was determined for each time point (T1–T6). If a variable was clearly shown to be related to drug use in the last 30 days for four or more of the six time points, then this variable was selected for the second part of data analysis. The criterion of "4" time points is used because this shows that the independent variable concerned was significantly related to drug use for more than half of the number of time points.

The second part of the data analysis involves the logistic regression of drug use in the last 30 days (dichotomy) for all of the sociodemographic, psychosocial, and drug use history variables for not less than four time points. As the attrition rate increased substantially with the time points (e.g., the T6 sample size was only 38.2% that of T1), data collected from T2 to T6 were "stacked" to provide a larger number of subjects for the regression analysis. At some time points, the influence of independent variables from the previous time point might be present in the regressions. However, as the T1 data did not have any previous time point to compare it with, data from this time point were not included in the stacked pool of data.

13.3   Results

13.3.1   Characteristics of the sample

In the full baseline sample at T1 ($N = 757$), 65.8% of the subjects were male, and almost 72.7% were below the age of 21 (i.e., mean age = 20.6 years). The majority of the subjects (88.8%) were not married; 63.2% had received only Secondary Form 3 or a lower level of education; 31.1% were students, and among those who were not students and who were 16 years or older, 53.8% had a job. More than 68.0% of the subjects did not have a religion and only 34.7% of the subjects lived in private housing, while the others mostly lived in public housing or rooms/quarters (Cheung 2012).

At T1, 51.7% of the subjects had used drugs in the last 30 days. Among these subjects, 80% had used ketamine, 19.9% had used ecstasy, 10.0% had used ice (amphetamine), 11.3% had used cocaine, 7.9% had used cannabis, and 1.3% had used heroin.

In summary, our drug user cohort was largely composed of young unmarried men, with lower to upper secondary school education, who were living in public housing. They mostly used psychoactive drugs, of which ketamine was the most popular, and on the whole, they were not interested in taking heroin. This is consistent with the social image of young psychoactive drug users in Hong Kong today.

13.3.2   Bivariate associations

We first examined the bivariate relationship of each of the sociodemographic, psychosocial, and drug use history variables with drug use in the last 30 days. The results are summarized in Table 13.1.

For the sake of space, Table 13.1 reports only those sociodemographic variables that were significantly related to drug use in the last 30 days at each time point (based on χ^2 tests). The variable that was significant at four time points (T1 to T4) was "whether the subject was still a student." Gender was significant at one time point (i.e., T2); age was also significant at one time point (i.e., T4); and employment (for nonstudents) was significant at two time points (i.e., T1 and T2). Marital status, religion, and housing type were not significant at any time point. Overall, sociodemographic

Table 13.1 T1–T6 Bivariate Relationships between Drug Use in the Last 30 Days ("Yes" and "No") and Sociodemographic Variables (Only Significant Ones Are Reported)

Sociodemographic variables	Used drugs in the last 30 days		
	%	n	N
T1			
Whether still a student ($p < 0.001$)			585
Yes, always attending school	48.2	81	168
Yes, but not always attending school	76.5	39	51
No	74.0	271	366
Employment (nonstudent and ≥16 years) ($p < 0.05$)			319
No	78.8	130	165
Yes	68.2	105	154
T2			
Gender ($p < 0.05$)			497
Male	45.0	136	302
Female	54.4	106	195
Whether still a student ($p < 0.001$)			497
Yes, always attending school	36.4	47	129
Yes, but not always attending	73.7	28	38
No	50.6	167	330
Employment (nonstudent and ≥16 years) ($p < 0.01$)			305
No	60.1	92	153
Yes	42.8	65	152
T3			
Whether still a student ($p < 0.01$)			377
Yes, always attending school	24.2	22	91
Yes, but not always attending	47.4	18	38
No	43.5	108	248
T4			
Age ($p < 0.05$)			356
14–16	41.7	25	60
17–20	30.8	64	208
21+	44.3	39	88
Whether still a student ($p < 0.05$)			357
Yes, always attending school	23.4	15	64
Yes, but not always attending	47.1	8	17
No	38.4	106	276

Table 13.1 (Continued) T1–T6 Bivariate Relationships between Drug Use in the
Last 30 Days ("Yes" and "No") and Sociodemographic Variables
(Only Significant Ones Are Reported)

	Used drugs in the last 30 days		
Sociodemographic variables	%	n	N
T5			
No significant variables			
T6			
No significant variables			

Note: Significance levels are based on χ^2 tests.

variables were not good predictors of drug use. Unlike sociodemographic
variables, a much larger number of psychosocial variables were significantly related to drug use in the last 30 days (see Table 13.2).

Altogether, 6 of the 16 psychosocial and drug use history variables
were significant at all six time points. They were "permissiveness to the
use of drugs," "found goal in life," "satisfied with life," "thought that
doing extreme things shows vitality of young people," "self-esteem," and
"depression." The "sense of uncertain future" variable was significant
at five time points, whereas "stricken by traumatic events in the last 6
months" was significant at four time points and "hopelessness" was significant at four time points. The "number of years of drug use," "how disappointed if aspired education level not achieved," and "thought parents
did not know how to teach children" variables were significant at one
time point each. On the other hand, the "wished to have own family, job,
and normal life," "thought school did not know how to teach students,"
and "subjective weathering compared with same age group" variables
were not significant at any of the six time points.

13.3.3 *Logistic regression analysis*

The next step was to perform a regression analysis on the independent
and dependent variables, in order to test for spuriousness of each of the
above significant bivariate relationships and compare the strengths of
the effects of independent variables on the dependent variable. In select-
ing sociodemographic, psychosocial, and drug use history variables for
the analysis, we included those that are more significant. As mentioned
above, if a variable was found to be significantly related to drug use at
four or more of the six time points, then it was selected. From the bivariate
relationships presented above, the following 10 variables were selected:

Table 13.2 T1–T6 Bivariate Correlations between Drug Use in the Last 30 Days ("Yes" and "No") and Psychosocial and Drug Use History Variables

Psychosocial variables	T1	T2	T3	T4	T5	T6
				r		
No. of years of drug use	0.149[a]	0.062	0.073	0.056	0.020	0.092
Permissiveness to use drugs	0.306[a]	0.359[a]	0.416[a]	0.279[a]	0.322[a]	0.285[a]
Level of disappointment if aspired education level is not achieved	-0.095[b]	-0.050	0.047	0.037	0.007	0.017
Ever been discriminated by other people	-0.033	-0.049	0.060	0.059	0.039	0.156[a]
Found goal in life	-0.140[a]	-0.247[a]	-0.251[a]	-0.283[a]	-0.291[a]	-0.179[a]
Satisfied with life	-0.179[a]	-0.281[a]	-0.288[a]	-0.375[a]	-0.378[a]	-0.268[a]
Sense of uncertainty about future	0.035	0.146[a]	0.061	0.203[a]	0.094[b]	0.141[b]
Think that doing extreme things shows the vitality of young people	0.106[b]	0.100[b]	0.176[a]	0.220[a]	0.241[a]	0.127[b]
Wish to have own family, job, and normal life in future	-0.072	-0.023	-0.049	0.010	-0.100	-0.073
Think that parents don't know how to teach children	-0.024	-0.040	-0.119[b]	0.098	0.102	0.096
Think that school doesn't know how to teach students	0.053	0.014	0.027	0.029	0.153[a]	0.040
Self-esteem	-0.135[a]	-0.121[a]	-0.194[a]	-0.288[a]	-0.269[a]	-0.244[a]
Hopelessness	0.070	0.136[a]	0.068	0.106[b]	0.223[a]	0.211[a]
Subjective weathering compared with same age group	0.003	0.002	0.020	0.019	-0.094	-0.045
Depression	0.115[a]	0.162[a]	0.200[a]	0.370[a]	0.358[a]	0.360[a]
Stricken by traumatic events in the last 6 months	0.062	0.070	0.126[b]	0.170[b]	0.247[a]	0.162[a]

[a] $p < 0.01$.
[b] $p < 0.05$.

1. Student status (i.e., whether still a student, whether the student is regularly attending school)
2. Permissiveness to the use of drugs
3. Found goal in life
4. Satisfied with life
5. Uncertain about future
6. Thought that doing extreme things shows vitality of young people
7. Self-esteem
8. Hopelessness
9. Depression
10. Stricken by traumatic events in the last 6 months

The results of the logistic regressions are shown in Table 13.3.

There were four types of independent variables in the regression models. The first type was each of the time points from T3 to T6, which were classified as independent variables because we wanted to determine if there is any effect attributed to time. The second type was the selected sociodemographic and psychosocial variables; the drug use history variable of number of years of drug use was not included as its bivariate relationship with drug use was significant at only one time point. The third type was the same set of variables at the previous time point (–1), while the fourth type of independent variables pertains to those at two time points before (–2). Thus, for the T2 to T6 data, the time point variables were T3, T4, T5, and T6, and there were no (–2)'s, whereas for the T3 to T6 data, the time points were T4, T5, and T6, and there were *both* (–1)'s and (–2)'s.

Let us first examine the regressions pertaining to the data pooled from T2 to T6. None of the time points (T3, T4, T5, or T6) was a significant marker for drug use, suggesting that the relationships between the independent variables and drug use did not significantly differ across the time points. Second, the sociodemographic variable of student status remained significant when all other variables were controlled for. Those subjects who were students but not regularly attending school were more likely to be involved in drug use than active students ($\beta = 0.615$, $p < 0.05$, OR = 1.850).

Not all the psychosocial variables remained significant in the regressions. Permissiveness to the use of drugs, satisfaction with life, depression, and stricken by traumatic events in the last 6 months were significantly related to drug use, whereas being uncertain about the future, thoughts that doing extremely things shows vitality of young people, self-esteem, and hopelessness did not have a significant effect on drug use. Among the former, stricken by traumatic events and permissiveness to use drugs had stronger effects on drug use ($\beta = 0.337$, $p < 0.05$, OR = 1.401 for drastic events, and $\beta = 0.303$, $p < 0.001$, OR = 1.354 for permissiveness) than the other two (i.e., $\beta = -0.046$, $p < 0.001$, OR = 0.955 for satisfied with life, and $\beta = 0.038$, $p < 0.05$, OR = 1.039 for depression).

Table 13.3 Logistic Regressions of Drug Use in the Last 30 Days ("Yes and No") on Selected Sociodemographic and Psychosocial Variables

Independent variables	Data pooled from T2 to T6			Data pooled from T3 to T6		
	β	S.E.	OR	β	S.E.	OR
T3	−0.059	0.182	0.943			
T4	0.098	0.196	1.103	0.358	0.230	1.430
T5	−0.230	0.207	0.795	0.083	0.254	1.087
T6	−0.163	0.222	0.850	0.060	0.271	1.062
Student status: not active	0.615[a]	0.284	1.850	0.372	0.394	1.451
Student status: nonstudent	0.323	0.170	1.381	0.507[a]	0.246	1.660
Permissiveness to use drugs	0.303[b]	0.042	1.354	0.378[b]	0.062	1.459
Found goal in life	−0.069	0.082	0.933	−0.047	0.115	0.954
Satisfied with life	−0.046[b]	0.086	0.955	−0.573[b]	0.120	0.564
Uncertain about future	−0.100	0.086	0.905	−0.238	0.124	0.788
Doing extreme things shows vitality of young people	−0.220	0.080	0.978	0.043	0.114	1.044
Self-esteem	−0.020	0.019	0.980	−0.030	0.028	0.970
Hopelessness	0.033	0.031	1.034	0.011	0.044	1.011
Depression	0.038[a]	0.016	1.039	0.050[a]	0.022	1.051
Stricken by traumatic events in the last 6 months	0.337[a]	0.152	1.401	0.431[a]	0.214	1.539
Drug use in the last 30 days (−1)	1.963[b]	0.144	7.121	2.090[b]	0.204	8.085
Permissiveness to use drugs (−1)	−0.095[a]	0.041	0.909	−0.183[c]	0.063	0.833
Found goal in life (−1)	−0.216[c]	0.077	0.806	−0.252[a]	0.113	0.777

	β		OR	β		OR
Satisfied with life (−1)	0.119	0.082	1.126	0.206	0.120	1.229
Uncertain about future (−1)	0.022	0.086	1.022	0.029	0.121	1.029
Doing extreme things shows vitality of young people (−1)	0.099	0.077	1.104	0.012	0.115	1.012
Self-esteem (−1)	0.033	0.018	1.034	−0.001	0.027	0.999
Hopelessness (−1)	−0.097[c]	0.030	0.908	−0.092[a]	0.044	0.912
Depression (−1)	0.015	0.015	1.015	0.011	0.023	1.011
Stricken by traumatic events in the last 6 months (−1)	0.120	0.147	1.127	0.139	0.209	1.149
Drug use in the last 30 days (−2)				1.015[b]	0.202	2.759
Permissiveness to use drugs (−2)				−0.032	0.056	0.969
Found goal in life (−2)				0.178	0.109	1.195
Satisfied with life (−2)				−0.145	0.114	0.865
Uncertain about future (−2)				−0.025	0.116	0.975
Doing extreme things shows vitality of young people (−2)				0.186	0.103	1.204
Self-esteem (−2)				0.040	0.025	1.041
Hopelessness (−2)				−0.012	0.042	0.988
Depression (−2)				0.022	0.021	1.022
Stricken by traumatic events in the last 6 months (−2)				−0.143	0.200	0.867
Pseudo R^2	0.356			0.466		
Model χ^2	679.636[b]			565.726[b]		

Note: All βs are unstandardized logit coefficients. OR is odds ratio, which is the exponential value of the logit coefficient.

[a] $p < 0.05$.
[b] $p < 0.001$.
[c] $p < 0.01$.

As drug use at a particular time point might also be influenced by variables at the previous time point, we examined the effects of independent variables with (–1)'s in Table 13.3. This group of independent variables included drug use at the previous time point. The results showed that the variables at the previous time point that also cast significant influence on drug use at the time point for comparison were as follows: drug use in the last 30 days (–1), permissiveness to the use of drugs (–1), found goal in life (–1), and hopelessness (–1). The greatest effect came from drug use in the last 3 days (–1) (i.e., $\beta = 1.963$, $p < 0.001$, OR = 7.121). Found goal in life (–) also had a moderate effect (i.e., $\beta = -0.216$, $p < 0.01$, OR = 0.806), and hopelessness (–1) and permissiveness to use drugs (–1) had only weak effects ($\beta = -0.097$, $p < 0.01$, OR = 0.908 for hopelessness, and $\beta = -0.095$, $p < 0.05$, OR = 0.909 for permissiveness).

Turning to the data pooled from T3 to T6, time points T4, T5, and T6 did not significantly affect drug use. For the variable student status (–2), subjects who were not students were more likely to use drugs (i.e., $\beta = 0.507$, $p < 0.05$, OR = 1.660). Regarding the psychosocial variables, those that were significant in the regression for the T2–T6 pooled data were also significant in this regression. They were permissiveness to the use of drugs ($\beta = -0.378$, $p < 0.001$, OR = 1.459), satisfied with life ($\beta = -0.573$, $p < 0.001$, OR = 0.564), depression ($\beta = 0.050$, $p < 0.05$, OR = 1.051), and stricken by traumatic events ($\beta = 0.431$, $p < 0.05$, OR = 1.539).

Psychosocial variables and the variable drug use in the last 30 days at the previous time point (–1) that were significant in the regression for the T2 to T6 data were also significant in the regressions for the T3 to T6 data. They were as follows: drug use (–1) ($\beta = 2.090$, $p < 0.001$, OR = 8.085), permissiveness to the use of drugs (–1) ($\beta = -0.183$, $p < 0.01$, OR = 0.833), found goal in life (–1) ($\beta = -0.252$, $p < 0.05$, OR = 0.777), and hopelessness (–1) ($\beta = -0.092$, $p < 0.05$, OR = 0.912).

For the T3 to T6 data, we also tested psychosocial variables (–2) as well as the variable drug use in the last 30 days (–2). These variables were two time points earlier than the comparison time point. Only drug use in the last 30 days (–2) was significant ($\beta = 1.015$, $p < 0.001$, OR = 2.759). All the psychosocial (–2) variables were not significant, suggesting that the influence of psychosocial variables on drug use did not extend to two time points.

In summary, our regression analysis shows that the sociodemographic variable student status and the psychosocial variables permissiveness to use drugs, satisfied with life, depression, and stricken by drastic events in the last 6 months were the most significant variables affecting drug use in the last 30 days. Moreover, the variables that were found to affect drug use at the next time point were drug use (–1), permissiveness to the use of drugs (–1), found goal in life (–1), and hopelessness (–1). Regarding the variables that might continue to affect drug use after the next time point, the only variable that yielded such extended significant effect was drug use (–2).

13.4   Discussion and summary

On the basis of data collected from the 3-year longitudinal survey of psychoactive drug abusers in Hong Kong, which was conducted by the author between 2009 and 2012, this chapter examines the influence of sociodemographic, psychosocial, and drug use history variables on drug use in a sample of psychoactive drug users recruited from youth outreach agencies and treatment programs. Altogether, six series of interviews were conducted, spaced out at 6-month intervals. The baseline sample size at T1 was 754, and this decreased to 288 in the last series of interviews at T6, mainly because of attrition in the form of natural case closure or premature departure. A long list of sociodemographic, psychosocial, and drug use history variables were tested for their bivariate relationships with the dependent variable "drug use in the last 30 days." Altogether, one sociodemographic variable and nine psychosocial variables were found to have significant bivariate relationships with drug use at four or more time points. They were thus selected for a series of regression analyses. The results showed that student status, permissiveness to the use of drugs, satisfied with life, depression, and stricken by traumatic events in the last 6 months significantly affected drug use. Those variables that also exerted influence on drug use at the next time point included the following: drug use in the last 30 days (–1), permissiveness to the use of drugs (–1), found goal in life (–1), and hopelessness (–1). The effect of the variable drug use in the last 30 days (–2) even extended to the time point after next. How should these results be interpreted? What theoretical and practical implications do these results have?

First, nonstudents, and students who were always absent from school, were more likely to use drugs than students who regularly attended school. This finding suggests that schools have a protective effect against drug use among students who are already drug users. This does not sound like a new finding, as traditionally schools are places of discipline and good conduct. However, a series of incidents that occurred in the mid-2000s, involving secondary school students being caught carrying or using ketamine in public parks, and students being overdosed with ketamine, were a wake-up call indicating that the use of psychoactive drugs by secondary school students was much more serious than the public thought. All of a sudden, the school was perceived to have fallen victim to the drug attack and to have become a venue for the easy distribution and consumption of ketamine among students. In view of the alarming increase of ketamine abuse in young people, Mr. Donald Tsang, then Chief Executive, in his 2007 Policy Address, announced the formation of an interdepartmental task force headed by Mr. Wong Yan-lung, then Secretary for Justice, to study the youth's drug problem and recommend solutions, including better equipping schools and teachers with knowledge and skills for drug prevention and intervention (Task Force on Youth Drug Abuse 2008). The Trial Scheme on School Drug Testing in all secondary schools in the Tai Po District was launched in 2009 to 2011, with the primary objective of creating an antidrug atmosphere in schools not just in Tai Po but all over Hong Kong (Narcotics Division 2010, 2011). These efforts greatly empowered schools in the prevention of drug abuse by students. Keeping drug-using students in school rather than letting them spend time on the street with gangs is an essential strategy. The question for the teachers is "How?" Drug-using students tend to be low achievers who are not interested in school work. How can they be motivated to come to school more often?

Second, permissiveness to the use of drugs significantly affected drug use and had a causal effect on drug use at the next time point as well. Previous studies have ascertained the influence of prodrug attitude on drug use behavior. For example, a study of 504 marginal youths and 503 regular secondary students in Hong Kong in 2002–2004 found that the level of permissiveness to the use of drugs of the marginal youths (80% of them had taken drugs at some point) was six times higher than that of regular students (Cheung and Cheung 2006). In the present sample of drug users, a higher level of permissive attitude toward drug use was associated with a greater likelihood to continue drug use.

Our findings also showed that half of the subjects had a higher level of permissiveness (T1 data; median = 10, with lowest = 4 and highest = 16). As many as 79.5% disagreed that drug use was an inappropriate behavior, and 71.1% thought that the recreational use of drugs was acceptable. Why was the recreational use of ketamine so appealing to young drug users? One of the reasons given was their perception of the risk of ketamine vis-à-vis heroin, which, for many decades (until the late 1990s), had been the dominant drug used in Hong Kong. Most of young drug users today think that ketamine is not as harmful as heroin, because they perceive it to be less addictive and so easier to stop using it. According to information we collected from several focus group sessions, ketamine users believe that as long as they do not try heroin, then they are safe. Such a misconception reinforces their false assessment of the harm of using ketamine, thereby facilitating them using this drug as part of their everyday lives. Many of them do not consider using ketamine to be drug addiction. Rather, they view it to be a bad habit, one of many (such as drinking, smoking, gambling, excessively playing computer games, swearing, gang fighting, theft, and vandalism) that a lot of young people consider to be part of being young. This reduction in the conception of "drug abuse" as a "bad habit," or what I call "bad habitization of drug use" (Cheung and Zhong 2014), lowers the level of awareness in young drug users to the actual dangers involved in psychoactive drug abuse, which encourages their denial of drug abuse, and hinders their motivation to seek help. Efforts should be made to address the bad habitization tendency as well as the larger drug subculture of young drug users in developing effective intervention and preventive strategies targeting ketamine and other psychoactive drugs.

Another significant psychosocial variable was satisfaction with life. I showed that subjects with higher levels of life satisfaction are less likely to use drugs. This finding is consistent with many previous studies that have found a negative relationship between life satisfaction and high-risk behavior among young people (Thatcher et al. 2002; Valois et al. 2004). Incidentally, in a previous 3-year longitudinal study of chronic drug abusers (mainly heroin abusers with a mean age of 36.3 years) in Hong Kong that I conducted between 2000 and 2002, life satisfaction was also found to exert significant direct and indirect effects on drug use (Cheung 2009). The level of life satisfaction in the present sample of young psychoactive drug users was medium only, with mean life satisfaction scores ranging from 3.01 to 3.40 (1 = lowest, 5 = highest) at the six time points. Unlike chronic heroin abusers, who perceived that their drug addiction had resulted in wasted opportunities for building a successful career and raising a normal family, young psychoactive drug abusers were more concerned with their immediate lives than their future. Participants of our focus group sessions expressed that schooling was so boring that they had no motivation to study hard. They became low achievers in school, at risk of developing a deviant subculture that deemphasizes mainstream school values and promotes the involvement in risk behavior as a mechanism to cope with their dissatisfaction with school life. Moreover, most drug-using students were not happy with their families. Conflict with parents and the lack of care and support from parents did not make their families a source of security and gratification. How to increase the life satisfaction of student drug users in their school and family spheres so as to reduce their likelihood to continue to use drugs is a tough question for teachers, parents, youth professionals, policy makers, and other related parties.

Depression was also positively related to drug use. The mean score of this sample of drug users at T1 was 9.6, which, in the simplified version of Beck's Depression Scale (score range, 0–39) used in this study, represented a moderately serious degree of depression (Mok et al. 2008). This finding suggests that counseling or psychiatric services provided to young psychoactive drug users might help decrease their likelihood to continue to use drugs.

Subjects struck by traumatic events that occurred in the last 6 months were more likely to use drugs than those who had not. This is an example of strain produced by "exposure to negative stimuli" as discussed in Agnew's strain theory (Agnew 1992). While the occurrence of distressing events is out of the drug user's control, access to counseling and social support could be very helpful in reducing the traumatic impact of these events on the drug user, thus resulting in decreasing the likelihood of continuous drug use.

Some variables exhibited causal effects on drug use at the next time point. Permissiveness to the use of drugs affected not only drug use at the same time point but also drug use at the next time point. Finding a goal in life reduced the likelihood of drug use at the next time point but did not affect drug use at the same time point. Hopelessness also affected drug use at the next time point rather than drug use at the same time point.

Finally, the variable that was able to exhibit strong effects on drug use at the next two time points was drug use itself. This finding is hardly surprising, as drug use at one time point indicates the presence of physical and psychological dependence at that time point. Unless the dependence is effectively removed before the next time point, through the subject's own will or via a treatment program, then drug use will continue. On the other hand, successfully becoming drug free at one time point increases the likelihood of maintaining the drug-free status for the next two time points.

Before we end this chapter, several limitations of the study should be mentioned. First, the overall attrition rate of the sample was high (i.e., 38.2% from T1 to T6). This type of sample is always quite unstable, because during the 3-year study period, some subjects leave their agencies because of program completion or because of premature departure. The fact that the participating agencies tried their best to maintain contact with their clients actually guaranteed the best retention rate possible. Also, for the present analysis, only those subjects who were not in a residential treatment program at the time of interview, or had been admitted into a treatment program not more than 30 days at the time of interview, were selected. Thus, this selection process further reduced the number of subjects eligible for analysis.

Second, the target population of the study were people who had at some time used any type of psychoactive drug. This refers to a very broad category of drug users who had a diverse pattern of drug use and abuse. There is no way to determine the size of this population and obtain a sampling frame for drawing a random sample of subjects. Despite the disadvantage of using a nonprobability sample, efforts were made in the data collection process to improve the representativeness of the sample by soliciting the participation of as many agencies as possible and recruiting as many subjects as possible for the baseline T1 sample.

Third, the dependent variable "drug use in the last 30 days" was measured with a "yes" and "no" dichotomy. This measurement was not able to differentiate different levels of use. Unfortunately, the quantity of drugs taken during each drug use episode was not determined in the questionnaire. If information regarding these variables had been acquired in the questionnaire, then they could have been combined to form a variable that collects more in-depth information regarding drug use.

This chapter reports the findings of a pioneer longitudinal survey of ketamine users as well as other psychoactive drug users in Hong Kong, addressing the influence of sociodemographic factors, psychosocial factors, and drug use history among young drug users. These findings indicate the need to address the issues of school attendance, psychosocial conditions, and drug subculture of young drug users, in order for us to better understand the social aspects of ketamine use and improve prevention and intervention strategies.

Acknowledgment

The data presented here are based on those collected for the "Longitudinal Survey of Psychoactive Drug Abusers in Hong Kong" (Cheung 2012), which was part of the "Socioeconomic and Health Impacts of Substance Abuse in Hong Kong—A Longitudinal Study" (Lee 2012) funded by the Beat Drugs Fund from 2009 to 2012.

References

Agnew, R. 1992. Foundation for a general strain theory of crime and delinquency. *Criminology* 30:47–88.

Beck, A.T., Ward, C.H., Mendelson, M., Mock, J., and Erbaugh, J. 1961. An inventory for measuring depression. *Arch. Gen. Psychiatry* 4:561–571.

Beck, A.T., Weissman, A., Lester, D., and Trexler, L. 1974. The measurement of pessimism: The hopelessness scale. *J. Consult. Clin. Psychol.* 42:861–865.

Becker, H. 1963. *Outsiders: Studies in the Sociology of Deviance.* New York: Free Press.

Cheung, N.W.T., and Cheung, Y.W. 2006. Is Hong Kong experiencing normalization of adolescent drug use? Some reflections on the normalization thesis. *Subst. Use Misuse* 41:1967–1990.

Cheung, Y.W. 1998. The hidden challenges. *Spotlight* (The Hong Kong Council of Social Service) 28:7–9.

Cheung, Y.W. 2009. *A Brighter Side: Protective and Risk Factors in the Rehabilitation of Chronic Drug Abusers in Hong Kong.* Hong Kong: The Chinese University Press.

Cheung, Y.W. 2012. *A Longitudinal Survey of Psychoactive Drug Abusers in Hong Kong.* Report submitted to the Beat Drugs Fund, Narcotics Division, Security Bureau, The Government of the Hong Kong SAR, August. Available at: http://www.nd.gov.hk/pdf/Longitudinal%20Survey%20of%20Psychoactive %20Drug%20Abusers%20in%20Hong%20Kong%20-%20report.pdf (accessed on January 13, 2015).

Cheung, Y.W., and Cheung, N.W.T. 2003. Social capital and risk level of post-treatment drug use: Implications for harm reduction among male treated addicts in Hong Kong. *Addict. Res. Theory* 11:145–162.

Cheung, Y.W., and Ch'ien, J.M.N. 1996. Drug use and drug policy in Hong Kong: Changing patterns and new challenges. *Subst. Use Misuse* 31:1573–1597.

Cheung, Y.W., and Zhong, S.H. 2014. Official reactions to crime and drug problems in Hong Kong, in C. Liqun, I. Sun and B. Hebenton (eds.), *The Routledge Handbook of Chinese Criminology.* Oxon: Routledge, pp. 295–308.

Lee, K. 2012. *Socioeconomic and Health Impacts of Psychoactive Drug Abuse in Hong Kong—A Longitudinal Study.* Overall report submitted to the Beat Drugs Fund, Narcotics Division, Security Bureau, The Government of the Hong Kong SAR, August. Available at: http://www.nd.gov.hk/pdf/Introduction.pdf.

Liang, H.J., Lau, C.G., Tang, A., Chan, F., Ungvari, G.S., and Tang, W.K. 2013. Cognitive impairments in poly-drug ketamine users. *Addict. Behav.* 38:2661–2666.

Liang, H.J., Lau, C.G., Tang, K.L., Chan, F., Ungvari, G.S., and Tang, W.K. 2014. Are sexes affected differently by ketamine? An exploratory study in ketamine users. *Subst. Use Misuse* 49:395–404.

Merton, R. 1957. *Social Theory and Social Structure.* New York: Free Press.

Mok, P.W.K., Wong, W.H.S., Lee, P.W.H., and Low, L.C.K. 2008. Is teenage obesity associated with depression and low self-esteem?: A pilot study. *H.K. J. Paediatr.* (New Series) 13:30–38.

Narcotics Division. 2010. *Trial Scheme on School Drug Testing in Tai Po District (School Year 2009/10) Evaluation Research Report.* Hong Kong: Narcotics Division, Security Bureau, Government of the Hong Kong SAR, November.

Narcotics Division. 2011. *Trial Scheme on School Drug Testing in Tai Po District (School Year 2010/11) Evaluation Research Report.* Hong Kong: Narcotics Division, Security Bureau, Government of the Hong Kong SAR, November–December.

Narcotics Division. 2012. *Central Registry of Drug Abuse Sixty-First Report, 2002–2011.* Hong Kong: Narcotics Division, Security Bureau, Government of the Hong Kong SAR.

Poon, T.L., Wong, K.F., Chan, M.Y., Fung, K.W., Chum, S.K., Man, C.W., Yiu, M.K., and Leung, S.K. 2010. Upper gastrointestinal problems in inhalational ketamine abusers. *J. Dig. Dis.* 11:106–110.

Rosenberg, M. 1965. *Society and the Adolescent Self-Image.* Princeton, NJ: Princeton University Press.

Shek, D.T.L. 2005. Hopelessness in Chinese adolescents in Hong Kong: Demographic and family correlates. *Int. J. Adolesc. Med. Health* 17:279–290.

Tang, W.K., Liang, H.J., Lau, C.G., Tang, A., and Ungvari, G.S. 2013. Relationship between cognitive impairment and depressive symptoms in current ketamine users. *J. Stud. Alcohol Drugs* 74:460–468.

Task Force on Youth Drug Abuse. 2008. *Report of the Task Force on Youth Drug Abuse.* Hong Kong: Government of the Hong Kong SAR.

Thatcher, W.G., Reininger, B.M., and Drane, J.W. 2002. Using path analysis to examine adolescent suicide attempts, life satisfaction, and health risk behavior. *J. Sch. Health* 72:71–77.

Valois, R.F., Zullig, K.J., Huebner, E.S., and Drane, J.W. 2004. Life satisfaction and suicide among high school adolescents. *Soc. Indic. Res.* 66:81–105.

Wang, C., Zheng, D., Xu, J., Lam, W., and Yew, D.T. 2013. Brain damages in ketamine addicts as revealed by magnetic resonance imaging. *Front. Neuroanat.* 7:23. doi: 10.3389/fnana.2013.00023.

Wei, Y.B., Yang, J.R., Yin, Z., Guo, Q., Liang, B.L., and Zhou, K.Q. 2013. Genitourinary toxicity of ketamine. *Hong Kong Med. J.* 19:341–348.

Wong, G.L., Tam, Y.H., Ng, C.F., Chan, A.W., Choi, P.C., Chu, W.C., Lai, P.B., Chan, H.L., and Wong, V.W. 2014. Liver injury is common among chronic abusers of ketamine. *Clin. Gastroenterol. Hepatol.* 12:1759–1762.e1. doi: 10.1016/j.cgh.2014.01.041.

chapter fourteen

Mechanisms of ketamine-induced neuroplasticity
Potential effects on brain and behavior

Vincenzo Tedesco, Ginetta Collo, and Cristiano Chiamulera

Contents

14.1 Introduction

The recreational use of ketamine has increased over recent years in many parts of the world, and physical harm and addiction have been reported in heavy users (Morgan and Curran 2012; Schifano et al. 2006). Initially confined to certain subcultures, ketamine is now the fourth most popular drug among UK clubbers after cannabis, ecstasy, and cocaine, which suggests a high potential of abuse (Morgan and Curran 2012). This abuse potential can be related to its complex psychoactive profile, observed even at low doses, which include analgesic, psychotropic, and antidepressant effects. Recent research has shown that ketamine can induce long-lasting effects after just a single low-dose acute administration. For example, clinical studies have shown that one infusion of ketamine induced a rapid antidepressant response in subjects with major depressive disorder that lasted for up to 7 days (Berman et al. 2000; Zarate et al. 2006). Remarkably, ketamine acutely induces effects similar to those observed after chronic administration of currently used antidepressants. These effects might be related to the capacity of ketamine to trigger molecular mechanisms of neuroplasticity, resulting in a different arrangement of neuronal circuits involved in behavioral control. Interestingly, when ketamine is taken by subjects with no symptoms of depression, it induces elation and the dazed behavior observed with other similar addictive drugs. The neurobiological and molecular events associated with ketamine exposure have been extensively investigated in areas of the brain related to depression and mood changes (Duman et al. 2012). However, similar mechanisms are known to be involved in areas of the brain that respond to reward and are involved in motivation. This therefore supports a possible role for neuroplasticity in the development of ketamine dependence and abuse (Dietz et al. 2009; Russo et al. 2010; Tedesco et al. 2013). This rapid induction of neuroplasticity could determine the establishment of a liability for ketamine dependence since the very first exposure, that is, a fast-onset acquisition of ketamine-taking behavior, its maintenance, and resistance to extinction and vulnerability to ketamine-seeking relapse.

In this chapter, we describe the effects of ketamine on the molecular and neurochemical mechanisms known to be involved in the drug-related neuroadaptation of the brain as determined in preclinical models. Ketamine-induced changes in neurotransmitters, their receptors, neurotrophic factors, and neuronal morphology both in vitro and in vivo will be discussed as correlates of behavioral phenomena associated with addiction, such as locomotor sensitization, and drug self-administration. This overview will be discussed in the context of ketamine abuse as well as the implications for basic research on psychiatric disorders.

14.2   Receptor profile of ketamine

14.2.1   Glutamate ionotropic receptors

Ketamine neuropharmacology is complex. Ketamine acts on glutamate binding sites of the ionotropic N-methyl-D-aspartate receptor (NMDAr) and on non-NMDArs (Kohrs and Durieux 1998). It is a glutamate noncompetitive antagonist at the NMDAr. Like other NMDAr noncompetitive antagonists (e.g., phencyclidine, MK-801, dextromethorphan, and memantine), it is fastened to an intrachannel site called the phencyclidine site (Lipton 2004). Ketamine intrachannel binding decreases the channel opening time and amplifies the response to repeated stimulation. Enhanced effects were observed when the NMDAr channel was previously opened after glutamate fixation (Arendt-Nielsen et al. 1995; Guirimand et al. 2000).

In addition to acting on glutamate binding sites, ketamine also binds to a site that is located in the hydrophobic domain of the NMDAr, where it reduces the frequency of channel opening (Schmid et al. 1999). Furthermore, it acts as an allosteric antagonist at the NR2B subunit of NMDAr. The NR2B subtype of the NMDAr is abundant not only in the limbic system and in the cingulate cortex but also in the hippocampus and amygdala, which are areas of the brain involved in emotion and memory (Mion and Villevieille 2013). The NMDAr is central to learning and memory because of its role in synaptic plasticity; thus, inhibition of the NMDAr is the mechanism responsible for the specific ketamine effects on synaptic plasticity.

Non-NMDA glutamate ionotropic receptors such as the α-amino-3-hydroxy-5-methyl-4-isoxazolepropionic acid receptor (AMPAr) and the kainate receptor interact with, and are inhibited by, ketamine (Gonzales et al. 1995). Activation of AMPAr, similar to NMDAr activation, stimulates nitric oxide (NO) synthesis, which in turn increases the production of intracellular cyclic guanosine monophosphate (cGMP) (Garthwaite 1991; Marin et al. 1993; Wood et al. 1990). Therefore, the effects of ketamine are mediated via the glutamate/NO/cGMP system, a mechanism suggested to underlie the neuroprotective and neurotrophic properties of ketamine (Gordh et al. 1995). However, recent data suggest that the involvement of the ketamine/AMPAr interaction in neuroplasticity may be attributed to an increase in the ketamine-induced glutamate levels observed in some brain regions and consequently to the release of AMPAr-mediated neurotrophins (Duman et al. 2012) rather than via a direct effect on the NO/cGMP mechanism.

14.2.2   Dopamine and serotonin receptors

Dopamine receptors are a class of G protein–coupled receptors categorized into two subfamilies, the D1-like dopamine receptors (D1r and D5r) and the D2-like dopamine receptors (D2r, D3r, and D4r) (Beaulieu and Gainetdinov 2011; Tiberi et al. 1991; Vallone et al. 2000). Serotonin receptors, also known as 5-hydroxytryptamine receptors (5-HTr), are a group of G protein–coupled receptors (5HT1r, 5HT2r, 5HT4r, 5HT5r, 5HT6r, and 5HT7r) and ligand-gated ion channels (5HT3r). Some reports suggest that ketamine can bind to dopamine D2r and serotonin 5-HT2r, in particular to the high-affinity states of these receptors, suggesting an agonist-like profile. Indirect evidence of 5HT2r binding was obtained with the ketamine analog PCP (Kapur and Seeman 2002), but its relevance in vivo and in humans in particular is still under scrutiny.

14.2.3 Dopamine and serotonin transporters

Ketamine has been shown to enhance the endogenous levels of dopamine in rodents when measured in vivo with microdialysis (Lorrain et al. 2003). Ketamine has also been shown to enhance serotonin transmission in monkeys (Yamamoto et al. 2013), and this effect on both dopamine and serotonin is probably mediated via a direct action on monoamine reuptake transporters (Nishimura et al. 1998). Indeed, it is well documented that ketamine has an affinity for the dopamine transporter and this effect shows stereo-selectivity with *S*-ketamine being more potent than *R*-ketamine (Nishimura and Sato 1999). Furthermore, ketamine shows relevant affinity for the serotonin transporter (Martin et al. 1988, 1990).

14.3 Neurochemical effects

14.3.1 Effects of ketamine on glutamate levels

The mechanism by which ketamine produces its effects has been partially attributed to the blockade of NMDAr located on inhibitory GABAergic neurons in the limbic and subcortical brain regions (Moghaddam et al. 1997; Nakao et al. 2003). As a result of this disinhibitory action, glutamate release and dopamine release in the prefrontal cortex and limbic striatal regions increase dramatically owing to augmented neural activity (Duncan et al. 1998; Gass et al. 1993; Lorrain et al. 2003). The glutamate content increases particularly in the cortex region after acute and chronic ketamine administration, which leads to an increased glutamatergic input on AMPAr relative to NMDAr. It is also possible that ketamine has direct effects also on the pyramidal neurons, enhancing synaptogenesis and maturation of the spines (Li et al. 2010a). On the other hand, there seems to be a different effect of ketamine in the hippocampus. Acute (but not chronic) ketamine administration reduces glutamate levels in the hippocampus (Chatterjee et al. 2012). Furthermore, ketamine at subanesthetic doses induces desensitization of AMPAr in the hippocampus (Wang et al. 2011), an effect that is observed in the cortex and striatum only after an anesthetic dose (Snyder et al. 2007).

14.3.2   Effects of ketamine on dopamine levels

Dopamine plays a critical role in modulating synaptic plasticity (Bai et al. 2009; Jay 2003). Most of the dopamine in the central nervous system is derived from projections of dopaminergic neurons located in the ventral tegmental area/substantia nigra and raphe nucleus (Ikemoto 2007). In vivo, dopaminergic neurons in the midbrain are regulated by cortical glutamatergic projections through a facilitatory pathway that is mediated via the NMDAr and AMPAr and an inhibitory pathway controlled by the NMDAr via GABA interneurons (Kegeles et al. 2000). As an NMDAr antagonist on GABA interneurons, ketamine disinhibits glutamate neurotransmission at the AMPAr and also reduces the inhibitory pathway activity leading to an increased excitatory input and rapid dopamine efflux in the cortex (Moghaddam et al. 1997). In addition, the uptake of dopamine is inhibited by ketamine interaction with the dopamine transporter, which contributes to the increased central dopaminergic activity (Nishimura and Sato 1999; Nishimura et al. 1998). The prefrontal dopaminergic system, which is critically involved in working memory and executive function, is particularly vulnerable to the chronic effects of ketamine (Narendran et al. 2005). While an acute subanesthetic dose of ketamine rapidly increases dopamine release in the prefrontal cortex, chronic ketamine application results in a further increase in the dopamine content that persists for at least 10 days after drug withdrawal (Chatterjee et al. 2012; Lindefors et al. 1997; Moghaddam et al. 1997), which suggests that there might be a longer duration effect of chronic ketamine use on the increase of dopamine levels in the cortex. In the striatum, a significant increase in the dopamine level is observed with both acute and chronic ketamine treatment, but, unlike the cortex, the dopamine level decreases during the drug withdrawal phase in this region. In contrast, the concentration of dopamine in the hippocampus did not change significantly after either acute or chronic ketamine treatments, whereas an increase was observed after drug withdrawal (Chatterjee et al. 2012). Interestingly, in the hippocampus, the mechanism of ketamine-induced AMPAr desensitization could be mediated by dopamine. In addition, it has recently been shown that ketamine induces AMPAr endocytosis-dependent synaptic depression via D1/D5 receptors at Schaffer collateral-CA1 synapses in this region of the brain (Duan et al. 2013).

14.3.3   Effects of ketamine on serotonin levels

The uptake of serotonin is also inhibited by ketamine (Martin et al. 1988, 1990). In the cortex, the level of serotonin does not change after acute ketamine administration, but it does increase after chronic ketamine administration, and this effect is further exacerbated upon withdrawal of the drug. Similarly, in the striatal region, chronic ketamine application results in an increase in the level of serotonin level; however, unlike in the cortex region, the serotonin levels return to normal after drug withdrawal. In the hippocampus, as for dopamine, there are no significant changes in the level of serotonin after either acute or chronic ketamine treatment (Chatterjee et al. 2012).

The data acquired regarding ketamine modulation of dopamine and serotonin in the cortex and striatal regions suggest that the ketamine effect on dopamine release is more rapid, and this might be responsible for early cellular events, which subsequently leads to the modulation of the serotonin levels at a later stage.

14.3.4   Effects of ketamine on noradrenaline content

The pattern of noradrenaline release after acute or chronic ketamine treatment or ketamine withdrawal is similar to that observed for dopamine across the different brain regions. An increase in noradrenaline levels is observed after acute and chronic ketamine treatment in the cortex and striatal regions, with a higher increase occurring during drug withdrawal in the cortex and hippocampus (Chatterjee et al. 2012). Ketamine, in particular *R*-ketamine, also inhibits the neuronal uptake of noradrenaline, which induces a prolonged synaptic response (Kress 1994). With the stimulation of noradrenergic neurons and an inhibition of uptake of catecholamines, ketamine provokes a hyperadrenergic state (i.e., with the release of noradrenaline, dopamine, and serotonin).

14.3.5   Effects of ketamine on the acetylcholine level

Cortical acetylcholine (ACh) regulates the detection, selection, and processing of stimuli, and it is known to be involved in cognitive functions (Hasselmo 2006). Administration of ketamine in acute or chronic doses leads to an increase in the level of ACh in the cortex, with a higher increase occurring after chronic treatment. On the other hand, no change in the level of ACh level has been found in striatal or hippocampal regions. Ketamine also increases acetylcholinesterase (AChE) activity in the cortex, with a higher level of activity occurring after acute administration than after chronic administration (Chatterjee et al. 2012). These findings are consistent with the fact that acute ketamine treatment might result in the dramatic activation of AChE, which likely leads to low ACh content. On the other hand, chronic ketamine treatment results in a lower level of AChE activation, which leads to an increase in ACh levels. As expected, ketamine inhibits NMDAr-mediated ACh release (Kohrs and

Durieux 1998). Other findings indicate that, at clinical concentrations, keta-mine acts as an antagonist of nicotinic and muscarinic ACh receptors (Kress 1994).

14.3.6   Imaging studies

Ketamine action in the cortex and limbic brain regions has also been exten-sively investigated by preclinical and clinical imaging studies. Functional magnetic resonance imaging studies conducted in rats, primates, and humans have shown that both acute and chronic subanesthetic doses of ketamine change metabolic activity in various regions of the brain. Acute ketamine use increases metabolic activity particularly in the limbic corti-cal regions (including the medial prefrontal, ventrolateral orbital, cingu-late, and retrosplenial cortices), in the hippocampal formation (i.e., in the dentate gyrus, CA3 stratum radiatum, stratum lacunosum moleculare, and presubiculum), and in the basolateral amygdala (BLA) (Duncan et al. 1998, 1999). On the other hand, chronic ketamine use increases metabolic activity in the right lentiform nucleus, fusiform gyrus, and entorhinal cor-tex, whereas it reduces activity in the substantia nigra, ventral tegmental area, posterior cingulated cortex, visual cortex, and somatosensory cortex (Yu et al. 2012). White matter changes in the cortex of mice and humans after chronic ketamine use have also been confirmed by recent imaging analyses that highlight bilateral frontal and left temporoparietal reduc-tions in fractional anisotropy in patients with ketamine dependence (Li et al. 2010b; Liao et al. 2010). Together, these findings suggest that ketamine, whether acutely or chronically administered, leads to widespread func-tional anomalies in the brain circuits that are relevant to drug addiction.

14.4   Signal transduction

From a molecular perspective, the acute ketamine antidepressant effect seems to be mediated by the activation of the mammalian target of rapa-mycin (mTOR), a protein kinase involved in translation control and long-lasting synaptic plasticity, via brain-derived neurotrophic factor (BDNF), a key neurotrophin modulated by chronic antidepressant administration (Duman and Monteggia 2006; Krishnan and Nestler 2008), and involved in the neuronal remodeling that occurs after chronic drug administration (Russo et al. 2010).

The increased extracellular glutamate induced by ketamine in the prefrontal cortex of rats shows a time course similar to the rapid induction of mTOR (Moghaddam et al. 1997). One of the two multiprotein complexes of mTOR (i.e., mTOR complex 1 or mTORC1) is involved in neuroplastic-ity via activation of p70 ribosomal S6 kinase (p70S6K) and repression of eukaryotic initiation factor 4E binding proteins (4E-BPs); which in turn lead to increased phosphorylation of ribosomal protein S6 (rpS6P) and release of the activated forms of 4E-BP1, respectively (Klann and Dever 2004). An increased level of mTOR, 4E-BP1, and p70S6K found in a preparation enriched in synaptoneurosomes (Li et al. 2010a) after acute subanesthetic ketamine administration confirms that ketamine rapidly activates the mTOR signaling pathway in the prefrontal cortex of rats. Furthermore, rpS6P expression increases in the prelimbic and infralimbic cortices, nucleus accumbens core, and BLA of rats after a single subanesthetic dose of ketamine, suggesting that acute ketamine administration also induces neuroplasticity in drug addiction–related brain areas (Tedesco et al. 2013). Since neuroplasticity is the key mechanism of addiction development, this effect of acute ketamine administration might underlie an early development of addictive behavior.

Stimulation of mTOR signaling and synaptic protein synthesis is dependent on glutamate activation of the AMPAr (Hoeffer and Klann 2010). Glutamate binding leads to activation of the AMPAr, which in turn modulates the opening of L-type voltage-dependent calcium channels, leading to an influx of intracellular calcium and the subsequent activity-dependent release of BDNF, which results in the activation of protein kinase B, and extracellular signal–related kinase (ERK), as well as stimulation of mTOR and synaptic protein synthesis (Hoeffer and Klann 2010; Jourdi et al. 2009; Slipczuk et al. 2009; Takei et al. 2004).

Increased phosphorylation of glycogen synthase kinase-3 (GSK-3), a target of mood-stabilizing agents (Li and Jope 2010), has been proposed as another possible molecular mechanism that is engaged after acute ketamine administration. Since mice containing a knock-in mutation that blocks the phosphorylation of GSK-3 do not respond to ketamine in a behavioral model of depression (Beurel et al. 2011), GSK-3 is a new candidate for molecular investigations of the effects of acute ketamine use. The mechanisms underlying the induction of GSK-3 phosphorylation by ketamine are not yet well understood, but it is speculated that Akt, a major regulator of GSK-3, and a key molecule in the synaptic protein expression triggered by ketamine, may be involved (Li et al. 2010a).

14.5   Structural plasticity induced by ketamine

Ketamine is known mostly for its neurodegenerative effects both in vivo and in vitro (Vutskits et al. 2006; Yu et al. 2012), and yet it is particularly important when exposure takes place during neurodevelopment (Ibla et al. 2009). Surprisingly, at lower doses, ketamine appears to have an opposite effect, busting structural plasticity (Li et al. 2010a). Structural plasticity can be defined as the ensemble of morphological changes that occur in neurons of specific regulatory neural systems when engaged in learning, in adaptive responses to external stressors, or during recovery after damage. It is tempting to suggest a role of structural plasticity in ketamine treatment.

A single subanesthetic dose of ketamine is associated with an increase in synapse formation and maturation in the medial prefrontal cortex of adult rodents. Ketamine increases spine densities and size (i.e., mushroom-like functional spines) in the distal and proximal dendrite segments of glutamatergic pyramidal neurons of the cortical layer V (Li et al. 2010a); these effects were observed 24 h after administration, suggesting a rapid action of ketamine.

Currently, there is no evidence to suggest that structural plasticity occurs in other brain regions of adult animals after ketamine application. The fact that cocaine, amphetamine, and nicotine all increase synaptic spine density and dendrite length in pyramidal neurons of the prefrontal cortex, in medium spiny neurons of the nucleus accumbens (Janson et al. 1988; Robinson and Kolb 2004), and in dopaminergic neurons of the mesencephalon (Collo et al. 2012, 2013; Mueller et al. 2006; Sarti et al. 2007), suggests that ketamine might also affect structural plasticity in other neuronal systems relevant for addictive behavior.

Recent in vitro data indicate neuroplasticity effects of ketamine on dendrite length and branching of mesencephalic dopaminergic neurons isolated from mouse embryos (Cavalleri et al. 2013). Interestingly, the neuroplasticity effects of ketamine were also observed during neural development in vivo. Ketamine exposure for 5 h was shown to increase the number of functional dendritic spines in the somatosensory cortex and hippocampus of 15-day-old mice (De Roo et al. 2009). While these effects could be seen as transient and related to the developmental stages, their relevance as potential mechanisms in addiction is an intriguing possibility that deserves some attention.

Recent molecular findings indicate that ketamine activates intracellular pathways associated with cell growth, such as Raf–MEK–ERK, Akt–PI3K–mTORC1 or GSK3, and IP_3–Ca^{2+}–CaMK, which are all relevant for the generation of synaptic spines and dendrite elongation (Duman et al. 2012; Tedesco et al. 2013). Interestingly, these pathways are primarily activated by neurotrophins such as BDNF or NT4, which bind to the TrkB receptor and trigger structural changes in most neurons (Duman and Bhavya 2012; Li et al. 2008). A reasonable and commonly accepted view indicates that ketamine effects are mediated by BDNF since ketamine increases the levels of BDNF in the hippocampus, cerebral cortex, and midbrain (Duman et al. 2012; Garcia et al. 2008; Ibla et al. 2009; Li et al. 2010; Tan et al. 2012). BDNF is known to increase the expression of several synaptic proteins, including the spine scaffold PSD95 and the density of AMPArs, leading to synaptic maturation (Chen et al. 2009; Duman et al. 2012). This effect appears to be mediated by mTORC1 since rapamycin blocks dendritic arborization or functional spine formation induced by either ketamine (Li et al. 2010a) or BDNF (Kumar et al. 2005). Interestingly, some findings indicate that ketamine also increases the AMPAr in the prefrontal cortex of Sprague–Dawley rats (Li et al. 2010a) and in the hippocampus of Wistar–Kyoto rats (Tizabi et al. 2012). Moreover, Jourdi et al. (2009) showed that positive modulation of the AMPAr leads to dendritic growth via BDNF secretion and TrkB activation. Mechanistically, the enhanced AMPAr signaling associated with TrkB intracellular pathway activation might result in a feed-forward system that is necessary to maintain the cellular growth machinery needed to produce structural changes. However, to be functional, this model requires an increase in glutamate release in the synaptic cleft. It was proposed that ketamine is specifically active in blocking NMDAr located in the GABA-releasing interneurons of the cortex, leading to a disinhibition of the presynaptic glutamatergic neurons and an increase in glutamate release (Duman et al. 2012; Moghaddam et al. 1997). Therefore, according to this hypothesis, in the presence of ketamine, glutamate (which is increased) will be acting primarily on the AMPAr and metabotropic glutamate receptors since the NMDArs are blocked. In one experiment, Li et al. (2010a) showed that the AMPAr antagonist NBQX blocks the activation of ERK, Akt, and mTOR pathways in the prefrontal cortex of adult rats administered with ketamine (but the structural plasticity effects of ketamine were not shown). Conversely, when NBQX was administered to a nonstimulated hippocampal organotypic preparation from young rats, an increase in both the size and number of dendritic spines was observed, a phenomenon also observed after NMDA blockade with MK801, and GABA enhancement with midazolam (De Roo et al. 2009). While these data on the glutamate system are promising, they do not rule out the possibility that some of the structural plasticity effects of ketamine might be mediated via the increased release of other neurotransmitters affecting the excitatory/inhibitory balance, such as dopamine and serotonin. In conclusion, there is evidence that ketamine produces structural plasticity, but the functional relevance of these changes is still a matter of exploration.

14.6 Behavioral correlates of neuroadaptation

14.6.1 Drug addiction as a form of behavioral adaptation

Several phases characterize the addictive experience (Kalivas and O'Brien 2008). The "social use" of drugs is compatible with a normal lifestyle, and it does not permeate all aspects of life. On the other hand, the "regulated relapse" use is more severe; nevertheless, it still involves some behavioral control. The most severe phase of drug use is termed "compulsive relapse." The shift from declarative to automatic behaviors—owing to chronic exposure to the drug—is an expression of neuroadaptation in areas of the

prefrontal cortex (Goldstein and Volkow 2002) toward a general reduction of activity, that is, a reduced inhibitory control over the striatal brain
areas. Over time and with continued chronic exposure to the drug, there
is a shift from ventral to dorsal striatum control on behavior, that is, from
goal-directed action to stimulus–response. That means that the flexible,
reversible motor behavior to get the drug becomes a habit (Belin et al.
2009). The shift from social use to compulsive relapse does not necessarily indicate that there is an increase in the frequency of taking the drug.
Some types of drug addiction (ketamine included) develop on a circadian
or even on a weekly basis. Therefore, additional nonpharmacological
factors may also play a role in the onset of addictive behavior—drug
taking—or on drug-seeking relapse. In fact, drug effect–associated stimuli
and context changes are stored in brain areas such as the amygdala (for
the incentive value of stimuli) and in the prefrontal cortical regions where
the executive control of behavior allows adaptation to the modified environment. Behavior becomes less conscious, more automatic, and is characterized by action schemata and motor pattern generators (Everitt and
Robbins 2005).

The current paradigm of drug addiction as a form of learning and
memory, that is, of neuroadaptation, is supported by several pieces of evidence at the molecular, cellular, and behavioral level. Historically, drug
addiction research has developed experimental procedures to study the
behavioral effects of drug of abuse and to measure variables that may
define the validity of the addictive disorder in humans. More recently,
neurobiological investigations on drug addiction as a form of neuroadaptation showed correlations between mechanisms and behaviors as
assessed in laboratory animal models. Table 14.1 shows the main experimental procedures that have been used to study the behavioral effects of
drugs of abuse, including ketamine. The data presented in the following describe the effects of ketamine in these models and are discussed in
terms of ketamine-induced neuroadaptation as it emerges at the behavioral level. First of all, a brief overview of the acute effects of ketamine in
laboratory animals is given.

14.6.2   Behavioral effects of ketamine in rodents

Ketamine has a biphasic effect on motor behavior in rodents. At low
doses (i.e., 1–10 mg/kg intraperitoneal delivery), ketamine induces a dose-
related increase in exploration, rearing, and ambulation. These effects are
more evident at moderate doses (i.e., 20–30 mg/kg intraperitoneal delivery), when they are associated with the onset of stereotypic movements.
At higher doses (i.e., close to anesthetic doses of ~100 mg/kg), motor coordination and muscle tone decrease. Ataxia and loss of posture are typical
signs. Locomotion is still activated at these doses, but the total distance

Table 14.1 Experimental Procedures and Definitions of the Behavioral Paradigms for the Investigation of Chronic Effects of Drugs of Abuse

Procedure	Definition
Drug-induced sensitization	An assay whereby a response is amplified by repeated drug administration. Drug-induced motor sensitization is a nonassociative learning process that is observed with repeated administration of drugs of abuse such as amphetamine and cocaine.
Self-administration	An operant assay in which a response (i.e., lever press or nose poke) is followed by a reinforcer (e.g., drug). Self-administration is thought to measure the reinforcing effects of drugs. A drug is self-administered when the probability of a response increases over time. Drugs can be self-administered by different routes (i.e., intravenously, intranasally, or orally).
Conditioned place preference	An assay in which administration of a drug is paired with a specific context, whereas administration of vehicle is paired with a different one. Drug-pairing sessions are alternated with vehicle-pairing sessions. After a sufficient number of sessions, the animals are tested in a drug-free state in which both contexts are presented. If animals spend significantly more time in the context previously paired with the drug, then the drug may have rewarding effects.
Reinstatement	An assay whereby a previously extinguished response (i.e., lever press, nose poke, conditioned place preference) is reestablished by administering the drug, which was used previously for conditioning (i.e., drug-induced reinstatement).

covered is lower because of incoordination, loss of equilibrium, and "crawling-like" deambulation. Interestingly, Castagné et al. (2012) showed that ketamine at a low to moderate range of doses also stimulates eating and drinking behaviors. This stimulant effect is also seen at high doses, but only after a longer period. In fact, this temporal profile of behavioral effects of ketamine is due to the fact that at high doses, the earliest effect is the sedative one; this then fades away and reveals the stimulant effect.

14.6.3 *Motor sensitization*

Sensitization was for a long time defined as reverse tolerance, or as an increased biological response to chronic treatment with a substance. The conceptualization of the phenomenon of sensitization is not intuitive, as it is supposed that the body puts in place adaptive responses contrary to the action of the drug, which are neither additive nor synergistic. In general, a

physiologically enhanced biological response corresponds to an improvement of function—for example, the hypersensitivity acquired owing to a
tactile stimulus pain reliever may prompt the anticipation of the perception and therefore avoidance. In the case of pharmacological sensitization,
however, there is a risk of toxicity. Sensitization is a phenomenon that only
in recent years has been proposed as one of the main features of the psychological addiction to drugs of abuse (Robinson and Berridge 1993). The
increasing response of a chronic dose of addictive drug has been observed
at different levels of analysis, that is, neurochemical, electrophysiological,
and behavioral. Chronic treatment with amphetamine, cocaine, nicotine,
and other substances of abuse in laboratory animals has been shown to
induce an increase in neuronal processes (dendrites) and in the number of
synapses (Robinson and Kolb 2004). This modification is similar to those
induced in other phenomena of neuroplasticity, such as memory, thus
suggesting that the sensitized response of substances of abuse may gain
stability over time. The sensitization to drug response might explain the
increased motivation, salience to environmental stimuli, and the higher
risk of relapse observed.

Surprisingly, there are relatively few reports describing ketamine-
induced motor sensitization (Meyer and Phillips 2003; Trujillo et al. 2008;
Uchihashi et al. 1993; Wiley et al. 2008). Motor sensitization is the behavioral phenomenon that has been observed after chronic administration
of cocaine, amphetamine, nicotine, and several other drugs of abuse.
Although it is not observed in humans, it is a valid measure of neuroadaptation to chronic treatment since it correlates with neuroplasticity
events at the molecular and cellular level. Trujillo et al. (2008) investigated
the motor sensitization effect of ketamine in rats at a moderate dose
(20 mg/kg), which was given once a week. During the first test, the effect
of this moderate dose was not significantly different from that induced
by the drug delivery vehicle alone. Ketamine was then given as a single
intraperitoneal injection once a week for the following 5 weeks. During
this time, the effect of the drug became more obvious, such that the same
dose induced an increase in motor hyperactivities, that is, time of activity,
ambulation, and fine movements, which was up to twofold greater than
the initial activity measured.

An increase in motor sensitization is demonstrated not only when
there is an escalation effect of the same dose but also when a dose that is
noneffective per se is able to induce an effect when given to a "sensitized"
subject, such as to an animal previously exposed to repetitive drug administration. In the Trujillo et al. (2008) study, for example, ketamine at 20 mg/
kg induced an increase of total activity counts in both saline-injected (controls) rats and in animals that had been sensitized with the drug for 6
weeks; however, a greater effect was observed in the previously sensitized group when compared with the vehicle-alone group. The occurrence

of motor sensitization was also shown in the 7th week when a ketamine challenge, a single, high dose of ketamine (at 50 mg/ml), was administered. Trujillo et al. (2008) also performed a second experiment on motor sensitization with the highest dose of ketamine (50 mg/kg) given in an "experimental cage." Motor sensitization for all the motor signs described above was observed after 6 weeks of a once-a-week treatment. Another group received the same dose regimen of ketamine but they remained in the "home cage." Interestingly, the motor sensitization observed in the group that received ketamine in the experimental cage was not observed in the group that received repetitive ketamine in the home cage. These data suggest that motor sensitization is a form of learning that—as seen with other drugs—is attributed to neuroadaptive processes (Anagnostaras et al. 2002). The association between the context where the motor activity took place and the ketamine effect enhances the initiation and expression of motor sensitization.

14.6.4 Ketamine self-administration

The effects of ketamine addiction have been described by using a rat model of addictive behavior and is called the drug self-administration paradigm (Ahmed 2010). This conditioning task provides behavioral measures that have been shown to possess predictive validity for drug reinforcing properties (Markou and Paterson 2009). Briefly, rats were trained to get a reward if they produced a specific behavioral response. The reward increased the probability of responding occurrence, therefore acting as a "reinforcer." This process is physiological and it sustains motivated behavior for seeking and obtaining natural rewards such as food, water, and sex. Drugs of abuse may act as primary reinforcers, as similar behavior is observed in human addicts and as it has been characterized in laboratory studies in both humans and animals (Table 14.1).

Drug self-administration protocols may also include Pavlovian components that are the conditioning of neutral stimuli. These stimuli acquire a conditioned value upon repetitive associations to the unconditioned stimulus (i.e., drug infusion). The reexposure to conditioned stimuli previously associated with the drug is able to trigger cue reactivity. The cue reactivity is an individual adaptive response to salient information that is present in the internal and external environment of an individual. Salience is motivational information that informs the approach or withdrawal behavior for seeking and taking rewards. Environmental components that may trigger cue reactivity are of two types: proximal or distal stimuli. Proximal stimuli are those directly associated with the motivated behavior and are discretely defined in terms of structure and properties. Distal stimuli are often defined as a complex of various stimuli that own a conditioned value as a whole. For instance, a club or a social environment is generally speaking defined as a context or setting. The real experience of an individual—the situation—includes both proximal and distal stimuli, but researchers have investigated the features and the motivational values of these two categories of stimuli separately. Previous studies have characterized real-life smoking and craving situations (Dunbar et al. 2010), as previously reported for other drugs of abuse (Epstein et al. 2009; Sussman et al. 1998, 2001). In addition, studies by Badiani et al. (2011) have cross-validated the effect of a setting between preclinical and human studies for heroin and cocaine (Caprioli et al. 2009) and put it in the broader perspective of the role of the environment in drug addiction (Badiani et al. 2011).

Similar to the Trujillo et al. (2008) study described previously, De Luca and Badiani (2011) investigated the effect of the setting (i.e., in the context of ketamine self-administration) on the response for ketamine infusion. They trained two groups of rats to self-administer different doses of ketamine. One group resided permanently in an experimental operant conditioning box (residential rats), whereas the other group was placed in the experimental box only during the 3-h daily session of intravenous ketamine self-administration and then they spent the rest of the day in their home cage (nonresidential rats). The acquisition rate of responding to ketamine infusion and the number of infusions were then compared between the groups. There was a significant interaction between the unit doses of ketamine self-administered, ketamine intake, and lever press for infusion (all dependent variables of responding for ketamine as a reinforcer) with the setting. Nonresidential rats showed significantly higher values for these variables than their residential counterparts.

These findings confirmed that ketamine, similarly to other psychostimulants, is able to induce reinforcing properties that interact with the conditioned value of the context, in this case a novel nonfamiliar one. Similar to the motor sensitization study reported by Trujillo et al. (2008), De Luca and Badiani (2011) demonstrated that ketamine effects may be enhanced by a novel environment.

14.6.5 Conditioned place preference

The conditioned place preference model has been widely used to assess the reinforcing effect of addictive drugs (Tzschentke 2007). The effect of the drug is repeatedly paired with one distinct context, whereas a neutral event is paired with a different context. Allowing the animal to move between the two contexts and measuring the amount of time spent on each context determines preference (Table 14.1). Administration of ketamine at 3 and 10 mg/kg has been shown to produce conditioned place preference in mice (Suzuki et al. 1999), which is the ability to induce Pavlovian conditioning between its effects and a specific context.

Interestingly, the extinction of a previously established drug-induced place preference can be accomplished with either repeated place preference testing without any drug exposure or by pairing the two environments of the test apparatus with saline injections. Drug priming injections (Shaham et al. 2003) can then reinstate the extinguished place preference in a similar way that the drug priming effect might reinstate the response for drug self-administration (see Table 14.1). The significance of this behavioral measure is the face validity with the effect of the drug-associated context described above. It is well known, for example, that ex-alcoholics should refrain from taking substances that contain even small quantities of alcohol (such as liqueur chocolates), in order to avoid the conditioned effects that can lead to a relapse. The drug priming effect has been studied from both a behavioral and molecular perspective (Shaham et al. 2003). Different types of receptors are involved, including the glutamate NMDA subtype. It has been shown that the ketamine conditioned place preference may be extinguished and reinstated by a single ketamine injection. Li et al. (2008) showed that application of a dose of ketamine lower than the normal conditioning dose (i.e., 5 mg/kg as opposed to 10 mg/kg) was able to reinstate the conditioned place preference in mice. This is another piece of evidence to suggest that ketamine not only maintains self-administration, conditioned place preference, and motor activation but also exerts adaptive effects on behavior such as the ability to induce motor sensitization and create a conditioned association with context and ketamine priming injections. The implication of these effects for ketamine abuse and dependence is of fundamental importance.

14.7 Summing up and conclusions

On the basis of the data described in the above sections, the neuroadaptive processes induced by ketamine are characterized by an early and acute induction, with relatively long-lasting molecular, neuronal, behavioral changes that can be summarized in relation to neuroplasticity. These effects might help elucidate the mechanisms underlying the potential therapeutic effects of ketamine as a rapid-onset antidepressant as well as addictive drug (Duman et al. 2012; Moghaddam and Krystal 2012). Since the early 1990s, drug addiction is considered to be a maladaptive disorder owing to persistent exposure to the drug and to the consequent neuronal changes in the critical brain circuitry that are involved in the resulting dysfunctional behavior (Nestler et al. 1993). Addiction research has therefore provided a model for the study of drug-induced neuroadaptation that not only explains the neurobiological mechanisms of drug abuse and dependence but also might be able to unravel the mechanisms for CNS adaptation in psychiatric disorders. The case of ketamine is paradigmatic since the drug induces plasticity changes after acute administration.

Remarkably, ketamine has a vast array of effects; it is a dissociative anesthetic, but at higher doses, it has psychotomimetic effects; it is also an analgesic (Elia and Tramèr 2005) and an antidepressant. The relatively safe profile of ketamine and its well-characterized receptorial and neurochemical actions make it a specific and feasible investigation tool.

The data described in this chapter suggest that a specific feature of ketamine abuse is the strong association of its psychoactive effects with contextual and conditioned stimuli. This feature may explain the setting-dependent effects of ketamine shown by preclinical studies (as reviewed in Section 14.6) and the maintenance of intermittent use for long periods in ketamine users (Lankenau et al. 2010). Although ketamine impairs cognitive processes at the level of different tasks, it appears that the strengthening of information associated with its administration might be defined as a paradoxical drug-related memory enhancement. In fact, recent data on the effect of ketamine on prediction error tasks support the hypothesis that ketamine may enhance some cognitive processing in reward-related cognitive tasks (Corlett et al. 2013).

In conclusion, research on the effects of ketamine on neuroadaptation in a broader sense, that is, at different levels of analysis, may provide better knowledge for the rapid transfer to prevention and treatment.

References

Ahmed, S.H. 2010. Validation crisis in animal models of drug addiction: Beyond non-disordered drug use toward drug addiction. *Neurosci. Biobehav. Rev.* 35:172–184.

Anagnostaras, S.G., Schallert, T., and Robinson, T.E. 2002. Memory processes governing amphetamine-induced psychomotor sensitization. *Neuropsychopharmacology* 26:703–715.

Arendt-Nielsen, L., Petersen-Felix, S., Fischer, M., Bak, P., Bjerring, P., and Zbinden, A.M. 1995. The effect of N-methyl-D-aspartate antagonist (ketamine) on single and repeated nociceptive stimuli: A placebo-controlled experimental human study. *Anesth. Analg.* 81:63–68.

Badiani, A., Belin, D., Epstein, D., Calu, D., and Shaham, Y. 2011. Opiate versus psychostimulant addiction: The differences do matter. *Nat. Rev. Neurosci.* 12:685–700.

Bai, T.H.Y., Cao, J., Liu, N., Xu, L., and Luo, J.H. 2009. Sexual behavior modulates contextual fear memory through dopamine D1/D5 receptors. *Hippocampus* 19:289–298.

Beaulieu, J.M., and Gainetdinov, R.R. 2011. The physiology, signaling, and pharmacology of dopamine receptors. *Pharmacol. Rev.* 63:182–217.

Belin, D., Jonkman, S., Dickinson, A., Robbins, T.W., and Everitt, B.J. 2009. Parallel and interactive learning processes within the basal ganglia: Relevance for the understanding of addiction. *Behav. Brain Res.* 199:89–102.

Berman, R.M., Cappiello, A., Anand, A., Oren, D.A., Heninger, G.R., Charney, D.S., and Krystal, J.H. 2000. Antidepressant effects of ketamine in depressed patients. *Biol. Psychiatry* 47:351–354.

Beurel, E., Song, L., and Jope, R.S. 2011. Inhibition of glycogen synthase kinase-3 is necessary for the rapid antidepressant effect of ketamine in mice. *Mol. Psychiatry* 16:1068–1070.

Caprioli, D., Celentano, M., Dubla, A., Lucantonio, F., Nencini, P., and Badiani, A. 2009. Ambience and drug choice: Cocaine- and heroin-taking as a function of environmental context in humans and rats. *Biol. Psychiatry* 65:893–899.

Castagné, V., Wolinsky, T., Quinn, L., and Virley, D. 2012. Differential behavioral profiling of stimulant substances in the rat using the LABORAS™ system. *Pharmacol. Biochem. Behav.* 101:553–563.

Cavalleri, L., Bono, F., Tedesco, V., Di Chio, M., Merlo Pich, E., Spano, P.F., Missale, C., Chiamulera, C., and Collo, G. 2013. Ketamine produces structural plasticity of mouse mesencephalic dopaminergic neurons via activation of Akt-mTOR pathway: Role of dopamine D3 receptor. P061 presented at the *Meeting Dopamine 2013*, Alghero, Italy, May 24–28, 2013.

Chatterjee, M., Verma, R., Ganguly, S., and Palit, G. 2012. Neurochemical and molecular characterization of ketamine-induced experimental psychosis model in mice. *Neuropharmacology* 63:1161–1171.

Chen, W., Prithviraj, R., Mahnke, A.H., McGloin, E., Tan, J.W., Goock, A.K., and Inglis, F.M. 2009. AMPA GluR1 and GluR2 receptor subunits regulate dendrite complexity and spine mobility in neurons of the developing neocortex. *Neuroscience* 159:172–182.

Collo, G., Bono, F., Cavalleri, L., Plebani, L., Merlo Pich, E., Millan, M.V., Spano, P.F., and Missale, C. 2012. Pre-synaptic dopamine D3 receptor mediates cocaine-induced structural plasticity in mesencephalic dopaminergic neurons via ERK and Akt pathways. *J. Neurochem.* 120:765–778.

Collo, G., Bono, F., Cavalleri, L., Plebani, L., Mitola, S., Merlo Pich, E., Millan, M.J., Zoli, M., Maskos, U., Spano, P.F., and Missale, C. 2013. Nicotine induced structural plasticity in mesencephalic dopaminergic neurons is mediated by dopamine D3 receptor and Akt-mTOC1 signaling. *Mol. Pharmacol.* 83:1176–1189.

Corlett, P.R., Cambridge, V., Gardner, J.M., Piggot, J.S., Turner, D.C., Everitt, J.C., Arana, F.S., Morgan, H.L., Milton, A.L., Lee, J.L., Aitken, M.R., Dickinson, A., Everitt, B.J., Absalom, A.R., Adapa, R., Subramanian, N., Taylor, J.R., Krystal, J.H., and Fletcher, P.C. 2013. Ketamine effects on memory reconsolidation favor a learning model of delusions. *PLoS One* 8:e65088. doi: 10.1371/journal.pone.0065088.

De Luca, M.T., and Badiani, A. 2011. Ketamine self-administration in the rat: Evidence for a critical role of setting. *Psychopharmacology (Berl.)* 214:549–556.

De Roo, M., Klause, P., Briner, A., Nikonenko, I., Mendez, P., Dayer, A., Kiss, J.Z., Muller, D., and Vutskits, L. 2009. Anesthetics rapidly promote synaptogenesis during a critical period of brain development. *PLoS One* 4:e7043. doi: 10.1371/journa.pone.0007043.

Dietz, D.M., Dietz, K.C., Nestler, E.J., and Russo, S.J. 2009. Molecular mechanisms of psychostimulant-induced structural plasticity. *Pharmacopsychiatry* 42:S69–S78.

Duan, T.T., Tan, J.W., Yuan, Q., Cao, J., Zhou, Q.X., and Xu, L. 2013. Acute ketamine induces hippocampal synaptic depression and spatial memory impairment through dopamine D1/D5 receptors. *Psychopharmacology (Berl.)* 228:451–461.

Duman, R.L., and Monteggia, M. 2006. A neurotrophic model for stress-related mood disorders. *Biol. Psychiatry* 59:1116–1127.

Duman, R.S., and Bhavya, V. 2012. Signaling pathways underlying the patho-physiology and treatment of depression: Novel mechanisms for rapid-acting agents. *Trends Neurosci.* 35:47–56.

Duman, R.S., Li, N., Liu, R.J., Duric, V., and Aghajanian, G. 2012. Signaling pathways underlying the rapid antidepressant actions of ketamine. *Neuropharmacology* 62:35–41.

Dunbar, M.S., Scharf, D., Kirchner, T., and Shiffman, S. 2010. Do smokers crave cigarettes in some smoking situations more than others? Situational correlates of craving when smoking. *Nicotine Tob. Res.* 2:226–234.

Duncan, G.E., Miyamoto, S., Leipzig, J.N., and Lieberman, J.A. 1999. Comparison of brain metabolic activity patterns induced by ketamine, MK801 and amphetamine in rats: Support for NMDA receptor involvement in responses to subanesthetic dose of ketamine. *Brain. Res.* 843:171–183.

Duncan, G.E., Moy, S.S., Knapp, D.J., Mueller, R.A., and Breese, G.R. 1998. Metabolic mapping of the rat brain after subanesthetic doses of ketamine: Potential relevance to schizophrenia. *Brain Res.* 787:181–190.

Elia, N., and Tramèr, M.R. 2005. Ketamine and postoperative pain—A quantitative systematic review of randomised trials. *Pain* 113:61–70.

Epstein, D.H., Willner-Reid, J., Vahabzadeh, M., Mezghanni, M., Lin, J.L., and Preston, K.L. 2009. Real-time electronic diary reports of cue exposure and mood in the hours before cocaine and heroin craving and use. *Arch. Gen. Psychiatry* 66:88–94.

Everitt, B.J., and Robbins, T.W. 2005. Neural systems of reinforcement for drug addiction: From actions to habits to compulsion. *Nat. Neurosci.* 8:1481–1489.

Garcia, L.S.B., Comin, C.M., Valvassori, S.S., Réus, G.Z., Barbosa, L.M., Andreazza, A.C., Stertz, L., Fries, G.R., Gavioli, E.C., Kapczinski, F., and Quevedo, J. 2008. Acute administration of ketamine induces antidepressant-like effects in the forced swimming test and increases BDNF levels in the rat hippocampus. *Prog. Neuropsychopharmacol. Biol. Psychiatry* 32:140–144.

Garthwaite, J. 1991. Glutamate, nitric oxide and cell-cell signalling in the nervous system. *Trends Neurosci.* 14:60–67.

Gass, P., Herdegen, T., Bravo, R., and Kiessling, M. 1993. Induction and suppression of immediate early genes in specific rat brain regions by the non-competitive N-methyl-D-aspartate receptor antagonist MK-801. *Neuroscience* 53:749–758.

Goldstein, R.Z., and Volkow, N.D. 2002. Drug addiction and its underlying neuro-biological basis: Neuroimaging evidence for the involvement of the frontal cortex. *Am. J. Psychiatry* 159:1642–1652.

Gonzales, J.N., Loeb, A.L., Reichard, P.S., and Irvine, S. 1995. Ketamine inhibits glutamate-, N-methyl-D-aspartate-, and quisqualate-stimulated cGMP production in cultured cerebral neurons. *Anesthesiology* 82:205–213.

Gordh, T., Karlsten, R., and Kristensen, J. 1995. Intervention with spinal NMDA, adenosine, and NO systems for pain modulation. *Ann. Med.* 27:229–234.

Guirimand, F., Dupont, X., Brasseur, L., Chauvin, M., and Bouhassira, D. 2000. The effects of ketamine on the temporal summation (wind-up) of the R(III) noci-ceptive flexion reflex and pain in humans. *Anesth. Analg.* 90:408–414.

Hasselmo, M.E. 2006. The role of acetylcholine in learning and memory. *Curr. Opin. Neurobiol.* 16:710–715.

Hoeffer, C.A., and Klann, E. 2010. mTOR signaling: At the crossroads of plasticity, memory and disease. *Trends Neurosci.* 33:67–75.

Ibla, J.C., Hayashi, H., Bajic, D., and Soriano, S.C. 2009. Prolonged exposure to ketamine increases brain derived neurotrophic factor levels in developing rat brains. *Curr. Drug Saf.* 4:11–16.

Ikemoto, S. 2007. Dopamine reward circuitry: Two projection systems from the ventral midbrain to the nucleus accumbens-olfactory tubercle complex. *Brain Res. Rev.* 56:27–78.

Janson, A.M., Fuxe, K., Agnati, L.F., Kitayama, I., Harfstrand, A., Andersson, K., and Goldstein, M. 1988. Chronic nicotine treatment counteracts the disappearance of tyrosine-hydroxylase-immunoreactive nerve cell bodies, dendrites and terminals in the mesostriatal dopamine system of the male rats after partial hemitransection. *Brain Res.* 455:332–345.

Jay, T.M. 2003. Dopamine: A potential substrate for synaptic plasticity and memory mechanisms. *Prog. Neurobiol.* 69:375–390.

Jourdi, H., Hsu, Y.T., Zhou, M., Qin, Q., Bi, X., and Baudry, M. 2009. Positive AMPA receptor modulation rapidly stimulates BDNF release and increases dendritic mRNA translation. *J. Neurosci.* 29:8688–8697.

Kalivas, P.W., and O'Brien, C.P. 2008. Drug addiction as a pathology of staged neuroplasticity. *Neuropsychopharmacology* 33:166–180.

Kapur, S., and Seeman, P. 2002. NMDA receptor antagonists ketamine and PCP have direct effects on the dopamine D(2) and serotonin 5-HT(2)receptors-implications for models of schizophrenia. *Mol. Psychiatry* 7:837–844.

Kegeles, L.S., Abi-Dargham, A., Zea-Ponce, Y., Rodenhiser-Hill, J., Mann, J.J., Van Heertum, R.L., Cooper, T.B., Carlsson, A., and Laruelle, M. 2000. Modulation of amphetamine-induced striatal dopamine release by ketamine in humans: Implications for schizophrenia. *Biol. Psychiatry* 48:627–640.

Klann, E., and Dever, T.E. 2004. Biochemical mechanisms for translational regulation in synaptic plasticity. *Nat. Rev. Neurosci.* 5:931–942.

Kohrs, R., and Durieux, M.E. 1998. Ketamine: Teaching an old drug new tricks. *Anesth. Analg.* 87:1186–1193.

Kress, H.G. 1994. Actions of ketamine not related to NMDA and opiate receptors. *Anaesthesist* 43:15–24.

Krishnan, V., and Nestler, E.J. 2008. The molecular neurobiology of depression. *Nature* 455:894–902.

Kumar, V., Zhang, M.X., Swank, M.W., Kunz, J., and Wu, G.Y. 2005. Regulation of dendritic morphogenesis by Ras-PI3K-Akt-mTOR and Ras-MAPK signaling pathways. *J. Neurosci.* 25:11288–11299.

Lankenau, S.E., Bloom, J.J., and Shin, C. 2010. Longitudinal trajectories of ketamine use among young injection drug users. *Int. J. Drug Policy.* 21:306–314.

Li, N., Lee, B., Liu, R.-J., Banasr, M., Qwyer, J.M., Iwata, M., Li, X.-Y., Aghajanian, G., and Duman, R.S. 2010a. mTOR-dependent synapse formation underlies the rapid antidepressant effects of NMDA antagonists. *Science* 329:959–964.

Li, Q., Cheung, C., Wei, R., Cheung, V., Hui, E.S., You, Y., Wong, P., Chua, S.E., McAlonan, G.M., and Wu, E.X. 2010b. Voxel-based analysis of postnatal white matter microstructure in mice exposed to immune challenge in early or late pregnancy. *Neuroimage* 52:1–8.

Li, X., and Jope, R.S. 2010. Is glycogen synthase kinase-3 a central modulator in mood regulation? *Neuropsychopharmacology* 35:2143–2154.

Li, Y., Luikart, B.W., Birnbaum, S., Chen, J., Know, C.-H., Kernie, S.G., Bassel-Duby, R., and Parada, L.F. 2008. TrkB regulates hippocampal neurogenesis and governs sensitivity to antidepressive treatment. *Neuron* 59:339–412.

Liao, Y., Tang, J., Ma, M., Wu, Z., Yang, M., Wang, X., Liu, T., Chen, X., Fletcher, P.C., and Hao, W. 2010. Frontal white matter abnormalities following chronic ketamine use: A diffusion tensor imaging study. *Brain* 133:2115–2122.

Lindefors, N., Barati, S., and O'Connor, W.T. 1997. Differential effects of single and repeated ketamine administration on dopamine, serotonin and GABA transmission in rat medial prefrontal cortex. *Brain Res.* 759:205–212.

Lipton, S.A. 2004. Failures and successes of NMDA receptor antagonists: Molecular basis for the use of open-channel blockers like memantine in the treatment of acute and chronic neurologic insults. *NeuroRx* 1:101–110.

Lorrain, D.S., Baccei, C.S., Bristow, L.J., Anderson, J.J., and Varney, M.A. 2003. Effects of ketamine and N-methyl-ᴅ-aspartate on glutamate and dopamine release in the rat prefrontal cortex: Modulation by a group II selective metabotropic glutamate receptor agonist LY379268. *Neuroscience* 117:697–706.

Marin, P., Quignard, J.F., Lafon-Cazal, M., and Bockaert, J. 1993. Non-classical glutamate receptors, blocked by both NMDA and non-NMDA antagonists, stimulate nitric oxide production in neurons. *Neuropharmacology* 32:29–36.

Markou, A., and Paterson, N.E. 2009. Multiple motivational forces contribute to nicotine dependence. *Nebr. Symp. Motiv.* 55:65–89.

Martin, D.C., Adams, R.J., and Watkins, C.A. 1988. Inhibition of synaptosomal serotonin uptake by Ketalar. *Res. Commun. Chem. Pathol. Pharmacol.* 62:129–132.

Martin, D.C., Introna, R.P., and Aronstam, R.S. 1990. Inhibition of neuronal 5-HT uptake by ketamine, but not halothane, involves disruption of substrate recognition by the transporter. *Neurosci. Lett.* 112:99–103.

Meyer, P.J., and Phillips T.J. 2003. Sensitivity to ketamine, alone or in combination with ethanol, is altered in mice selectively bred for sensitivity to ethanol's locomotor effects. *Alcohol Clin. Exp. Res.* 27:1701–1709.

Mion, G., and Villevieille, T. 2013. Ketamine pharmacology: An update (pharmacodynamics and molecular aspects, recent findings). *CNS Neurosci. Ther.* 19:370–380.

Moghaddam, B., Adams, B., Verma, A., and Daly, D. 1997. Activation of glutamatergic neurotransmission by ketamine: A novel step in the pathway from NMDA receptor blockade to dopaminergic and cognitive disruption associated with the prefrontal cortex. *J. Neurosci.* 17:2921–2927.

Moghaddam, B., and Krystal, J.H. 2012. Capturing the angel in "angel dust": Twenty years of translational neuroscience studies of NMDA receptor antagonists in animals and humans. *Schizophr. Bull.* 38:942–949.

Morgan, C.J., and Curran, H.V. 2012. Ketamine use: A review. *Addiction* 107:27–38.

Mueller, D., Chapman, C.A., and Stewart, J. 2006. Amphetamine induces dendritic grow in ventral tegmental area dopaminergic neurons in vivo via basic fibroblast grow factor. *Neuroscience* 137:727–735.

Nakao, S., Nagata, A., Miyamoto, E., Masuzawa, M., Murayama, T., and Shingu, K. 2003. Inhibitory effect of propofol on ketamine-induced c-Fos expression in the rat posterior cingulate and retrosplenial cortices is mediated by GABAA receptor activation. *Acta Anaesthesiol. Scand.* 47:284–290.

Narendran, R., Frankle, W.G., Keefe, R., Gil, R., Martinez, D., Slifstein, M., Kegeles, L.S., Talbot, P.S., Huang, Y., Hwang, D.R., Khenissi, L., Cooper, T.B., Laruelle, M., and Abi-Dargham, A. 2005. Altered prefrontal dopaminergic function in chronic recreational ketamine users. *Am. J. Psychiatry* 162:2352–2359.

Nestler, E.J., Hope, B.T., and Widnell, K.L. 1993. Drug addiction: A model for the molecular basis of neural plasticity. *Neuron* 11:995–1006.

Nishimura, M., and Sato, K. 1999. Ketamine stereoselectively inhibits rat dopamine transporter. *Neurosci. Lett.* 274:131–134.

Nishimura, M., Sato, K., Okada, T., Yoshiya, I., Schloss, P., Shimada, S., and Tohyama, M. 1998. Ketamine inhibits monoamine transporters expressed in human embryonic kidney 293 cells. *Anesthesiology* 88:768–774.

Robinson, T.E., and Berridge, K.C. 1993. The neural basis of drug craving: An incentive-sensitization theory of addiction. *Brain Res. Brain Res. Rev.* 18:247–291.

Robinson, T.E., and Kolb, B. 2004. Structural plasticity associated with exposure to drug of abuse. *Neuropharmacology* 47:33–46.

Russo, S.J., Dietz, D.M., Dumitriu, D., Morrison, J.H., Malenka, R.C., and Nestler, E.J. 2010. The addicted synapse: Mechanisms of synaptic and structural plasticity in nucleus accumbens. *Trends Neurosci.* 33:267–276.

Sarti, F., Borgland, S.L., Kharazia, V.N., and Bonci, A. 2007. Acute cocaine exposure alters spine density and long-term potentiation in the ventral tegmental area. *Eur. J. Neurosci.* 26:749–756.

Schifano, F., Corkery, J., Oyefeso, A., Tonia, T., and Ghodse, A.H. 2006. Trapped in the "K-hole": Overview of deaths associated with ketamine misuse in the UK (1993–2006). *J. Clin. Psychopharmacol.* 28:114–116.

Schmid, R.L., Sandler, A.N., and Katz, J. 1999. Use and efficacy of low-dose ketamine in the management of acute postoperative pain: A review of current techniques and outcomes. *Pain* 82:111–125.

Shaham, Y., Shalev, U., Lu, L., De Wit, H., and Stewart, J. 2003. The reinstatement model of drug relapse: History, methodology and major findings. *Psychopharmacology (Berl.)* 168:3–20.

Slipczuk, L., Bekinschtein, P., Katche, C., Cammarota, M., Izquierdo, I., and Medina, J.H. 2009. BDNF activates mTOR to regulate GluR1 expression required for memory formation. *PLoS One* 4:e6007. doi: 10.1371/journal.pone.0006007.

Snyder, G.L., Galdi, S., Hendrick, J.P., and Hemmings, H.C. 2007. General anesthetics selectively modulate glutamatergic and dopaminergic signaling via site-specific phosphorylation in vivo. *Neuropharmacology* 53:619–630.

Sussman, S., Ames, S.L., Dent, C.W., and Stacy, A.W. 2001. Self-reported high-risk locations of drug use among drug offenders. *Am. J. Drug Alcohol Abuse* 27:281–299.

Sussman, S., Stacy, A.W., Ames, S.L., and Freedman, L.B. 1998. Self-reported high-risk locations of adolescent drug use. *Addict. Behav.* 23:405–411.

Suzuki, T., Aoki, T., Kato, H., Yamazaki, M., and Misawa, M. 1999. Effects of the 5-ht(3) receptor antagonist ondansetron on the ketamine- and dizocilpine-induced place preferences in mice. *Eur. J. Pharmacol.* 385:99–102.

Takei, N., Inamura, N., Kawamura, M., Namba, H., Hara, K., Yonezawa, K., and Nawa, H. 2004. Brain-derived neurotrophic factor induces mammalian target of rapamycin-dependent local activation of translation machinery and protein synthesis in neuronal dendrites. *J. Neurosci.* 24:9760–9769.

Tan, S., Lam, W.P., Wai, M.S.M., Yu, W.-H.A., and Yew, D.T. 2012. Chronic ketamine administration modulates midbrain dopamine system in mice. *PLoS One* 7:e43947. doi: 10.1371/journal.pone.0043947.

Tedesco, V., Ravagnani, C., Bertoglio, D., and Chiamulera, C. 2013. Acute ketamine-induced neuroplasticity: Ribosomal protein S6 phosphorylation expression in drug addiction-related rat brain areas. *Neuroreport* 24:388–393.

Tiberi, M., Jarvie, K.R., Silvia, C., Falardeau, P., Gingrich, J.A., Godinot, N., Bertrand, L., Yang-Feng, T.L., Fremeau, R.T., and Caron, M.G. 1991. Cloning, molecular characterization, and chromosomal assignment of a gene encoding a second

D1 dopamine receptor subtype: Differential expression pattern in rat brain compared with the D1A receptor. *Proc. Natl. Acad. Sci. U.S.A.* 88:7491–7495.

Tizabi, Y., Bhatti, B.H., Manaye, K.F., Das, J.R., and Akinfiresoye, L. 2012. Antidepressant-like effects of low ketamine dose is associated with increased hippocampal AMPA/NMDA receptor density ratio in female Wistar-Kyoto rats. *Neuroscience* 213:72–80.

Trujillo, K.A., Zamora, J.J., and Warmoth, K.P. 2008. Increased response to ketamine following treatment at long intervals: Implications for intermittent use. *Biol. Psychiatry* 63:178–183.

Tzschentke, T.M. 2007. Measuring reward with the conditioned place preference (CPP) paradigm: Update of the last decade. *Addict. Biol.* 12:227–462.

Uchihashi, Y., Kuribara, H., Morita, T., and Fujita, T. 1993. The repeated administration of ketamine induces an enhancement of its stimulant action in mice. *Jpn. J. Pharmacol.* 61:149–151.

Vallone, D., Picetti, R., and Borrelli, E. 2000. Structure and function of dopamine receptors. *Neurosci. Biobehav. Rev.* 24:125–132.

Vutskits, L., Gascon, E., Tassonyi, E., and Kiss, J.Z. 2006. Effect of ketamine on dendritic arbor development and survival of immature dopaminergic neurons in vitro. *Toxicol. Sci.* 91:540–549.

Wang, X., Yang, Y., Zhou, X., Wu, J., Li, J., Jiang, X., Qu, Q.C., Liu, L., and Zhou, S. 2011. Propofol pretreatment increases antidepressant-like effects induced by acute administration of ketamine in rats receiving forced swimming test. *Psychiatry Res.* 185:248–253.

Wiley, J.L., Evans, R.L., Grainger, D.B., and Nicholson, K.L. 2008. Age-dependent differences in sensitivity and sensitization to cannabinoids and 'club drugs' in male adolescent and adult rats. *Addict. Biol.* 13(3-4):277–286.

Wood, P.L., Emmett, M.R., Rao, T.S., Cler, J., Mick, S., and Iyengar, S. 1990. Inhibition of nitric oxide synthase blocks N-methyl-D-aspartate-, quisqualate-, kainate-, harmaline-, and pentylenetetrazole-dependent increases in cerebellar cyclic GMP in vivo. *J. Neurochem.* 55:346–348.

Yamamoto, S., Ohba, O., Nishiyama, S., Harada, N., Kakiuchi, T., Tsukada, H., and Domino, E.F. 2013. Subanesthetic doses of ketamine transiently decrease serotonin transporter activity: A PET study in conscious monkeys. *Neuropsychopharmacology* 38:2666–2674. doi: 10.1038/npp.2013.176.

Yu, H., Li, Q., Wang, D., Shi, L., Lu, G., Sun, L., Wang, L., Zhu, W., Mak, Y.T., Wong, N., Wang, Y., Pan, F., and Yew, D.T. 2012. Mapping the central effects of chronic ketamine administration in an adolescent primate model by functional magnetic resonance imaging (fMRI). *Neurotoxicology* 33:70–77.

Zarate, C.A., Singh, J., and Manji, H.K. 2006. Cellular plasticity cascades: Targets for the development of novel therapeutics for bipolar disorder. *Biol. Psychiatry* 59:1006–1020.

chapter fifteen

The influence of ketamine on our understanding of depression

Andrew M. Perez

Contents

15.1 Major depressive disorder

Major depressive disorder (MDD) is a ubiquitous, heterogeneous clinical syndrome, characterized by the core symptoms of low mood or loss of interest in the environment, in addition to other features ranging from changes in energy level, sleep, appetite, sexual interest, psychomotor function, and cognition (American Psychiatric Association, Task Force on DSM-IV 2000). Throughout the world, MDD is one of the most disabling medical illnesses. Estimates suggest that it accounts for 65.5 million disability-adjusted life years and is ranked third among the leading causes of global health burden (World Health Organization 2008).

The World Health Organization reports that in the West, depression overtook hypertension as the leading cause of disability, and it is presumed that this will be the case throughout the world by 2020 (Simon 2003).

Although treatment options for depression include psychotherapy, counseling, meditation, and electroconvulsive therapy (ECT), for the remainder of this chapter, "treatment" will refer to pharmaceutical medications used to manage the disorder. Dr. Shekhar Saxena, Director of the Mental Health and Substance Abuse at the World Health Organization states that less than 10% of people with the disease obtain or have access to treatment. Individuals privileged enough to have access to mental health providers and treatment must wait weeks to months from treatment initiation for response, and of these patients, an estimated 15%–30% do not respond and are termed "treatment resistant." The clinical significance of the delay in drug response after initiation of antidepressant therapy is the increased incidence of suicidal ideation and behavior during this period, especially in adolescents and young adults. As MDD continues to grow, the search for a therapeutic intervention with rapid onset and a low suicide risk is even more critical.

Ketamine was formulated during the "birth" of psychopharmacology and is classically described as a dissociative anesthetic. In recent years, ketamine has seen an increase in off-label uses, including chronic pain, such as complex regional pain syndrome and cancer-related pain, as well as acute pain experienced on military frontlines. It is in the field of psychology, however, that ketamine has had a remarkable rebirth. "Science" calls this new awareness of the effects of ketamine, "arguably the most important discovery in half a century" of research on depression. The antidepressant effects are seen rapidly, often within the first few hours after treatment, and the duration persists for days to as long as several weeks. Although traditional antidepressants have focused on normalizing monoamine neurotransmitter levels, ketamine's antidepressant effect via the N-methyl-D-aspartate (NMDA)/glutamatergic pathway is establishing a new platform for clinical and pharmacologic research and development.

This chapter will discuss the monoaminergic system and the current antidepressant medications that focus on this pathway and their inherent limitations. We will summarize the current neurobiological understanding of ketamine's antidepressant properties as seen through animal and human research. We will review the human studies and discuss the safety and efficacy of ketamine as an antidepressant, and we will detail where the research is headed in the area of depression.

15.2 *Monoaminergic hypothesis of depression*

Current antidepressant medications can be said to lay their foundation on the monoaminergic hypothesis of the disorder, theorizing that low synaptic levels of norepinephrine and serotonin, whether via rapid metabolism, impaired storage, or excessive reuptake, are causal factors in depression. The hypothesis began from the incidental scientific finding that drugs that affected the monoaminergic system, such as isoniazid and reserpine, could treat or cause depression, respectively. These findings lead to the development of drugs that alter the levels of serotonin, dopamine, and norepinephrine. Medications that focus on this system include monoamine oxidase inhibitors (MAOIs), tricyclic antidepressants (TCAs), and selective serotonin reuptake inhibitors (SSRIs).

15.2.1 Monoamine oxidase inhibitors

MAOIs were developed during the "explosive birth" of psychopharmacology during the 1950s. Originally formulated MAOIs would irreversibly bind and inhibit monoamine oxidase (both A and B isoforms), leading to an increase in monoamine availability (norepinephrine, serotonin, dopamine, etc.). Not surprisingly, these older MAOIs carried serious adverse effects related to the increased circulating levels of these monoamines. Patients who were prescribed MAOIs were required to adhere to a strict dietary plan (avoidance of tyramine-containing foods, such as cheeses, red wines, and certain meats) while taking the medication. MAOIs' hazardous reputation is further cemented by the case involving Libby Zion, where the drug interaction between the MAOI, phenelzine, and the opiate, meperidine, led to the development of serotonin syndrome, a hypertensive crisis that resulted in cardiac arrest, circulatory failure, and resultant death. The Libby Zion Law placed a regulation on physician-training work hours to 80 hours a week, and no more than 24 consecutive hours in New York State (Brody 2007). Despite newer selective MAOIs that bind reversibly and have less hypertensive and dietary cautions, concern regarding their use still exists among medical providers. For this reason, MAOIs are often limited to atypical depression and treatment-resistant depression (TRD).

15.2.2 Tricyclic antidepressants

TCAs' history first commences with the development of chlorpromazine, a synthetic antihistamine, later noted in a Parisian hospital ward to have antipsychotic effects, characterized by increased energy and mood in depressed patients with previous history of psychomotor retardation. Research chemists then began exploring derivatives of chlorpromazine, leading to the development of the first TCA for the treatment of depression, imipramine. The term "tricyclic" is used because of its molecular three-ring structure. TCAs act via inhibition of reuptake of serotonin and norepinephrine, blocking the serotonin transporter (SERT) and norepinephrine transporter (NET), resulting in an elevation of synaptic concentrations of the neurotransmitters. Although TCAs have largely been supplanted by SSRIs as the first-line treatment for depression, it is important to understand that this is not attributed to differences in efficacy. Rather, TCAs' antimuscarinic side effects, dry mouth, constipation, blurred vision, constipation, orthostatic hypotension, as well as sexual dysfunction, hypersensitivity, and akathisia, proved difficult for many individuals to tolerate. Additionally, in the setting of purposeful or unintentional drug overdose, TCAs carry a greater morbidity and mortality than most SSRIs. TCAs continue to be widely used for a host of disorders, including neuropathic pain, fibromyalgia, and migraine headache prophylaxis. TCAs are still considered the most effective treatment for melancholic depression.

15.2.3   Serotonin reuptake inhibitors

Revolutionizing the field of depression, and certainly helping reduce the negative stigma associated with mental illness, was the age of SSRIs. They are the first class of psychotropic drugs created by the process called "rational design," where the molecule formed is based on targeting a specific biological target and creating a drug to affect it.

These newer antidepressants are valued for their improved safety, reduced number of adverse side effects, and lower risk of lethality after an overdose. SSRIs block the SERT, increasing the levels of serotonin and enhancing its neurotransmission. Although the monoamine hypothesis gave way to the clinical research and development of pharmacologic agents used to understand and treat depression, it is not a full explanation. Despite the fact that neurotransmitter levels quickly adjust after the initiation of treatment with MAOIs, TCAs, and SSRIs, there is a prolonged lag period of several weeks between initiating the medication and the desired antidepressant effect. The lack of correlation between monoaminergic level and antidepressant effect led researchers to look into an alternative explanation for depression, such as nerve growth factors, synaptic formation and remodeling, and neural plasticity.

15.3   Measuring depression severity

There are numerous clinical tools from which to measure and rate the severity of an individual's depression. The Hamilton Depression Rating Scale (HDRS), however, is one such instrument used to assess the severity of depression. Mild to moderate depression is defined as an HDRS score ≤18, severe depression is defined as an HDRS score between 19 and 22, and very severe depression is defined as an HDRS score ≥23. For the purposes of determining clinical efficacy of a drug, an effect size of 0.2 or less is considered small, and an effect size of >0.2 to 0.5 is considered medium. The UK's National Institute of Clinical Excellence set a threshold for clinical significance at an effect size of 0.50 or a drug versus placebo difference of 3 on the HDRS.

In recent years, there have been a number of articles presented in the literature questioning the efficacy of standard antidepressant medications. A recent meta-analysis including more than 100 randomized, placebo-controlled trials (RCTs) of antidepressant agents and 27,127 patients with MDD demonstrated only a 54% response rate to US Food and Drug Administration (FDA)–approved medications versus 37% for placebo (Undurraga and Baldessarini 2012). The number needed to treat (NNT) was determined to be 8, and the response rate was 1.42 (CI, 1.38–1.48). Another meta-analysis of antidepressant efficacy that included 44,240 patients with MDD yielded a response rate similar to the previously mentioned study, with responses of 54.3% and 37.9% for patients who received the active drug and placebo, respectively (Levkovitz et al. 2011). One important point that is worth mentioning from the two large meta-analyses by Undurraga and Baldessarini is that they found a small but statistically significant advantage of older agents, such as TCAs, over newer agents.

With respect to reported data, it is important to consider the possibility of publication bias for RCTs and to note the influence of baseline depression severity on observed efficacy. Kirsch (2010) conducted a meta-analysis of placebo-controlled RCTs of FDA-approved antidepressants on patients with major and minor depressive disorders. The study questioned the efficacy of the newer-class antidepressants, such as fluoxetine and paroxetine (Kirsch et al. 2008). The study looked to determine the antidepressant efficacy on the basis of different levels of initial depression scores. It found that when looking at mild to moderate depression, antidepressants fared no better than placebo. This is based on the following results: the effect size for mild to moderate depression is +0.11 (95% CI, −0.18 to 0.41), the effect size for severe depression is +0.17 (95% CI, 0.22 to 0.71), and the effect size for very severe depression is +0.47 (95% CI, 0.22 to 0.71). The study stated that with respect to patients with mild to moderate depression, "The magnitude of benefit of antidepressant medication compared with placebo…may be minimal or nonexistent" (Kirsch 2010). Fournier et al.'s similar result determined that the NNT for patients with very severe depression is 4, versus an NNT of 16 for patients with mild to moderate depression.

The recent body of literature, however, may not correlate with the true effectiveness of antidepressants from a practical, nonclinical perspective. The most validated study involving nonclinical trial patients comes from the large Sequenced Treatment Alternatives to Relieve Depression (STAR*D) study, which evaluated 4041 outpatients with MDD over an extended period. Following up to 12 weeks of treatment with citalopram, an SSRI, the response and remission rate were 48.6% and 36.8%, respectively (Rush et al. 2006). For patients who did not remit at the previous step, they would undergo several augmentation or switching strategies, yielding response rates of 28.5%, 16.8%, and 16.3% at the second, third, and fourth treatment stages, respectively. However, despite the modest gains of these steps, one-third of patients remained ill after these aug-mentation strategies. Of the strategies utilized, none appeared to have a meaningful advantage over the other for patients who did not remit at a previous step.

A follow-up study to the STAR*D is the Combining Medications to Enhance Depression Outcomes study. It was designed to evaluate the difference between monotherapy with an SSRI versus two antidepres-sant combinations with respect to response and remission rates over the 12-week period. The study involved 665 outpatients with chronic or recur-rent MDD randomized to escitalopram and placebo, escitalopram plus bupropion SR, or venlafaxine XR plus mirtazapine. The study demon-strated that the 12-week response rates varied between 51.6% and 52.4%, not differing significantly between the three groups. Remission rates were also not significantly different, varying between 37.7% and 38.9%. If we scrutinize this information closely, it indicates that the currently available treatments for depression are only modestly effective for patients with severe and very severe depression, yet for patients with mild to moderate depression, they are largely ineffective.

As aforementioned, SSRIs may only offer a statistical benefit in patients with severe and very severe depression. Despite their efficacy in this pop-ulation, 20% of patients remain refractory to treatment with SSRIs. Recent research, however, has shed new light on an old medication and may in fact be as *TIME* magazine asks in the article titled "Ketamine for Depression: The Most Important Advance in Field in 50 Years?" (Szalavitz 2012). Scientists have long known that depression is not simply a problem related to the chemical balance of serotonin and norepinephrine. If this were the case, then SSRIs and TCAs, which rapidly augment the brain's neurotransmitter levels, would take effect immediately rather than weeks to months later. Research is now looking at the role of other pathways to treat depression rather than relying solely on monoamine dysregulation.

15.4   Molecular mechanisms involved in ketamine's antidepressant effects

15.4.1   Synaptogenesis

The mechanism by which ketamine ameliorates depressive symptoms is still unclear; however, preclinical science research is beginning to shed light on the molecular processes that are believed to be associated with the antidepressant effects of ketamine. Recent studies have demonstrated compelling evidence that ketamine enhances the formation of stronger and more numerous synaptic connections in the cortical regions of the brain and that such formation correlates with antidepressant behavior in rodents (Li et al. 2010).

Ketamine also elevates levels of synaptic proteins (postsynaptic density 95, glutamate receptor 1, and synapsin I), increases spine density, and augments the excitatory postsynaptic current responses. These results are also observed by Li et al., as his team sought to observe the effects of ketamine and selective NMDA receptor 2B antagonist Ro-25-6981 on an animal model of chronic unpredictable stress (CUS). Observations from the study demonstrated a decrease in both anhedonic and anxiogenic behavior as well as the biochemical changes mentioned above. Furthermore, the behavioral deficits in the CUS model were also blocked by pretreatment with rapamycin.

Low-dose ketamine increases glutamate signaling in prefrontal cortical regions, causing a watershed of molecular reactions that lead to synaptogenesis and enhanced synaptic functioning.

15.4.2 mTOR, BDNF, and AMPA

There are three metabolic pathways/factors that are believed to be critical for this antidepressant effect of ketamine:

1. The mammalian target of rapamycin (mTOR) signaling pathway
2. The brain-derived neurotrophic factors (BDNFs)
3. The AMPA receptor (AMPAR)

Li et al. reported that ketamine activates the mTOR signaling pathway, leading to a rapid increase in levels of proteins involved in neuron signaling at the synaptic level, in addition to an increase in number and function of new spine synapses in the prefrontal cortex of rodents (Duman et al. 2012). He provided evidence that protein changes at the cellular level correlate with new neural connections and that these changes are associated with behavioral changes in the animal model. It appears that activation of the mTOR signaling pathway also causes transient increases in the levels of phosphorylated eukaryotic initiation factor 4E-binding protein 1, p70S6 kinase, and mTOR. Other important mediators include extracellular signal regulated kinase (ERK and protein kinase B [PKB]/ Akt). Blockade of the mTOR pathway, which can be done by administering rapamycin 30 min before ketamine, seems to block all these effects. In addition, blockade of ERK and PKB abolishes the impact of ketamine on the mTOR signaling pathway, demonstrating the important interplay involved in neuronal plasticity signaling pathways, which may explain the clinical differences between those who respond and those who do not respond to ketamine as an antidepressant (Duman et al. 2012).

Another important receptor/pathway is BDNF. Chronic stress has been shown to result in lower levels of BDNF and is associated in animal studies with depression. BDNF is involved in the proliferation, differentiation, neurogenesis, and apoptosis of neuronal and nonneuronal cells in both the developing brain and the adult brain. With conflicting evidence, the "neurotrophin hypothesis" postulates that lower levels of BDNF are directly related to symptoms of depression and that the increase in BDNF after ketamine administration is responsible for the antidepressant effects. In animal studies, increasing the dose administered leads to higher levels of BDNF activity in the hippocampus, and it correlated clinically to less depressive symptoms (Garcia et al. 2008). Human studies also demonstrate the important role of BDNF, which will be discussed in Section 15.5.

15.5   Human studies

Via a revolutionary, proof-of-concept clinical trial, Berman et al. randomized nine patients to receive either 0.5 mg/kg of ketamine or normal saline, intravenously over 40 min, in a crossover design study. He reported a mean decrease of 14 points on the HDRS in the ketamine arm and a mean decrease of zero in the placebo arm at 72 h posttreatment. Interestingly, the antidepressant effects did not coincide with the neurocognitive (psychomimetic and dissociative) effects that occurred during the infusion, but rather they followed afterward. In a similar, yet larger study, Zarate et al. randomized 18 patients with TRD and reported a large response size when comparing ketamine to placebo, with a response rate of 71% at 24 h and 35% at 1 week. Zarate noted that antidepressant effects were seen as early as 110 min after drug administration and were sustained for approximately 7 days. These studies demonstrate that ketamine exerts its antidepressant effect long after the drug itself is metabolized ($t_{1/2}$ is 180 min in humans). It is believed that by initiating a cascade of events that culminates in increasing neural network density and function, ketamine exerts its antidepressant effect. In one of the largest studies to date on human subjects with TRD by Murrough et al., 73 patients were randomized to receive intravenous ketamine 0.5 mg/kg over 40 min or midazolam as an active placebo. At 24 h, the results demonstrated a 63.8% response rate in the ketamine group compared to 28% in the placebo group (Murrough et al. 2013). The NNT for the study was 2.8. For the sake of comparison, the NNT for placebo-controlled phase 3 FDA administration trials to determine the efficacy of antidepressants is 6–7, and this is for depressed patients who are not considered "treatment resistant." Following the observations of many in various fields, multiple case reports and several small-scale clinical trials of ketamine have added further evidence to the hypothesis that ketamine possesses antidepressant properties (Table 15.1) (Murrough et al. 2011). Furthermore, it is the rapidity with which ketamine treatment ameliorates depressive individuals, effective within 1–3 days as opposed to weeks to months, that has the psychiatric world intrigued. Current antidepressants, such as SSRIs, serotonin–norepinephrine reuptake inhibitors (SNRIs), and MAOIs, are slow to act and are associated with an

Table 15.1 Antidepressant Effects of Ketamine in Clinical Populations

Reference	Sample size	Design	Intervention	Primary finding
aan het Rot et al. 2010	TRD ($N = 10$)	Open label	Ketamine 0.5 mg/kg IV infusion (six doses over 12 days)	Well tolerated; 85% mean reduction in depressive symptoms after six infusions
Berman et al. 2000	MDD ($N = 6$) BD ($N = 1$)	Placebo-controlled, double-blind, crossover	Ketamine 0.5 mg/kg IV infusion (single dose)	Significant improvement in depressive symptoms within 72 h
Diazgranados et al. 2010a	TRD ($N = 33$)	Open label	Ketamine 0.5 mg/kg IV infusion (single dose)	Significant improvement in depression and SI within 4 h
Diazgranados et al. 2010b	TRBD ($N = 18$)	Placebo-controlled, double-blind, crossover add-on study	Ketamine 0.5 mg/kg IV infusion (single dose)	Significant improvement in depressive symptoms within 40 min up to 3 days; overall 71% response
Larkin et al. 2011	Depressed patients presenting in the ED	Open label	Ketamine 0.2 mg/kg IV infusion (single dose)	Significant improvement in depressive symptoms and SI within 2 h and up to 10 days
Machado-Vieira et al.	TRD ($N = 23$)	Open label	Ketamine 0.5 mg/kg IV infusion (single dose)	Significant improvement in depressive symptoms within 4 h; no change in BDNF plasma levels

(Continued)

Table 15.1 (Continued) Antidepressant Effects of Ketamine in Clinical Populations

Reference	Sample size	Design	Intervention	Primary finding
Mathew et al. 2010	TRD (N = 26)	Open-label ketamine followed by double-blind, placebo-controlled riluzole for relapse prevention	Ketamine 0.5 mg/kg IV infusion (single dose); riluzole 100–200 mg p.o. daily	Response at 24 h, 65%; at 72 h, 54% No effect of riluzole on time to relapse
Murrough et al. 2013	TRD (N = 73)	Placebo-controlled, double-blind study	Ketamine 0.5 mg/kg IV infusion or midazolam IV infusion	Response in treatment group, 63.8% Response in placebo group, 28%
Phelps et al. 2009	TRD (N = 26)	Open label	Ketamine 0.5 mg/kg IV infusion (single dose)	Significant improvement in depressive symptoms within 4 h; family history of alcohol dependence predicted greater improvement
Price et al. 2009	TRD (N = 26)	Open label	Ketamine 0.5 mg/kg IV infusion (single dose)	Significant reduction in depression and SI 24 h after ketamine administration
Zarate et al. 2006	TRD (N = 17)	Placebo-controlled, double-blind, crossover	Ketamine 0.5 mg/kg IV infusion (single dose)	Significant improvement in depressive symptoms up to 1 week; 71% response at 24 h

Note: The table describes clinical trials of low-dose ketamine in patients with major depression or bipolar depression (does not include case reports). ED, emergency department; MDD, major depressive disorder; p.o., oral; SI, suicidal ideation; TRD, treatment-resistant depression.

increase in suicidal ideation and action in the short interim period between drug initiation and maximal effect. Researchers observed that ketamine reduces a wide array of depressive symptoms, including anhedonia, low energy, impaired concentration, negative conditions, and sadness, and does not appear to increase the risk of suicidal ideation/behavior during this initial period.

Studies are being conducted with ketamine treatment for acute suicidal ideation in the both the psychiatric ward and the emergency department. Their results may alter the way patients with suicidal ideation are managed in the hospital setting. Investigators have been looking into the prolongation of ketamine's antidepressant effects. Two RCTs involved the use of riluzole, a glutamate release inhibitor, to determine whether it had a role in relapse prevention. Both studies demonstrated a benefit of ketamine on treating patients with TRD; however, they failed to show any statistical significance in patients who had received riluzole versus placebo with respect to time to relapse. The average time to relapse in one of the studies was 22–24.4 days (Mathew et al. 2010).

Using the model of ECT for the treatment of depression, one study observed the effects of repeated exposure to ketamine rather than a single dose. The report followed 10 patients who were treated with ketamine, 9 of whom demonstrated a response. Responders were then treated with five more courses of treatment on a Monday, Wednesday, Friday schedule, over the course of two consecutive weeks. The report demonstrated that the six courses of treatment were tolerated and safe for the participants (aan het Rot et al. 2010). Additionally, it demonstrated a considerably variable duration of effect, with relapse varying widely between subjects. There was a follow-up to the study, which added 14 more patients in addition to the original 10 ($N = 24$); the response rate at 24 h and at the end of the study was 70.8% for patients with TRD. The study demonstrated that response at 4 h was 94% sensitive and 71% specific for predicting the response at the end of the study (Murrough et al. 2012).

In humans, chronic stress is associated with decreases in neurotrophin factors and atrophic changes in the hippocampus, one of the principal areas of emotional response in the brain. Ketamine was found to reverse not only the neuronal changes associated with stress but also the behavioral changes, and the mechanism of these changes is believed to be attributed to activation of the mTOR signaling pathway and stimulation of BDNF (Li et al. 2011). Postmortem evidence in fact has demonstrated lower levels of BDNF in the hippocampus of depressed patients and higher levels in patients who were treated with antidepressants. SSRIs, MAOIs, and ECT have been demonstrated to cause increased blood levels of BDNF. Arguments against the neurotrophin hypothesis, however, are stacking up in the literature. Either the SSRI fluoxetine has no effect or it decreases mRNA for BDNF protein in the rat hippocampus. Other animal studies have demonstrated increased depressive behavior with increased exogenous BDNF, and other studies have demonstrated no correlation with endogenous BDNF and reduction in depressive-like behavior (Groves 2007). Further investigation on the link between depression and BDNF is required.

A recent study of depressed patients utilized magnetoencephalography and found evidence of increased cortical excitability after ketamine that was specific to antidepressant responders, which is potentially consistent with synaptic potentiation from AMPAR-mediated glutamatergic neurotransmission (Cornwell et al. 2012). However, more human studies are needed to determine the antidepressant mechanism of ketamine.

15.6 Current direction of pharmaceutical research

The current state of antidepressants, that being of limited effectiveness, highlights the need for new research and new approaches to treatment and management of MDD. With respect to the new drugs being developed for the treatment of depression, the majority of them seem to focus on the monoaminergic system. These include triple reuptake inhibitors (TRIs), which alter the neurotransmitter levels of serotonin, norepinephrine, and dopamine by blocking the SERT, the NET, and the dopamine transporter, respectively. In the era of SSRIs and SNRIs, TRIs can easily be seen as the next logical sequence of treating depression by engaging the monoamine systems. Unfortunately, there is scarce evidence that SNRIs are more efficacious at treating depression than SSRIs. As of now, the efficacy of TRIs is still yet to be determined. Other agents that are in phase III development include the dual reuptake inhibitor (SERT, NET) levomilnacipran (Forest Laboratories), the selective NET inhibitor LY2216684 (Eli Lilly), and a D2 receptor partial agonist (OPC-34712, Lundbeck).

As ketamine has demonstrated a novel therapeutic way of treating depression, pharmaceutical companies are now turning to the glutamate system as a novel mode of treating depression. Glutamate acts on metabotropic and ionotropic receptors in the brain. The latter group includes the NMDA, AMPA, and kainate receptors. Ketamine is a known antagonist at the NMDA receptor; however, its long-term therapeutic efficacy is hindered by its psychomimetic side effects. Lanicemine (AZD6765), a drug produced by AstraZeneca, is a low trapping NMDAR antagonist, which in phase 2 trials demonstrated efficacy and tolerability with minimal psychomimetic effects when compared to ketamine, in patients with TRD. Ketamine and lanicemine are nonselective antagonists with respect to the various NR2 subtypes, of which there are four (NR 2A, B, C, and D). Because of serious adverse effects, the makers of lanicemine have had to stop further trials. GLYX-13 and NRX-1074, made by Naurex, are partial antagonists at the NR2B receptor. GLYX-13 is a monoclonal antibody–derived peptide and is selected because it is an NMDAR glycine site-specific partial agonist.

GLYX-13 is currently in phase IIb clinical trials, after phase 2a trials demonstrated safety, efficacy, and a rapid onset of antidepressant activity (within hours). Their second-generation molecule is currently in phase I trials and is not only bioavailable after oral administration but also several thousand times more potent than their first-generation molecule. Researchers believe that GLYX-13 via NR2B activation leads to intracellular calcium influx and expression of AMPA, which is responsible for increased synaptic formation between neurons (Burgdorf et al. 2013). These two compounds modulate glutamate metabotropic receptors (mGluR). Traxoprodil is a medication that antagonizes the NMDA 2B receptor specifically. Two other products in development are looking at targeting the mGluR. RG7090 (Roche) is an mGluR5 antagonist and BCI-632 (BrainCells) is an mGluR2/3 antagonist (Murrough and Charney 2012). The momentum for drugs focusing on the glutamate system is based on the recent evidence that glutamate-mediated neuroplasticity may be the final common pathway of action for antidepressants. This will be the subject of the next section; however, for the sake of completeness, other sites of interest for pharmaceutical research on depression besides the glutamate system include the hypothalamic–pituitary–adrenal axis, the galanin neuropeptide system, the melatonin system, and the inflammatory and the hippocampal neurogenesis.

15.7 Conclusions

The discovery of alterations in the density and function of neural and spine synapses represents a sharp divergence from the traditional understanding of depression based solely on the monoaminergic amines. The mechanism that ketamine is able to rapidly increase synaptogenesis and reverse the behavioral deficits seen after exposure to CUS serves as evidence of ketamine's effect, as well as the mechanism by which neuronal atrophy after stress exposure can lead to symptoms of depression. This interplay between genetic predispositions and environmental stress may be a key to understanding how depression develops. Currently in the United States, the FDA has not approved the use of ketamine for the treatment of depression, as there has yet to be a phase 2a trial evaluating the effects of chronic ketamine use for the treatment of depression. There is substantial evidence of the adverse effects of habitual ketamine use, both short term and long term. Animal and human data demonstrate adverse neurologic, cardiac, and genitourinary effects in chronic ketamine use. Whether these effects balance against the potential benefit of treating MDD is still to be determined.

Further research is needed to determine how ketamine's antidepressant effect can be best maintained, whether with daily, weekly, or
monthly administration, and the safety and tolerability of this course
of treatment. The safety of ketamine in certain scenarios is yet to be
studied, including in people younger than 18 years, psychotic forms of
depression, and elderly patients with genitourinary, neurologic, or cardiovascular disease. Further research is needed to determine an alternative to the intravenous drug delivery method, whether transdermal,
intranasal, or oral. Undoubtedly, ketamine's resurrection from obscurity
has been a major breakthrough in the area of depression. The monoaminergic hypothesis is now being questioned, and the glutamatergic
system may be a key component in the development and treatment of
depression.

References

aan het Rot M, Collins KA, Murrough JW et al. 2010. Safety and efficacy of
repeated-dose intravenous ketamine for treatment resistant depression. *Biol
Psychiatry*. 15;67(2):139–145.

American Psychiatric Association, Task Force on DSM-IV. 2000. *Diagnostic and
Statistical Manual of Mental Disorders*, 4th edn, text revision. American Psychiatric Association, Washington, DC.

Berman RM, Cappiello A, Anand A et al. 2000. Antidepressant effects of ketamine
in depressed patients. *Biol Psych*. 15;47(4):351–354.

Brody JE. 2007. A mix of medicines that can be lethal. *New York Times* February 27.

Burgdorf J, Zhang X-L, Nicholson KL et al. 2013. GLYX-13, a NMDA receptor
glycine-site functional partial agonist, induces antidepressant-like effects
without ketamine-like side effects. *Neuropsychopharmacology*. 38:729–742.

Cornwell BR, Slavadore G, Furey M et al. 2012. Synaptic potentiation is critical
for rapid antidepressant response to ketamine in treatment resistant major
depression. *Biol Psychiatry*. 72(7):555–561.

Diazgranados N, Ibrahim LA, Brutsche NE et al. 2010a. Rapid resolution of suicidal ideation after a single infusion of N-methyl-D-aspartate antagonist in
patients with treatment-resistant major depressive disorder. *J Clin Psych*.
71(12):1605–1611.

Diazgranados N, Ibrahim LA, Brutsche NE et al. 2010b. A randomized add-on
trial of an N-Methyl-D-Aspartate antagonist in treatment-resistant bipolar
depression. *Arch Gen Psych*. 67(8):793–802.

Duman RS, Li N, Liu RJ et al. 2012. Signaling pathways underlying the rapid antidepressant actions of ketamine. *Neuropharmacology*. 62(1):35–41.

Garcia LS, Comim CM, Valvassori SS et al. 2008. Acute administration of ketamine induces anti-depressant like effects in the forced swimming test and
increases BDNF in the rat hippocampus. *Prog Neuropsychopharmacol Biol
Psychiatry*. 32(1):140–144.

Groves JO. 2007. Is it time to reassess the BDNF hypothesis of depression? *Mol.
Psychiatry*. 12:1079–1088.

Kirsch I. 2010. Review: Benefits of antidepressants over placebo limited except in
very severe depression. *Evid Based Mental Health*. 13(2):49.

Kirsch I, Deacon BJ, Huedo-Medina TB et al. 2008. Initial severity and anti-depressant benefits: A meta-analysis of data submitted to the Food and Drug Administration. *PLoS Med.* 5(2):e45.

Larkin GL, Beautrais AL. 2011. A preliminary naturalistic study of low-dose keta-mine for depression and suicidal ideation in the emergency department. *Int J Neuropsychopharmacol.* 14(8):1127–1131.

Levkovitz Y, Tedeschini E, Papakostas GI. 2011. Efficacy of antidepressants for dysthymia: A meta-analysis of placebo-controlled randomized trials. *J Clin Psychiatry.* 72:509–514.

Li N, Lee B, Liu RJ et al. 2010. mTOR-dependent synapse formation underlies the rapid antidepressant effects of NMDA antagonists. *Science.* 329(5994):959–964.

Li N, Liu RJ, Dwyer JM et al. 2011. Glutamate N-methyl-D-aspartate recep-tor antagonists rapidly reverse behavioral and synaptic deficits caused by chronic stress exposure. *Biol Psychiatry.* 69:754–761.

Mathew SJ, Murrough JW, aan het Rot M et al. 2010. Riluzole for relapse prevention following intravenous ketamine in treatment resistant depression: A pilot ran-domized, placebo-controlled trial. *Int J Neuropsychopharmacol.* 13(1):71–82.

Murrough JW, Charney DS. 2012. Is there anything really novel on the antidepres-sant horizon? *Curr Psychiatry Rep.* 14(6):643–649.

Murrough JW, Iosifescu DV, Chang LC et al. 2013. Antidepressant efficacy of keta-mine in treatment-resistant major depression: a two site randomized con-trolled trial. *Am J Psychiatry.* 170(10):1134–1142.

Murrough JW, Perez AM, Mathew SJ et al. 2011. A case of sustained remission fol-lowing an acute course of ketamine in treatment resistant depression. *J Clin Psychiatry.* 72:414–415.

Murrough JW, Perez AM, Pillemer S et al. 2013. Rapid and longer term anti-depressant effects of repeated ketamine infusions in treatment-resistant major depression. *Biol Psychiatry.* 15;74(4):250–256.

Phelps LE, Brutsche N, Moral JR et al. 2009. Family history of alcohol dependence and initial antidepressant response to an N-methyl-D-aspartate antagonist. *Biol Psychiatry.* 15;65(2):181–184.

Price RB, Nock MK, Charney DS, Mathew SJ. 2009. Effects of intravenous keta-mine on explicit and implicit measures of suicidality in treatment-resistant depression. 1;66(5):522–526.

Rush AJ, Trivedi MH, Wisniewski SR et al. 2006. Acute and longer term outcomes in depressed outpatients requiring one or several treatment steps: A STAR*D report. *Am J Psychiatry.* 163:1905–1917.

Simon GE. 2003. Social and economic burden of mood disorders. *Biol Psychiatry.* 43(3):208–215.

Szalavitz M. 2012. "Ketamine for depression: The most important advance in field in 50 years"? *Time,* October 5. Available at http://healthland.time.com /2012/10/05/ketamine-for-depression-the-most-important-advance-in-field -in-50-years/ (accessed on December 3, 2012).

Undurraga J, Baldessarini RJ. 2012. Randomized, placebo-controlled trials of anti-depressants for acute major depression: Thirty-year meta-analytic review. *Neuropsychopharmacology.* 37:851–864.

World Health Organization. The global burden of disease: 2004, update 2008.

Zarate CA Jr, Singh JB, Carlson PJ et al. 2006. A randomized trial of an N-methyl-D-aspartate antagonist in treatment-resistant depression. *Arch Gen Psychiatry.* 63(8):856–864.

chapter sixteen

Clinical testing for ketamine

How it inspires the need to develop emerging drugs-of-abuse analysis in a clinical laboratory

Magdalene H.Y. Tang, Calvin Y.K. Chong,
Doris C.K. Ching, and Tony W.L. Mak

Contents

16.1   Introduction

Ketamine is a general anesthetic that is used in both humans and animals. Developed in the 1960s as an analogue of phencyclidine, ketamine was considered a promising anesthetic with a shorter duration of action and better safety profile compared with phencyclidine.[1] However, because of its dissociative and hallucinogenic effects, ketamine soon became a target of abuse. The first report of ketamine abuse appeared in 1965; by the 1990s, its recreational use had become widespread.[2]

Like other drugs of abuse, the analysis of ketamine has a wide range of applications. In the clinical setting, especially in acute intoxications, the identification of the offending agent assists in diagnosis and management of the patient. In such cases, the analysis needs to be rapid, giving results within minutes, and is often performed on the bedside (so called point-of-care testing [POCT]). Known abusers in drug treatment programs also require regular testing in order to monitor their rehabilitation progress; although the speed of analysis is not as critical as that for acute clinical cases, POCT may also be utilized in order to aid the counseling process.

Another large area of application is workplace drug testing, where pre- and post-employment checks help the organization monitor drug use of staff and potential employees. Workplace testing is especially important for certain professions, for example, pilots and train drivers, whose drug use and accompanying impaired mental status may have catastrophic results. On a similar note, school drug testing aims at identifying young abusers for early intervention; moreover, it serves as a preventive tool to discourage students from drug use.

Drug testing also plays a key role in legal and forensic investigations. Roadside testing (driving under the influence of drugs, or drug-driving) is often performed on-site with point-of-care devices, since immediate law enforcement actions are required for presumptive positive results; where screening is positive, samples will be referred to laboratories for further confirmatory testing and legal proceedings. Needless to say, forensic testing demands an extremely high level of standard before, during, and after analytical processes owing to the potential legal consequences. Highly sensitive and accurate methodologies are hence required.

In addition, forensic investigations involve a large variety of biological matrices such as organ tissues, vitreous humor, and muscle. The sampling site and time often greatly influence the interpretation of results and pose great challenges to the forensic toxicologist.

The remainder of this chapter discusses drug testing, in particular ketamine testing, in the clinical context. It should be noted that drug analysis is a complex topic involving different techniques and a wide range of sample types—each serving different purposes—as well as specialized result interpretation expertise and chain-of-custody considerations. Importantly, a positive drug result may have severe legal and social consequences. Hence, results should always be interpreted with proper understanding of the context and purpose for which the drug testing was undertaken.

16.2   Testing methods

16.2.1   Point-of-care testing

16.2.1.1   Application

POCT is generally utilized where the analytical result affects certain critical decisions and rapid results are required. For ketamine, the major clinical application is in the emergency room or in the ambulance, where the presence or absence of the drug can help determine the cause of intoxication and, more importantly, the treatment of the patient. Patients with overdose of ketamine are typically given benzodiazepines or other sedatives. For drugs with antidotes that may be harmful, POCT is especially important.[3]

Another area where POCT is used for clinical purpose is in drug rehabilitation centers or substance abuse clinics, where POCT may be used for on-site assessment of drug abstinence and compliance such that the optimal treatment and counseling may be provided to the patient.

Ketamine analysis by POCT devices is also useful under other contexts. Roadside testing for drug-driving requires rapid results such that the driver with suspected ketamine intoxication can be immediately prohibited from further vehicle control, since ketamine is known to alter the mental and physical state of the driver.[4] In addition, POCT may be utilized in places that lack easy access to testing service but where a drug test result is nevertheless useful for various reasons such as intervention and management, for example, health clinics, schools, or workplace.[5] Many POCT devices are currently available on the market for ketamine. The usual specimen type is urine, although oral fluid is also becoming a popular testing matrix.[4]

16.2.1.2 Principle

POCT devices currently available for ketamine are in the lateral-flow immunochromatography format. The results can be read visually and are produced within a few minutes. Figure 16.1 shows a typical design of a ketamine POCT device.

POCT devices work on the principle of competitive antigen–antibody binding. Within the device are three essential components:

1. Labeled ketamine-specific antibody: typically monoclonal or polyclonal mouse antibodies raised against ketamine. This antibody is usually "tagged" with a visible label such as colloidal gold.
2. Immobilized ketamine: ketamine that is immobilized on the reaction pad, which competes against the ketamine that is present in the subject's urine for binding to ketamine-specific antibody.
3. Control antibody: anti-mouse antibody that serves as a procedural control.

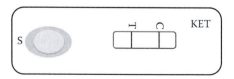

Figure 16.1 A typical design of ketamine POCT device. S, sample area where urine is added; C, control line; T, test line. The result is interpreted according to whether the "T" line is visible. The "C" line serves as a procedural control.

Figure 16.2 illustrates the components of a POCT device and how it works. Urine is added to the sample area; it carries the labeled ketamine-specific antibodies and flows laterally toward the test area, which consists of immobilized ketamine and the control antibodies (Figure 16.2a). In the absence of ketamine in the subject urine, the labeled ketamine-specific antibodies are free to bind to immobilized ketamine in the test area. Together with the control band, where the ketamine-specific antibody is captured by anti-mouse antibody, two bands are visible. This indicates a negative result (Figure 16.2b).

If ketamine is present in the subject urine at a concentration above the cutoff, these drug molecules capture the ketamine-specific antibodies, such that they can no longer bind to the immobilized ketamine in the test area. Therefore, only the control band is visible, indicating a positive result (Figure 16.2c).

The control band serves as an important procedural check and the test is regarded as invalid if the control band cannot be seen, which may indicate that the urine sample is insufficient or has failed to reach the test area.[3]

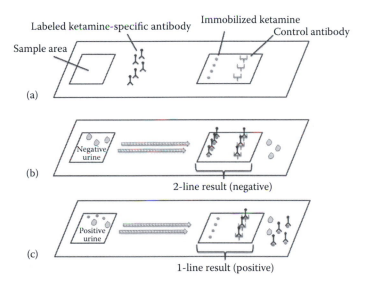

(a)

(b) 2-line result (negative)

(c) 1-line result (positive)

Figure 16.2 Diagram showing the components of a ketamine POCT device and how it works. (a) POCT device typically consists of a sample area, labeled ketamine-specific antibodies, plus immobilized ketamine and control antibodies in the test area. (b) In the absence of ketamine in the subject urine, the ketamine-specific antibodies bind to the immobilized drug. Together with the control line, two bands are visible in the test area, indicating a negative result. (c) In the presence of ketamine, the ketamine-specific antibodies are no longer free to bind to the immobilized ketamine. Only the control band is visible in the test area, indicating a positive result.

POCT devices for ketamine yield qualitative results (a "yes or no" answer) and no quantitative measurements are made on the drug. The qualitative result is based on a "cutoff" concentration, defined by each device manufacturer and below which the drug is regarded as being absent. A commonly used cutoff of ketamine is 1000 ng/mL. This means that a ketamine concentration of >1000 ng/mL in the subject urine will saturate the binding sites of the ketamine-specific antibodies and yield a positive result.[6,7] Since this cutoff is relatively high compared with other analytical techniques (discussed in Sections 16.2.2 to 16.2.5), POCT devices are primarily used to detect acute overdose or recent consumption of ketamine.

As with other immunotechniques, the drug-specific antibodies have varying cross-reactivity toward other compounds. A compound bearing structural similarity to ketamine, for example, phencyclidine, may be recognized by the antibody and produce a "false-positive" result.

Hence, it is important for the POCT device to be assessed for interference by other chemicals, such as commonly used drugs or structurally related compounds, to avoid giving inaccurate results. On the other hand, cross-reactivity allows the antibody to bind to ketamine metabolites, thus enhancing the sensitivity of the device since metabolites are often present in urine at higher levels than the parent drug.[8]

POCT devices for ketamine come in different forms. The device may either be dipped directly into the urine held in a container, or the urine sample may be transferred onto the device by a dropper. In the former case, special containers that detect substitute or "adulterated" urine sample may be used (see Section 16.3.1.5). In addition, devices are also able to detect either single or multiple analytes. A single-analyte device detects only one drug, that is, ketamine, whereas POCT cassettes may be used to detect multiple classes of drug simultaneously, each at their respective cutoff concentrations.

16.2.1.3   Advantages and limitations

The major advantage of POCT is the speed of analysis and improved turnaround time, which allows critical decisions to be made in a timely manner, typically within a few minutes. The analysis is performed on-site, reducing the time needed for transport of samples and results.

Additionally, since the target users come from all walks of life and often do not possess specialized technical skills (e.g., nurses, social workers, and policemen), POCT devices are often designed to be easy and robust to use. The urine sample is directly applied to the device and no sample pretreatment is required. Results are also relatively easy to interpret without the need of expert knowledge in the field.

POCT is an invaluable tool as an alternative testing means to central laboratory analysis. In comparison to the latter, POCT does not require a high capital investment in equipment or instrumentation. It often involves a much smaller amount of paperwork and administrative procedures typically required for test request or reporting in central laboratories. In the context of forensic testing, the chain-of-custody documentation is also reduced. As a relatively accurate result can be obtained rapidly, samples that are screened as negative can be excluded from confirmatory testing, thus reducing the workload of the central laboratory.

Despite its speed of analysis, POCT suffers from the major drawback of providing only preliminary positive results that often require further confirmatory testing. As mentioned previously, cross-reactivity owing to structurally related compounds may yield false-positive results; hence, a more definitive method is needed to confirm the presence of a drug. Moreover, only qualitative results are obtained and further testing is again required for measurement of drug level.

Ideally, POCT yields a simple "yes or no" answer; however, at times, the result may be difficult to interpret, for instance, the ambiguity of whether a "line" is present in the test area. Hence, POCT often involves subjective judgment that may give rise to inconsistencies and errors in result interpretation.

Although POCT does not require capital investment, its consumable cost is high compared with other non-immunoassay techniques. Moreover, the test menu is also limited by the availability of commercial devices. Owing to the reduced requirement on test request and reporting procedures, POCT results may not be properly documented in patient records. In the forensic setting, POCT results are also not accepted by court because of the preliminary nature of the test result and the lack of chain-of-custody documentation. Table 16.1 gives a summary of the advantages and limitations of POCT.

Table 16.1 Advantages and Limitations of POCT

Advantages	Limitations
Reduced turnaround time	Preliminary results that require confirmatory testing
Improved patient management	Qualitative results only
Easy and robust to use; low expertise level required	Inconsistent and possible erroneous interpretation of results
No capital investment	High consumable cost
Reduced amount of paperwork for test request, reporting, and chain-of-custody	Test menu limited by commercially available devices
Screens out negative specimens to reduce confirmatory analyses	Lack of proper documentation of results

16.2.2   Laboratory-based immunoassays

16.2.2.1   Application

Laboratory-based immunoassays for ketamine have not been commercially available until recent years. They are used in settings that require high-throughput and relatively short turnaround times. Although working on a similar immunochemical principle as POCT, laboratory-based immuno-assays differ in that they require more advanced instruments and hence are usually performed in laboratories. Clinical units that require ketamine testing by immunoassays include the psychiatric department to monitor abstinence and compliance of drug abusers, the obstetrics department for evaluation of potential drug abuse of pregnant women, the pediatrics unit to assess the *in utero* exposure of neonates to drugs, and pain management programs to identify possible misuse of drugs.[9] Additionally, workplace and forensic testing may also utilize immunoassays as frontline drug screening. Due to the high volume of samples that can be analyzed each time, immu-noassay is one of the most useful tools in high-throughput drug screening.

16.2.2.2   Principle

Immunoassays utilize an antibody that specifically recognizes and binds to a molecule, or a group of molecules with similar chemical struc-tures. Due to the high specificity and sensitivity of this binding action, antibodies can be used to identify and quantify an antigen of interest when it is present above a particular threshold concentration.[5] Various types of immunoassays are available for different forms of testing. In a competitive immunoassay (limited reagents), the analyte in the subject's sample competes with the labeled analyte for antibody binding; hence, the analyte concentration is inversely proportional to the test signal. On the other hand, in a non-competitive immunoassay (excess reagent, "sandwich" assay), the test signal reflects the analyte concentration in a directly proportional manner.

Heterogeneous assays require the physical separation of bound and unbound antigen in order to measure the amount of analyte present; examples of such assays include radioimmunoassay and enzyme-linked immunosorbent assay (ELISA). A homogeneous assay, on the other hand, does not require this separation ("washing") step and is thus more convenient to use. Examples of homogeneous assays include enzyme-multiplied immunoassay technique, cloned enzyme donor immunoas-say, fluorescence polarization immunoassay, and kinetic interaction of microparticles in solution.[5,10]

Laboratory-based immunoassays for ketamine are mostly in the ELISA format, which is a heterogeneous competitive immunoassay tech-nique. Figure 16.3 illustrates how ELISA works.

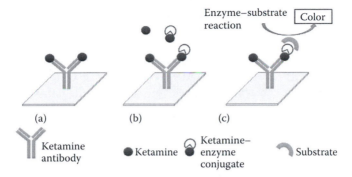

Enzyme–substrate reaction

Color

(a)   (b)   (c)

Ketamine antibody

Ketamine

Ketamine–enzyme conjugate

Substrate

Figure 16.3 Diagram showing how an ELISA for ketamine works. (a) Ketamine antibodies are pre-coated on the sample well. Test sample containing ketamine is added to the well, and the ketamine binds to the antibodies. (b) Ketamine–enzyme conjugate is added and the reaction is incubated. Ketamine present in the sample and the ketamine–enzyme conjugate compete for binding to the antibodies. (c) The wells are washed to remove unbound ketamine and ketamine–enzyme conjugate. Substrate is then added, and the reaction between enzyme/substrate causes color formation, which is read by a spectrophotometer.

Typically, an ELISA for ketamine involves the following:

1. Sample wells are pre-coated with purified antibodies to ketamine or its metabolite, norketamine.
2. The test sample containing free ketamine is added to the wells.
3. Ketamine–enzyme conjugate (ketamine labeled with the enzyme horseradish peroxidase) is then added to the wells.
4. The above reaction mixture is incubated for a period, during which free ketamine present in the sample and the ketamine–enzyme conjugate compete for binding to the antibodies. The more free ketamine present in the test sample, the less ketamine–enzyme conjugate that can be bound.
5. After incubation, the wells are washed to remove any unbound ketamine or ketamine–enzyme conjugate.
6. The substrate is added and incubated in the wells. Reaction between the substrate and the enzyme on the ketamine–enzyme conjugate causes a color to form.
7. After incubation, the reaction is stopped by addition of an acidic stop solution.
8. The resulting color intensity is read by a spectrophotometer. The reading is inversely proportional to the concentration of ketamine in the test sample.

9. For qualitative results, the absorbance of the sample is compared against the "cutoff calibrator"—a higher absorbance indicates a negative result, and vice versa. Specific cutoff levels may be established by individual testing laboratories to define a positive finding. For quantitative results, a calibration curve should be constructed and the sample concentration should be determined using the calibration curve.

Immunoassays operate on a "molecular recognition" principle; hence, compounds with similar chemical structures may also be recognized by the same antibody. Although this cross-reactivity imposes a risk of generating false-positive results, it also has the advantage of being able to recognize metabolites of the specific drug. In the case of ketamine, different ELISA brands have differing cross-reactivity toward the major metabolites of ketamine, which are norketamine and dehydronorketamine. Since metabolites often appear at a later time point than the parent drug and are usually present for longer periods inside the body, assays that cross-react with ketamine metabolites will be able to detect ketamine use even when ketamine is only present at very low levels. These assays are termed to have a longer "detection window." On the other hand, assays that react specifically with ketamine and not its metabolites will give a better estimation of ketamine concentration in the sample, since the measurement is not influenced by the concurrent detection of its metabolites.[8,11,12]

16.2.2.3 Advantages and limitations

One of the major advantages of laboratory-based immunoassays is its high throughput and rapid turnaround time. ELISA is typically performed on 96-well plates, which allows the simultaneous analysis of large numbers of samples. Additionally, the results are simple to interpret and are ready typically within a few hours. Together with reasonable sensitivity and accuracy, these features make immunoassay an excellent screening method to rapidly identify those specimens that require further confirmatory testing. Immunoassays for ketamine have been successfully applied as a screening method in oral fluid and urine before confirmatory testing by mass spectrometry (MS).[13,14]

In addition to high throughput, laboratory-based immunoassays also offer the capability for automation, in particular the homogeneous immunoassays. Large-scale automatic analyzers can be utilized for providing prompt results throughout the day, which is especially important in the emergency setting where results are required within a short period to facilitate rapid diagnosis and treatment decisions. Automation also reduces labor-intensive steps and decreases the chance of human error. In recent years, biochip arrays like the Evidence Investigator® from Randox are able to perform simultaneous and semi-automated analysis of multiple drugs of abuse, including ketamine, using a single sample.[13]

Such approach offers cost- and time-effective means of producing quality results with only minimal sample required.

In comparison to POCT, laboratory-based immunoassays provide more accurate and properly documented results. Since testing is done in the central laboratory, adequate quality controls are normally performed within each batch, thus ensuring the accuracy of results. Tests are performed by qualified laboratory staff and the results are interpreted by trained personnel. Test results are stored in a central database and can usually be retrieved easily upon the need to review data. Test request and reporting are also performed and documented according to the laboratory guidelines. In addition, the laboratory has the capability to perform other tests that verify the authenticity of the urine sample (see Section 16.3.1.5), for example, whether the urine is a substituted or adulterated sample, which is especially important in forensic testing.

Akin to POCT, laboratory-based immunoassays suffer from the following limitations:

- Relatively poor specificity: compounds with similar structures can bind to the antibodies and interfere with the assay. It is often not feasible for a manufacturer to assess the cross-reactivity of every single drug. In particular, designer drugs are continually emerging, many of which bear structural resemblance to conventional drugs of abuse.
- Qualitative results: the results generated are often only qualitative in nature and, at best, semi-quantitative. To measure accurately the concentration of ketamine, more sophisticated methods are required, such as chromatography- or MS-based analysis.
- Commercial availability: the test menu depends solely upon the availability of commercial kits. It is not feasible for a laboratory to develop its own immunoassay method owing to the complexity of generating antibodies.

The major drawback of laboratory-based immunoassays compared with POCT is the complexity of the testing procedure, which delays clinical action. Beginning with test request, sample transport, sample registration through to actual analysis, test reporting, and result documentation, this form of testing involves a much larger amount of time, resources, and manpower. A higher level of expertise is also required. Additionally, testing is affected by the performance of the analyzers, the breakdown of which may result in suspension of service.

16.2.3   Chromatographic techniques

The accurate identification of ketamine requires its physical separation from interfering components in the sample matrix.[10] Separation can be

achieved in different dimensions—by means of sample cleanup to physi-
cally remove the interfering components, as well as by means of separation according to the chemical properties of the compound such as
lipophilicity, boiling point, charge, mass, and so on. An example of the
latter is a technique called chromatography.

Chromatography, as defined by the International Union of Pure and
Applied Chemistry, is a "physical method of separation in which the components to be separated are distributed between two phases, one of which
is stationary (stationary phase) while the other (the mobile phase) moves
in a definite direction."[15] A diversity of chromatographic techniques exists,
including liquid–liquid (partition) chromatography, paper chromatography, thin-layer chromatography, and so on.[10,15] The two most predominant
forms employed for ketamine analysis are gas chromatography (GC) and
liquid chromatography (LC).

After chromatographic separation, ketamine may be detected by a diode
array detector (ultraviolet [UV] absorption),[16–18] nitrogen phosphorus,[19] or
flame ionization detectors.[20] However, by far the most frequently utilized
detection of ketamine is by MS, a universal technique that can be used to
detect a wide variety of compounds as well as provide detailed structural
information. Chromatography in gas (GCMS) and liquid (LCMS) phase
coupled to MS share some common applications and features, which will
be discussed in this section. However, the principles and advantages/
limitations of each methodology are nevertheless method specific and
will be separately discussed in detail in Sections 16.2.4 and 16.2.5.

16.2.3.1 Application

The use of chromatography in the analysis of ketamine dates back to the
1970s[21–23]; since then, it has been used in numerous clinical, forensic, and
research applications.[11,12,17–19,24–70] Chromatographic techniques coupled with
MS are highly specific, sensitive, and definitive. They are often used as a
means of confirmation. Full scan analysis in GCMS, with its highly reproducible and definitive spectral results, is considered a "gold standard" and
the results it yields are accepted by court as forensic evidence. LCMS was
developed relatively recently compared with GCMS but is rapidly gaining
popularity because of its versatility and easy sample preparation. High-
performance liquid chromatography (HPLC), LCMS, and GCMS methods
are commonly employed in clinical and forensic laboratories, where accurate and traceable results are required. On the other hand, because of the
longer time needed for sample preparation and analysis, these technologies are rarely used in settings where rapid results are needed.

Detection by chromatography/MS methods offers several features
that enable the accurate quantitation of ketamine: specificity, accuracy,
precision, and linearity. As such, LCMS and GCMS methods are commonly used for the quantification of ketamine and its metabolites. A range

of quantitative methods has been published for various clinical applications including overdose cases and pharmacokinetic studies.[26,46,49,52,54,57,70] Quantitative analyses are also useful in workplace testing and drug monitoring programs that have cutoff drug concentrations, as well as forensic cases that also require the accurate measurement of the drugs present in the biological specimens of a subject.[11,12,24,25,29,31,34,35,41,42,44,45,47,48,50,52,54,56,62,64,66–69]

Ketamine is metabolized in the body to its active metabolite norketamine. In addition, it is biotransformed to other metabolites such as dehydronorketamine.[5,71] These compounds share similar structures and are difficult to differentiate by molecular recognition techniques. However, this can be achieved by chromatography/MS methods. The differentiation of ketamine from its metabolites is particularly important in pharmacokinetic studies as well as in the elucidation of the metabolic profile in pharmacological evaluations.

LCMS and GCMS are highly versatile methods that enable the simultaneous analysis of a large number and a wide variety of drugs. In drug-of-abuse testing, it is desirable to screen for as many candidates as possible for accurate diagnosis and effective treatment. Immunoassays are typically for single-compound analysis; at best, a group of analytes of similar structures can be detected by the same device (e.g., opiates, benzodiazepines). While multidrug POCT devices are available, their coverage is still far more limited compared with MS methods. A voluminous amount of chromatography/MS methods have been published for the simultaneous detection of ketamine and a large number of drugs/pharmaceuticals. This enables the screening of ketamine in systematic toxicological analyses (or general unknown screening) and the study of coingestion with other drugs of abuse.[25,30,32–34,46–48,50,52,55,56,60–63,65,66,69,72]

16.2.3.2   Principle

The principles involved in LC or GC coupled with MS may be found in detail in Sections 16.2.4 and 16.2.5.

16.2.3.3   Advantages and limitations

The major advantage of chromatographic techniques is the accuracy of its results, in particular when coupled with MS. Unlike immunoassay techniques, LCMS or GCMS methods rarely suffer false positivity because of the separation of ketamine from its interfering components by sample cleanup and chromatography. In MS, the measurement of molecular mass, especially exact mass measurement by high-resolution MS, is much more compound specific compared with molecular recognition techniques or UV detection. In particular, the study of the unique fragmentation pattern in tandem MS offers near unequivocal identification of a compound.[73] Moreover, laboratories utilizing chromatographic/MS methods for analysis often have adequate internal quality control as well as participate regularly in external proficiency testing so as to ensure the accuracy of the method at all times.[74]

The analysis by MS methods does not depend upon the availability of commercial kits. Where reference standards are available, a method can be established relatively easily for the detection of the compound on the chromatograph/mass spectrometer.

Existing methods are easily adapted to incorporate new analytes. In the instance where reference materials are unavailable, it is also possible to identify compounds tentatively by its mass spectrometric properties and comparison with published data.[57]

Compared with immunoassay, chromatography/MS techniques require a higher capital investment cost in the acquisition of the chromatograph and mass spectrometer. More sophisticated mass spectrometers, such as high-resolution or ion trap MS, are particularly expensive. On the other hand, the consumable cost of this technique is lower than that of immunoassay, which requires the costly manufacture of antibodies. The reagents required for LCMS or GCMS analysis are commonly used laboratory chemicals such as solvents, gas, and buffers.[75]

One of the limitations of chromatography/MS methodologies is the long turnaround time. Compared with the simple and fast procedures for POCT and immunoassays, LCMS and GCMS analyses often require lengthy sample preparation steps and long analysis times on the chromatograph. Recent developments have seen the emergence of rapid LCMS methods such as direct injection[47] or dilute-and-shoot[34] of urine samples, although it should be noted that such methods may reduce the lifespan of the analyzer owing to the introduction of complex matrix and contaminants into the system. Ultra-performance LC (UPLC) permits shorter chromatographic run times and is becoming increasingly popular in the analysis of ketamine.[12,39,62]

LCMS and GCMS are highly sophisticated analytical systems that require a high level of expertise in method development, daily operation, and result interpretation. The establishment of the method requires optimization of various chromatographic and MS parameters. Daily maintenance, calibration, and suitability checks are often required to ensure optimal operating condition of the high-end analyzers. Moreover, the results generated require expert review of chromatographic and mass spectrometric findings. All these necessitate ample training for staff and higher operation costs.

16.2.4   *Liquid chromatography–mass spectrometry*

16.2.4.1   *Principle*

The analysis by chromatography and MS requires clean sample material. Hence, before LCMS analysis, a biological sample typically needs to undergo a cleanup process to remove matrix components such as proteins, lipids, and electrolytes that may interfere with the assay. Various cleanup procedures have been utilized for the analysis of ketamine, including liquid–liquid extraction,[17,18,50,51,53,57,63,68–70] solid-phase extraction,[12,28,30,32,33,49,50,59,64,65] protein precipitation or filtration,[26,35,46,56,60] and direct injection with or without prior dilution.[1,47,62] Internal standards are often added during the sample preparation step to compensate for extraction loss and for more accurate quantitation. Deuterium ketamine[35,37,49,53,57] and other structurally related compounds such as bromoketamine[59] have been used as internal standards.

LC involves the partitioning of an analyte between the stationary phase and liquid mobile phase. The stationary phase comprises a cylindrical analytical column packed with particles that "trap" the analytes of interest—the particles can be broadly classified into hydrophobic, hydrophilic, or ionic in nature, or their combination, for analysis of different compounds. Sub-2 μm particles are used in UPLC for faster analysis and higher resolution. The mobile phase typically consists of an aqueous solution and an organic solvent—the combination of which, either in consistent ("isocratic") or varying ("gradient") composition, carries the analyte onto the stationary phase where retention and separation are achieved according to the chemical properties of the compound. Pumps are employed to control the flow and composition of the mobile phases.[75,76]

In reversed-phase LC, which is the most commonly employed form of LC for ketamine, the stationary phase takes the form of an analytical column packed with particles containing hydrophobic alkyl chains. Analytes are injected onto the column by an auto sampler and interact with the stationary phase. Under the flow of the mobile phase, which in "gradient" composition involves increasing solvent strength, the analytes elute at variable time points according to their hydrophobicity, interaction with the solvent, and other physicochemical properties. The time at which an analyte elutes off the column is its "retention time." The higher the hydrophobicity of a compound, the longer its retention time in reversed-phase LC.[75,76,77] After separation by LC, the analytes enter the mass spectrometer for analysis. Different types of mass spectrometers working on various principles are available, including quadrupole mass filters, ion traps, time-of-flight (TOF), orbitrap, and Fourier transform mass analyzers.[10,73,75,78] Those that are most commonly used for the detection of ketamine include the single- or triple-quadrupole analyzers[12,32,46,49–51,57,59–62,64,68,70] and the ion trap analyzer.[26,28,30,37,53,57,63] Other forms of MS that have been utilized for ketamine analysis include the orbitrap, which allows exact mass measurement by high-resolution MS[35,56] as well as the hybrid linear–ion-trap triple-quadrupole (QTrap) analyzer.[33,34]

Before entering the mass spectrometer, the eluent from the LC undergoes nebulization of the liquid to form a fine spray of droplets, removal of solvent, and ionization to form charged analyte ions. This occurs at atmospheric pressure in the interface between the LC and MS. Ionization of analytes is achieved in the ion source. One form of ionization method is atmospheric pressure chemical ionization (APCI). However, by far the majority of ketamine analytical methods utilize electrospray ionization (ESI) technology.

In ESI, the liquid sample flows through a capillary to which is applied a high voltage (~1–5 kV). At the tip of the capillary, the liquid is nebulized into a fine spray of charged droplets that travels toward a counter electrode. The solvent in the sample is further removed by drying gas and heat, ultimately resulting in the formation of charged analyte ions. These analyte ions are transported from atmospheric pressure, through a low-pressure region, toward the mass spectrometer that operates under a high vacuum by the action of vacuum pumps (Figure 16.4). The mass spectrometer measures the mass of a compound by mass filter, ion trap, or TOF mechanisms. The measured mass of a compound is expressed as a mass-to-charge (m/z) ratio.[10,73–75,79] A diagram of the major components of an LCMS system is shown in Figure 16.5.

The mass spectrometer may be used in different modes for the identification of ketamine and its metabolites. The simplest form is the direct

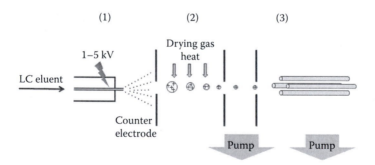

Figure 16.4 Schematic diagram showing the principles of ESI in LCMS analysis. (1) The liquid eluent is passed through a capillary to which is applied a high voltage (1–5 kV). At the tip of the capillary, the liquid forms a fine spray of charged droplets that travel toward a counter electrode. This occurs under atmospheric pressure. (2) Solvent in the sample is removed by drying gas and heat; charged analyte ions are formed and enter the mass analyzer under increasing vacuum. (3) Mass analysis is performed under a high vacuum to avoid the collision of analyte ions with other molecules. A quadrupole analyzer is shown here.

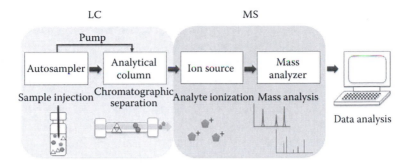

Figure 16.5 Schematic diagram of an LCMS system. A sample is introduced into the liquid chromatograph by autosampler injection. Through the action of the pumps, the analytes in the sample are carried to the analytical column, where chromatographic separation occurs. The separated analytes are then ionized at the ion source, and the charged analyte ions enter the mass spectrometer for analysis. Data are recorded and analyzed by computer and software systems.

monitoring of the nominal mass (m/z) of the compound,[26,49,53,59] although this method potentially suffers from lack of specificity owing to interferences with the same nominal mass. High-resolution MS by TOF and orbitrap analyzers improves the specificity of the detection by measurement of the exact mass, thereby confirming the elemental composition of the compound.[56]

A more advanced form of identification utilizes the unique "fragmentation" pattern of the compound by collision-induced dissociation to form "product ions." A triple-quadrupole mass spectrometer is commonly used for the specific monitoring of the transition between parent ion and its product ion after fragmentation. This technique is termed "multiple reaction monitoring" (MRM) (or "selected reaction monitoring") and is frequently adopted for the identification and quantitation of ketamine and its metabolites.[28,32,46,47,50,51,57,60,63,64,68–70,79] In addition, the relative intensities of the MRM transitions ("ion ratio") may be taken into account for further confirmation. The ion ratio is calculated by using the response of the most abundant transition as base and calculating the response of the next most abundant transition(s) as a percentage of the base.[12,61,62] The Clinical and Laboratory Standards Institute and other authorities have published guidelines for the maximum permitted tolerance of ion ratios in the confirmation of a compound.[74,80,81]

Besides the specific monitoring of parent-to-product transition, the fragmentation of the parent ion produces a "product ion mass spectrum," which may also be used as a means of identification. Upon fragmentation, all the product ions are captured in the product ion spectrum as a "finger-print" of the compound. Reference product ion spectra can be stored in a database and used for compound identification via "library matching" by comparing the sample spectrum with the reference spectrum.[73,74] The library match algorithm has been frequently utilized for the identification of ketamine and its metabolite.[30,33,34,56]

16.2.4.2   Advantages and limitations

The pros and cons of chromatography/MS techniques in general have been discussed in detail in Section 16.2.3.3. In comparison with GCMS, LCMS methodologies have the following advantages:

1. LCMS methods often involve simpler sample preparation steps. Dilute-and-shoot, simple filtration, and direct injection methods have been described for ketamine.[26,34,35,47,56,62] Online solid-phase extraction that yields clean extracts with minimal manual labor is also becoming increasingly popular for drug analysis.[82–84]
2. In LCMS analysis, the mobile phase may be modified in various ways for the optimal detection of different compounds, for example, the pH, organic solvent, buffer salt used and its strength, and the isocratic/gradient composition of the mobile phases.[73] This flexibility allows the analysis of a diverse range of drugs using this technology.
3. LCMS technology has seen the advancement to UPLC with shorter analysis times while retaining resolution and sensitivity. Typical run times for the analysis of ketamine and its metabolites with UPLC range from 2 to 8 min.[12,39,62]

On the other hand, LCMS also suffers from drawbacks in compari-son with GCMS technology. First, competition for ionization between the analyte and matrix components may lead to a decrease in analyte sig-nal (termed "matrix effect") and imprecision of results—ESI is particu-larly susceptible to such phenomenon. Matrix effects may be reduced by improvements in sample cleanup and chromatographic separation. Alternatively, its detrimental effects may be minimized by modifying mass spectrometric parameters, for example, the ionization technique (APCI appears to suffer less matrix effects) or ionization polarity, or by the introduction of a deuterium internal standard that compensates for ion suppression or enhancement effects.[35,37,49,53,57,73,75,77] Second, the product ion spectra produced by LCMS analyzers are often lacking in reproduc-ibility[85]; different analyzer types and collision energies used by labora-tories yield variable spectra, which make library standardization and establishing a universal database difficult. Recent advances have led to the development of sophisticated search algorithms such as SmileMS that may resolve the issue.[86]

16.2.5   Gas chromatography–mass spectrometry

16.2.5.1   Principle

GCMS has long been utilized for the analysis of ketamine in biological samples.[11,24,25,29,31,36,38,41–45,48,52,54,55,58,66,67] GC is used for the analysis of volatile, thermal-stable compounds. Since the analytes are in gas phase, GC is more easily coupled to MS (which also requires analytes to be in gas phase) compared with LC. Akin to LC, GC also comprises stationary and mobile phases. The stationary phase is typically a long coil of capillary column consisting of silica with a coating of a viscous liquid.[87] The main discrepancy between the two techniques is that GC utilizes gaseous mobile phase while LC utilizes liquids.

As with most chromatography methods, samples require a cleanup procedure before analysis. This typically involves liquid–liquid or solid-phase extraction to remove interfering components in the sample matrix. In GCMS analysis, a more recently developed technique has also been used for the analysis of ketamine—solid-phase microextraction, which is an efficient and solvent-less method that utilizes a coated fiber for the extraction (from matrix) and deposition (onto the chromatograph) of analytes.[19,24,36,55,76]

As mentioned above, a compound must be volatile and thermally stable for GCMS analysis. For those that are not, a reaction called derivatization may be used to increase the volatility, thermal stability, and amenability to GCMS analysis of the compounds. Derivatization methods typically include silylation, acylation, alkylation, and formation of cyclic or diastereomeric derivatives.[10,73] Derivatizing agents that have commonly been utilized for ketamine analysis include HFBA (heptafluorobutyric anhydride), PFBC (pentafluorobenzoyl chloride), TFAA (trifluoroacetic anhydride), and MBTFA [N-methyl-bis(trifluoroacetamide)].[31,38,42–45,48,52,66]

The gas chromatograph consists of three essential components—the injector, the gaseous mobile phase, and the stationary phase that is housed in a temperature-controlled oven.

First, the sample is introduced into the system by the injector. Liquid (direct injection) or gaseous (headspace) sampling are possible with different injector devices. Subsequently, the analytes are vaporized at the injection port by high temperatures. They are then swept onto the GC column by an inert carrier gas (the mobile phase) such as helium, nitrogen, or hydrogen. Chromatographic separation occurs in the column where analytes interact with the stationary organic phase under specific column temperatures. Such interactions continue as the carrier gas flushes the analytes down the length of the column, until each analyte elutes at its specific retention time. Similar to LC, the parameters of the GC may be modified for different chromatographic conditions. The commonly manipulated parameters include the oven temperature and carrier gas pressure.[79,87]

Analytes that elute off the GC column enter the mass spectrometer through an interface. They are first ionized at the ion source. The predominant ionization method for ketamine analysis is by electron ionization (EI, also known as electron impact).[11,24,29,31,36,38,42–45,52,54,55,58,66,67] As the gas-phase analytes travel through the ion source, they are bombarded by a stream of 70-eV electrons, which fragment the analyte and cause the loss of an electron from the molecule, thereby forming a positively charged "molecular ion." The ions are then propelled into the mass spectrometer as an ion beam through the action of a repeller and focusing lens. Due to the high energy (70 eV) exerted on the molecules, the analyte often fragments substantially and the parent molecular ion may not be observed in the mass spectrum.[10,87]

Another ionization technique that has been used for the analysis of ketamine, although to a much smaller extent, is chemical ionization (CI).[31,41] This is a "soft" ionization method compared with EI. In a CI source, a reagent gas such as methane is initially ionized; as the ionized gas and sample travel through the CI source together, they collide and the sample molecule becomes charged through energy transfer. As the energy involved is much lower in CI, the analyte does not fragment to a great extent and often the molecular ion can be observed in the mass spectrum to assist in determining the mass of the parent compound.[73,79]

The mass spectrometer coupled to a GC system is typically a quadrupole MS. This type of analyzer, similar to that of the LCMS, contains four rods arranged diagonally. Electric fields are generated among the rods by control of direct current and radio frequency voltages. Consequently, only compounds of a specific m/z value can travel in a stable trajectory toward the detector.[75] The quadrupole MS may be operated in selected ion monitoring (SIM) mode, full scan mode, or both simultaneously. SIM mode, which involves monitoring of certain characteristic ions of a compound throughout the run, has been extensively utilized in the analysis, particularly quantitation, of ketamine.[11,25,29,31,36,38,42–45,48,52,54,55,58,66,67] Full scan mode obtains a complete mass spectrum of a compound for identification purposes. In EI, the constant energy applied (70 eV) produces a highly reproducible mass spectrum of a compound. Similar to the "product ion scan" in LCMS, the mass spectrum in EI mode may be used for spectral matching and library search analysis.[87]

16.2.5.2   Advantages and limitations

The major advantage of GCMS over LCMS analysis is the reproducibility of the mass spectrum. Owing to the universally constant energy (70 eV) applied in EI analysis, the pattern and relative quantity of ions produced for a particular compound are consistent across many brands and models of GCMS analyzers and across different laboratories. This has two distinct advantages. First, the highly consistent and reproducible mass spectrum offers relatively easy and near unequivocal identification of a compound. This feature allows GCMS analysis to continue being the current "gold standard" method, the results of which are generally accepted by court in forensic investigations.[79] Second, the consistency in the mass spectrum obtained at 70 eV also allows universal databases and spectral libraries to be established and used in laboratories around the world. This is particularly useful in helping to identify compounds for which the laboratory does not have the reference standard. However, one should be aware that the spectrum of a given compound is drastically different with or without derivatization, as well as using different derivatization agents.

GCMS analysis also suffers some drawbacks. It may only be used for the analysis of volatile and thermally stable compounds. Although this can be somewhat improved by derivatization, this creates yet another obstacle owing to the time and labor required for such a process.

Derivatization often involves initial extraction of the analytes into an organic medium, which then has to be dried before addition of the derivatizing agent. Careful maneuvers are required to keep moisture to a minimum during the derivatization process, since the derivatizing agents readily reacts with the hydrogens in water.[87] In general, the derivatization process increases exposure to hazardous chemicals and is also difficult to automate.

In summary, a variety of analytical methods are currently available for the determination of ketamine and its metabolites in biological samples. Each methodology has its own strengths, weaknesses, and applications. In order to determine the appropriate method to use and, more importantly, to use the method appropriately, the user should have an understanding of the resources that are available (financial or staff expertise), the setting under which a compound is being tested, and the strengths and limitations of the method applied. Table 16.2 provides a summary of the above for each methodology discussed in this chapter.

16.3   Specimen types for testing

Laboratory testing of ketamine begins with the collection of specimen. Whereas urine has been the de facto standard of routine drug testing for decades, other body fluids such as serum, plasma, whole blood, dried blood spot, saliva, and sweat have been used as matrices for the

Table 16.2 Comparison of the Methodologies for the Analysis of Ketamine and Its Metabolites in Biological Samples

	Applicable settings	Sensitivity	Specificity	Quantitation	Turnaround time	Expertise requirement	Capital requirement	Running cost requirement
Point-of-care testing (POCT)	A&E departments, ambulance, drug rehabilitation clinics, roadside drug testing	+	+	–	Rapid	Low	Low	High
Laboratory-based immunoassay	Psychiatric departments, pediatric testing, pain management programs	++	++	++	Moderate	Medium	Medium	Medium
Liquid chromatography–mass spectrometry (LCMS)	Various clinical and forensic settings	+++	+++	+++	Slow to moderate	High	High	Low
Gas chromatography–mass spectrometry (GCMS)	Various clinical and forensic settings	+++	+++	+++	Slow	High	High	Low

Note: –, Not available; +, least desirable; ++, moderately desirable; +++, most desirable.

detection of drugs of abuse.[57,88-91] Technological advances, most notably the availability of MS in drug testing laboratories, have allowed the use of specimen types such as hair and nail for testing of drugs of abuse including ketamine.[44,67]

The different types of specimens listed above have their advantages and limitations in terms of the ease of collection, risk of adulteration or substitution, handling and transport, stability, drug levels, analytical methods applicable, detection window, and so on. The choice of specimen type could be a balance of these factors depending on the setting requiring drug analysis. Before looking into each individual specimen type, a brief account of ketamine pharmacokinetics is given in the following.

After consumption, ketamine is metabolized into norketamine (through demethylation by the cytochrome P450 system), 6-hydroxynorketamine (through hydroxylation), and dehydronorketamine.[57,92,93] CYP3A4 and CYP2B6 are reported to be the major enzymes responsible for the N-demethylation activity at analgesic, anesthetic, recreational, and toxic doses. CYP3A4 is the high-capacity, low-affinity system, whereas CYP2B6 is the high-affinity, low-capacity one. As a result, the two major enzymes contribute equally at analgesic doses, while at toxic levels, CYP3A4 becomes the major enzyme responsible for demethylation.[94] These metabolites, together with a small amount of parent drug, are excreted renally with or without conjugation.

The pharmacokinetic profile of ketamine has been thoroughly investigated and is described as a two-compartment model.[95,96] The reported terminal half-life of ketamine varied and ranged from 1.67 to 3.33 h, as reviewed by Malinovsky et al.[97] However, in patients with coexisting pathologies such as chronic alcoholism or with the use of enzyme inhibitors, the terminal half-life of ketamine can be up to 4.98 h.[98]

The bioavailability of ketamine has been evaluated for oral, intramuscular, intranasal, and rectal routes. Intramuscular route has the highest bioavailability at 85.9%–97.2%, whereas that of oral route is reported to be 14.5%–24.5%.[99] The bioavailability for ketamine in rectal and intranasal routes evaluated in one study in children were 25% and 51.5%, respectively.[97] For ketamine use through intravenous, intramuscular, or intranasal routes, the level of norketamine remains lower than that of ketamine within the first hour and then rises and peaks later compared with ketamine, whereas after oral use and, to a lesser extent, rectal use, norketamine rises with ketamine and peaks only slightly later, suggesting a significant first-pass metabolism.

16.3.1   Urine

Urine is often considered the most widely used type of specimen in drug testing. As described in the introduction in Section 16.3, ketamine and its metabolites are mostly excreted renally, with well-defined pharmacokinetics for both oral and parenteral administration routes.[96,99] Approximately 90% of a single dose of ketamine is excreted within 72 h, among which a large majority (80%) as conjugated hydroxyl metabolites, 16% as dehydronorketamine, and 2% each as norketamine and ketamine.[100]

16.3.1.1   Application

Urine has been used for ketamine testing in various settings including forensic,[67] drug-driving,[89] and school and clinical contexts, mostly because of its noninvasiveness and ease of collection.

16.3.1.2   Advantages and limitations

Urine as a specimen type for testing of ketamine and other drugs of abuse offers a number of advantages. Urine collection is easy and noninvasive. Besides, the ample volume produced in most patients (with the possible exception of those who have end-stage renal failure) and the relatively high drug concentrations, usually at parts per billion to parts per million levels, allow easy analysis using various techniques. In a study evaluating impairment symptoms with reference to ketamine concentration in oral fluid and urine in subjects attending disco, the urine ketamine concentrations in positive cases ranged from 112 to 14,160 ng/mL.[89] Most drugs and metabolites, including ketamine, are reasonably stable in urine when kept refrigerated or frozen.[101] However, urine collection and testing is not without its limitations. The specimen is relatively easily tampered with by urine adulteration or substitution. Direct observation of urine collection could minimize the risk but would intrude on one's privacy. Furthermore, the correlation between urine drug level and concomitant dosage or pharmacological effect was weak, if any, and difficult to interpret.

The detection window of ketamine and its metabolites in urine depends on a number of factors such as dosage, analytical method used, and individual variability, but is usually in terms of a few days. Ketamine can be detected in urine for 48–72 h after a single dose for most methods. In one study, ketamine and norketamine were measured in urine samples of six children after receiving a single intravenous dose of ketamine as anesthetic using sensitive GCMS and LCMS methods—ketamine and norketamine were detected up to 2 and 14 days, respectively, using GCMS, and up to 11 and 6 days, respectively, using LCMS.[92]

16.3.1.3   Methods of analysis

Ketamine can be detected in urine specimens using various methods including ELISA,[11] HPLC-UV,[16] GCMS,[102] LC-MS/MS,[28] and CE (capillary electrophoresis) coupled with UV or MS.[103] For mass spectrometric applications, depending on the desired sensitivity, different sample preparation procedures like dilute-and-shoot, liquid–liquid extraction, and solid-phase extraction approaches have been described.[104] Unlike other common drugs of abuse like opiates, amphetamines, cocaine, and so on, urine immunoassay kits in automated laboratory analyzers are not readily available for ketamine. ELISA and point-of-care tests for urine ketamine have been developed. The typical cutoff concentration of urine ketamine point-of care tests is 1000 ng/mL[6,7]; false positives and false negatives do exist and confirmatory testing is required.

Ketamine is stable in urine for at least three freeze–thaw cycles and is stable for up to 12 h in urine at 25°C.[26] External quality assurance program of urine toxicology has been made available by many organizations, including the RCPA Quality Assurance Program administered by the Royal College of Pathologists of Australasia.

16.3.1.4   Adulteration

Collection of urine is most often carried out without direct observation, and thus chance of adulteration exists.[105–110] Adulteration can be in vivo, meaning that substances are taken by the subject before the collection of a genuine urine specimen, or ex vivo (in vitro), meaning that substances are added into the collected urine specimen before submission, or a substitution urine specimen is submitted directly.

In vivo adulteration, in its most primitive way, is by consumption of large amounts of water. However, dilute urine specimens can be easily detected by the measurement of urine creatinine concentration.[109] More elaborate methods involve taking products containing creatine as well as riboflavin in addition to consumption of water, so as to produce a dilute urine but with a yellowish realistic-looking appearance together with a falsely normal creatinine level.[110] Limited data were available for the effectiveness of such in vivo adulteration strategies.

Ex vivo (or in vitro) adulteration can involve either adulteration of urine after voiding, or substitution of urine specimen by chemically formulated ones or clean urine of another person. Devices for delivering substituted urine specimen while avoiding detection in routine observation of sampling, together with heating devices to maintain temperature, have been reported to be commercially available.[110] Recatheterization, the act of filling bladder with clean urine via a Foley catheter, is another way in which urine specimen could be substituted. However, such act is risky and can introduce infection. Depending on the time spent between recatheterization and voiding, the temperature of the urine may not equilibrate well, or enough amounts of endogenous urine may result in a positive test, particularly when a sensitive detection method is utilized for analysis.

The addition of adulterants represents another way one can attempt to avoid detection. Adulterants used in avoiding urine drug detection can be divided into those that interfere with immunoassays by means of changing the chemical nature of the matrix, including acidity, ionic strength, presence of surfactants, and those that chemically modify the drug metabolites by oxidation. Whereas data regarding the effect of such adulterating agents for traditional drugs of abuse such as marijuana, amphetamines, opioids, cocaine, or even phencyclidine (a drug chemically related to ketamine) are available,[110–113] no published information exists for the robustness of immunoassays in detecting ketamine in the presence of such adulterants.

16.3.1.5 Detection of adulteration

There are multiple ways of preventing or detecting in vivo or in vitro urine adulteration, including direct observation of urine collection, on-site analysis, and laboratory analysis. Observation of urine collection is an effective means of preventing in vitro adulteration, but this may not be applicable to all settings of drug testing as privacy of subjects is a concern.

Procedures like turning off tap water and adding coloring agent to flushing water in the sites of sample collection have also been recommended. The United States mandatory guidelines for federal workplace testing program specifies requirements on temperature, creatinine concentration, specific gravity, acidity, and detection of oxidizing adulterants in urine specimens.[114] A urine specimen with creatinine concentration of equal to or greater than 0.18 mmol/L but less than 1.8 mmol/L is often considered dilute, whereas urine creatinine concentration of less than 0.18 mmol/L often indicates an invalid sample. Urine temperature is representative of core body temperature, and when measured shortly after voiding, 99% of measurements lie between 32.5°C and 36.7°C.[115]

16.3.2 Serum, plasma, and whole blood

Serum, plasma, and whole blood are specimens that are most commonly used for the determination of pharmacokinetic parameters. As a dissociative anesthetic agent, the pharmacokinetic and pharmacodynamic parameters of ketamine have been described in the 1980s.[95,96,116] Compared with urine, blood levels of ketamine likely better reflect the tissue levels of ketamine and provide superior correlation with dosage and pharmacological/toxicological effects.[117] Blood specimens are second only to urine specimens for drug testing in most clinical laboratories.

16.3.2.1 Application

Blood specimen for drug testing is mainly used in clinical and forensic settings. At times, blood specimens are taken in clinical settings when urine samples are not available, for example, in patients with end-stage renal failure, a not-uncommon complication of ketamine uropathy. Blood specimen is not commonly used in nonclinical contexts such as workplace drug testing because of its inherent invasiveness.

16.3.2.2   Advantages and limitations

One of the advantages of using a blood specimen is that the blood level represents drug concentration at the time of collection, rather than an integrated measure over a period as for urine specimens. As described earlier, levels in blood-derived specimens better correspond to the tissue levels (e.g., brain) of ketamine and provide the best correlation with drug dosage and also pharmacological and toxicological effects among other types of specimens. Besides, sample integrity can usually be ensured as blood specimens are usually taken by healthcare professionals, and it is uncommon to have adulterated blood specimens.

On the other hand, collection of blood specimens is relatively invasive, limiting its application in drug testing under nonclinical settings. Besides, as with most drugs that are renally excreted, the levels of ketamine and its metabolites in blood specimens are lower (usually in parts per billion levels) and short lived compared with those in urine. One study showed that the blood ketamine concentrations in 14 drugged drivers taking ketamine ranged from 170 to 850 ng/mL.[118] The window of detection of ketamine and its metabolites in blood is much shorter in comparison with urine.

16.3.2.3   Methods of analysis

Commercially available ELISA kits for the detection of ketamine in serum[119,120] have been available. Sample preparation is not required for most immunoassay-based kits, with dilution in buffer being integrated into the assay procedures.

Various methods for measuring ketamine in blood samples using HPLC-UV, GCMS, LC-MS/MS, and CE have been described. Due to the large protein content of serum, plasma, and whole blood, sample cleanup utilizing solid-phase extraction,[121] liquid–liquid extraction,[122] or protein precipitation[60] is necessary for chromatography-based methods.

Serum ketamine, norketamine, and dehydronorketamine are stable for at least 2 days when stored at 4°C. Plasma ketamine and norketamine levels remain stable when centrifugation of blood sample is delayed for 120 minutes both at ambient temperature and at 4°C. However, plasma dehydronorketamine level has a significant decrease when centrifugation is delayed while the blood sample is kept at 4°C, with only 53% remaining after a delay of 120 minutes.[123]

16.3.3   Oral fluid

Oral fluid is principally secreted by the three salivary glands: parotid gland, submandibular gland, and sublingual gland, with minor constituents such as secretion from the labial, buccal, and palatal glands, as well as gingival fluid, sloughed epithelial cells, and others.[124] The secretion rate is 0–6 mL/min with a daily production of 500–1500 mL.[125]

Oral fluid has been considered an ultrafiltrate of blood.[126] However, this is not an accurate description of oral fluid. As reviewed by Langman,[124] oral fluid is formed by ultrafiltration, active transport, and passive diffusion. The result is that nonionized as well as basic drugs are much better represented in oral fluid compared with acidic drugs.[124,127] The protein binding characteristic of the drug also determines the partition of the drug between saliva and blood.

Although the saliva-to-plasma drug ratio, commonly used to assess the suitability of analysis using oral fluid, has not been thoroughly assessed for ketamine, it follows that ketamine, as a slightly basic drug (pKa 7.5) with only 20%–50% protein bound fraction,[128] can likely be reliably determined in oral fluids.

Various methods of collecting oral fluid for drug testing including ketamine have been described, such as expectoration, gravitational draining from the buccal cavity, and the use of absorbent material put inside the buccal cavity.

16.3.3.1 Application

The testing of oral fluid, the collection of which is easy and noninvasive, has been applied in various clinical and nonclinical settings, including workplace drug testing, testing of drivers suspected to drive under the influence of drugs,[89] and monitoring of illicit drug use during treatment.[129] Commercially available collecting devices have been described.[127]

16.3.3.2 Advantages and limitations

Oral fluid collection is simple and noninvasive, and can be directly observed without interfering with the privacy of the subject. As a result, the risk of specimen substitution or adulteration is minimal compared with urine collection. In addition, as saliva drug concentration in general correlates with plasma-free drug concentration, it can provide a better measure than urine in the assessment of pharmacologic effect and level of impairment in the subject.

Whereas the secretion rate of oral fluid after consumption of drugs of abuse such as cannabis is very slow, ketamine is known to cause increased salivary secretion, as shown in studies when ketamine was used as an anesthetic drug.[128] Numerous techniques involving either mechanical or chemical stimuli have been applied to increase the flow rate of saliva.[130–132]

The mechanical methods stimulate saliva secretion by providing a substance (e.g., paraffin wax, chewing gum) for the subject to chew upon, whereas chemical stimuli (e.g., sour candies, a drop of citric acid) provide a gustatory response. A sixfold difference in oral fluid drug concentration has been reported between nonstimulated and stimulated oral fluid secretion.[127] The effect of the use of mechanical stimulus on the oral fluid levels of ketamine, however, has not been reported in the literature. Adsorption of the analyte onto the chewable material is the major issue for mechanical stimulation, whereas chemical stimulus, commonly acidic, may affect the kinetics of drug transfer into the oral fluid. However, as ketamine is a neutral or weakly basic drug, the oral fluid-to-plasma ratio of ketamine could be relatively independent of the rate of saliva formation.

Ketamine is commonly administered through nasal insufflation or oral ingestion. Another problem with oral fluid analysis of ketamine is that the level could be affected by recent drug use through these routes, as contamination of the oral cavity from these sources can lead to gross overestimation of the ketamine level.[89] The relatively low specimen volume and low drug levels, leading to more technically demanding analysis, are other limitations of using oral fluid as a specimen type for drug testing, including ketamine. The median ketamine levels in oral fluid and urine were 271 and 2125 ng/mL, respectively, in 21 positive cases in a study evaluating impairment symptoms in subjects attending disco.[89] Similar to blood sample, oral fluid has a short detection window, reflecting mainly recent drug use.

16.3.3.3 Methods of analysis

As a result of the low specimen volume, drug testing in oral fluid demands the use of sensitive methods such as immunoassays and chromatographic–mass spectrometric methods.[127] GCMS[133] and LC-MS/MS[89,127,133] methods have been described in the literature.

ELISA kits for oral fluid ketamine are commercially available.[119,134] A point-of-care immunoassay device for oral fluid ketamine has also been described. The sensitivity and specificity of the device were 87.5% and 97.9% at a cutoff of 15 ng/mL when evaluated against an LC-MS/MS method, with a false-positive rate of 4.5% and a false negative rate of 6%.[4] For chromatographic–mass spectrometric assays, sample preparation procedures like solid-phase extraction and liquid–liquid extraction are usually required,[4,62] although direct injection analysis after dilution with buffer has also been reported.[62] Apart from proteins, mucin is also a ubiquitous compound in oral fluid, which can be removed by one freeze–thaw cycle.

16.3.3.4 Adulteration

Adulteration in oral fluid testing is less commonly identified, as collection is often under direct observation. Thus, attempts in adulteration or decreasing the amount of drugs in oral fluids rely mostly on cleansing the mouth with or without dilution of the oral fluid sample. A crude but experimentally proven method is to use mouthwash to cleanse the mouth before oral fluid testing. In a study evaluating the effect of two commercially available adulteration kits on oral fluid cannabinoid and opiate levels, it was found that the two kits, together with the usual commercially available mouthwash, produced similar effects of decreasing the drug concentrations in oral fluid, probably through cleansing of the oral cavity and partially diluting the sample.[135]

16.3.4   Hair

Hair is an outgrowth from hair follicles, a characteristic organ in the skin of mammals. Most of the human body is covered in hair, but most commonly, scalp hair is used for analysis, despite it being the most manipulated and treated group of hair. Chemically, hair is mostly keratin (65%–95% by weight), with water and lipid being other major constituents.[136] The size of hair ranges from 15 to 120 μm in diameter.

There are three phases of hair growth: anagen (growth), catagen (transitional), and telogen (resting). The growth rate is approximately 1 cm/month (0.6–3 cm/month) during the anagen phase.[137] With such slow growth rates, hair testing can be seen as an indicator of drug use in the medium to long term, depending on the length of hair tested.

Drugs in hair represent a dynamic equilibrium between drug incorporation and drug removal. Drugs are incorporated into hair by multiple routes: transfer from blood during hair formation in the follicle, deposition on hair via sweat and sebum after the formation of hair, and environmental exposure (e.g., passive smoking, dissolution of drugs present in dirty hands into sebum or sweat, and incorporation as with ingested drugs). Drugs may be removed from hair during hair washing. A study showed that among ketamine, norketamine, and dehydronorketamine, the parent drug ketamine had the highest levels in hair specimens, with an average norketamine-to-ketamine ratio of 0.32 in 15 specimens.[67]

Hair contains melanin, a dark pigment produced by melanocytes in skin and hair follicles. Melanin produces binding sites for drugs, particularly amines and heavy metals. It has been shown that drug level in hair after a single dose and melanin content are strongly correlated for many drugs,[138] including ketamine, with significant difference found between patient groups, as well as guinea pigs (cavies), of different hair color.[67] This has triggered discussion as to whether hair testing results in racial bias.[137] Attempts in compensating the difference in melanin content included measuring permeability of dyes in the sample, as well as performing aqueous wash with drug homologue and adjusting for the uptake.[139] However, these methods are cumbersome and are not commonly used.

Hair specimens are best collected from the back of the head (vertex posterior), where hair grows with homogeneity with the highest anagen-to-telogen ratio (proportion of actively growing hair).[132] To perform segmental analysis, the hair should be cut as close as possible to the scalp, with the scalp end labeled. Hair specimens are commonly stored in envelopes with or without aluminum foil wrapping at room temperature to protect them from light and moisture.[132,137] Drugs in hair are stable for years when stored in a dry place without exposure to light or heat.

16.3.4.1   Application

Owing to the noninvasive nature of hair testing, as well as the ability of hair testing to reveal long-term drug use patterns, hair testing has been applied to various clinical and nonclinical settings, such as workplace drug testing, evaluation of driving privileges, drug abstinence monitoring (especially during pregnancy), child custody hearings, and postmortem forensic studies.[137,140,141]

16.3.4.2   Advantages and limitations

The principal advantage of using hair as a specimen of drug testing is the long detection window, thus allowing the monitoring of substance use over a period of weeks to months, as compared with only hours to days for other body fluid specimen types. The complementary use of body fluid specimens and hair specimens allows the profiling of substance use in both short and long term. Segmental analysis of hair specimens can also provide a rough historical profile of drug use, assuming a hair growth rate of approximately 1 cm/month. Other advantages of using hair specimens include the noninvasiveness and simplicity of sample collection, the relatively low risk of sample adulteration or substitution, and the long-term stability of drugs present in hair if properly stored.

Environmental contamination represents the most significant problem for hair testing. Procedures to control or detect environmental contamination of hair, such as sample washing, comparison of drug concentrations in washes and hair samples, use of cutoff values, and determining the metabolite-to-parent drug concentration ratio, have been described.[142] The Society of Hair Testing recommended washing of the sample as well as the use of metabolite-to-parent drug ratio to report positive results.[137] For hair analysis of ketamine, the norketamine-to-ketamine ratio can be used. Besides, dyeing or bleaching may also have deleterious effects on hair drug levels to different extents. As previously described, the possibility of hair color bias is another concern in hair testing. The low levels of drugs present in hair, generally in the range of picogram per milligram to nanogram per milligram, also constitute a significant analytical challenge. For example, the ketamine and norketamine concentrations in hair samples of four chronic ketamine abusers were found to be at low levels in a study, ranging from 0.2 to 5.7 ng/mg and from 0.1 to 1.2 ng/mg, respectively.[37] Tedious sample preparation procedures and chromatography–mass spectrometric analysis are usually required. Moreover, hair testing service is not commonly available in most clinical laboratories.

16.3.4.3 Methods of analysis

In contrast to blood and urine specimens, hair is a solid sample, which requires extraction of analyses before analysis by most techniques. Buffer extraction with micropulverization techniques, acid hydrolysis, alkaline hydrolysis, and enzymatic hydrolysis have all been reported in the literature.[39,132,137] Further sample cleanup procedures such as solid-phase extraction and liquid–liquid extraction are generally required for most chromatography-based methods.[132] GCMS and LC-MS/MS are the most common analytical methods for hair specimens,[39,67,137,140,143] in part owing to the low concentration of drugs found in hair. The use of other methods such as immunoassay[134] and CE or related techniques[103] has also been reported.

A novel technique in hair drug testing is the use of matrix-assisted laser desorption/ionization technique coupled with high-sensitivity and high-resolution MS, in which a single hair is cut open and mounted, and laser is used to cause desorption and ionization of analyses in the hair sample.[144] In contrast to segmental analysis, this technique allows profiling of drug use over the whole period and is essentially nondestructive.

16.3.4.4 Adulteration

Hair is often exposed to agents that cause drugs to leach out. This includes mild agents such as shampoo and conditioner, as well as more elaborate hair treatment such as bleaching, perming, and coloring.[132,137] These treatments not only cause drugs to leach out from hair but also increase the porosity of hair, resulting in increased leaching on further treatment (risk of false negative), as well as increased contamination from environmental sources (risk of false positive).[137] Further to cosmetic treatments, "detox" shampoo products marketed toward subjects of hair drug testing have been commercially available. However, the effect of such treatment has not been thoroughly studied.

16.3.5 Nail

Nail is a keratinic product of the nail matrix. The proximal part of the nail is embedded in skin, the proximal nail fold, below which is where the germinal matrix lies. The germinal matrix and the nail bed are supplied by the two paired digital arteries.

The detection window for nails can be considered in a number of different ways. Whereas at autopsy, forensic specimens can include all whole fingernails and toenails, clinical specimens include only nail clippings, of which the length available depends on the frequency of nail clipping. The reported average growth rate of fingernail and toenail are 3.47 and 1.62 mm/month, respectively.[145] With a commonly quoted great toenail length of 20 mm,[146] this represents a detection window of 12.3 months for the great toenail, if the whole toenail is considered. However, if one considers the length of nail clippings, which often ranges from 2 to 5 mm, the period covered would be much shorter.

The incorporation of drugs into nail, unlike hair, is not as straightforward. This has been analyzed by Madry et al. with single-dose zolpidem as a tracer—it was shown that the incorporation of drug taken orally can be detected in nail specimens that were collected monthly for up to 3 months after intake, with three peaks corresponding to 24 h, 2–3 weeks, and 3 months (with decreasing peak size) after ingestion of drug.[147] According to the author, this represents incorporation via sweat, nail bed, and germinal matrix, respectively.[147] Thus, nail specimen, while representing medium-term drug exposure, is useful in determining the temporal sequence of drug intake only if serial nail clippings or if the entire nail could be analyzed.

Permeation of drugs into nail has been demonstrated to be independent of the hydrophobicity of the drug, but is related to the molecular weight of the drug when the hydration of ionic drugs has been taken into account.[148] This suggests that ketamine, with a molecular mass of 237.7 Da, would have intermediate permeation into nail. In addition, as ketamine is often supplied as powder for insufflation, the handling of drugs by hands may result in significant contamination, which could compromise the validity of quantitation.

16.3.5.1   Application

Despite reports of nail toxicology being applied to forensic, workplace, and prenatal exposure of drugs,[149] the use of nail is still mostly restricted to forensic settings.

16.3.5.2   Advantages and limitations

Nail as a toxicology matrix has the advantage of being an intermediate-term indicator of drug use, situated between hair (months to years) and body fluid (hours to weeks) matrices. Besides, the date of ingestion can be estimated, likely to a better resolution than hair analysis, according to the appearance of three peaks with decreasing size after a single dose of medication. Similar to hair, nail can also be stored under room temperature in a dry environment almost indefinitely.

Cumbersome sample preparation procedures and chromatography–MS analysis are required for nail drug testing. It is still mainly in the research stage and is not commonly available in most clinical laboratories.

16.3.5.3   Methods of analysis

Nail is a solid keratinic matrix and requires dissolution before analysis. Similar to hair, washing of nail is necessary to minimize possible surface contamination, and this is often done with water and methanol.

Table 16.3 Sample Matrices and Applicable Analytical Techniques

Matrix	Detection window[a]	Analysis techniques
Urine	Days	Immunoassays, LC-MS/MS, GCMS, HPLC-UV, CE-UV, CE-MS
Blood	Hours	Immunoassays, LC-MS/MS, GCMS, HPLC-UV, CE-UV, CE-MS
Oral fluid	Hours	Immunoassays, LC-MS/MS, GCMS
Hair	Months to years	LC-MS/MS, GCMS, MALDI-IMS
Nail	Weeks to months	LC-MS/MS, GCMS

[a] Strongly dependent on limit of detection of the methodologies used. Typical values are given.

Sample preparation by alkaline hydrolysis, followed by liquid–liquid extraction and derivatization, has been reported in a study utilizing GCMS for the detection of ketamine in fingernail.[44] Another method employing acid hydrolysis followed by liquid–liquid extraction has been reported with LC-MS/MS detection of ketamine in nail.[150] Other sample preparation methods like micropulverization have also been reported.[42]

In summary, the use of various specimen types in the analysis of ketamine depends upon different factors including the purpose of the testing, the desired detection window, and the technologies available. Table 16.3 shows a comparison of the application of different matrix types.

16.4   Emerging drugs-of-abuse analysis

As discussed in Sections 16.2 and 16.3, there are a number of analytical techniques for the detection of ketamine in various biological specimens. The different combinations of analytical technique and specimen type offer a myriad of analytical tools to suit different needs: POCT on a salivary specimen would be ideal for roadside testing for suspected drug-driving; confirmation of exposure to ketamine for prosecution purposes would require GCMS analysis on a urine or blood specimen; and analysis by various methods on a hair specimen can reveal a much longer period of exposure to drugs of abuse than testing in biological fluids.

Although analytical methods for ketamine are discussed in this chapter, in reality, ketamine is seldom an isolated analytical target. Polysubstance abuse, deliberately or inadvertently, is a rule rather than an exception.[151] For this reason, most of the analytical techniques were developed to detect multiple substances of abuse.[72,152–155] Clinically, finding out which and how much drugs a user has consumed is important for management for obvious reasons.

While it is important to know the different substances used by a user, analytical techniques can only detect a finite number of drugs. Analytical techniques are developed to detect certain targets—the choices would be different for different geographical locations determined by the local drug abuse pattern. If something "new" to the analytical system were consumed by a user, their presence would not be detected regardless of the specimen type and sensitivity of the instrument. Such "new" substances are called "emerging drugs of abuse" or "novel psychoactive substances," which are becoming a significant healthcare problem and pose a great challenge to the clinical laboratory professionals.[156–159]

The last decade has seen a rapid and continual growth of "emerging" drugs of abuse. These substances bear chemical and pharmacological resemblance to conventional drugs of abuse and pose a threat to public health, but are often (initially) not controlled by law. In 2008, 13 emerging drugs were reported for the first time to the European Monitoring Centre for Drugs and Drug Addiction; by 2012, 73 new drugs were reported within a year.[160] These novel substances are often sold in disguise as "bath salts," plant food, chemical standards, spice mix, herbal incense, and so on, while being explicitly labeled as "Not for human consumption." They are readily available in head shops, from street dealers and over the Internet.[156,157,159] With particular reference to ketamine, its analogue methoxetamine has recently gained popularity on the illicit drug market.[161–163]

As new drugs emerge continually, some disappear while some remain. Ketamine was once an "emerging" drug of abuse in Hong Kong in the early 1990s. It has remained and continues to be a tremendous burden to society nowadays. When such emerging drugs of abuse first appear, the parties concerned—healthcare professionals, social workers, teachers, law enforcement officials, legislators, and so on—know very little about them. It often takes several years for the society to understand the nature and learn the impact of such novel substances.

During this period, the clinical and harmful effects are poorly understood. There is no analytical tool to detect and confirm their presence, and there is a lack of specific legislation to control these substances. This period of poor preparedness has to be shortened for the society to fight the battle more effectively. In the case of ketamine, if the society had realized its presence earlier and controlled it more promptly and vigorously, can history be modified?

Current anti–drug abuse strategies commonly focus on the conventional drugs of abuse. The foremost challenge in combating emerging drugs is that such substances are not covered by current routine drug testing methodologies. As a result, they may have penetrated deeply into society by the time they are recognized. If penetration of an emerging drug into society is recognized early, subsequent actions on education, clinical management, and legislation may be initiated in a much timelier manner before the damage becomes permanent. To achieve this, a system to monitor the emergence of novel drugs is essential.

Owing to its technical complexity, immunoassay-based methods— including POCT and laboratory-based immunoassays—are often developed commercially. There is often a time lag of years between a substance appearing in the abuse circle and its immunoassay-based detection method becoming available. Facing the rapid, exponential growth in the number of emerging substances, immunoassay-based methods are not suitable for their detection. On the other hand, chromatography-based methods, often coupled to different MS techniques, have been successfully applied to detect these emerging drugs of abuse.[65,72,158,164,165]

In the past decade, the use of these instruments has become much more widespread in routine clinical laboratories, and in-house methods can be developed readily on these platforms to detect emerging drugs. In the clinical setting, the detection of novel drugs combined with information on the toxicity profile will help identify highly dangerous compounds that require immediate regulatory actions.

A recent method employed a chromatography/MS-based analytical system to detect both conventional and emerging drugs of abuse in urine and hair.[166,167] Solid-phase extraction is used for urine sample cleanup and micropulverization for drug extraction from hair. The method covered 47 conventional drug analyses and more than 40 emerging substances, including methoxetamine, which is an analogue of ketamine. Emerging drugs of other classes were also detected by the method, including amphetamines, phenethylamines, tryptamines, piperazines, cannabinoids, cathinones, and so on. This method has been adopted as a routine drug of abuse screening within the authors' laboratory and served as a continual surveillance of novel drugs in the population. Indeed, after its implementation, the method has been successfully applied, leading to the identification of several emerging compounds.[166–168]

Drugs of abuse are notorious for their protean nature; slight chemical modifications are constantly being introduced to bypass the current legislation. As such, an analytical system for emerging substances must be flexible for expansion to cover "newer" drugs. Combining routine analysis of conventional drugs with surveillance of emerging substances in a single analytical process allows monitoring of the emergence of novel compounds at minimal cost.

The results may guide future directions in terms of legislation as well as social and clinical management of novel drugs. Hopefully, such efforts can minimize the damage of novel psychoactive substances as ketamine had imposed on the society in the past two decades.

16.5   Conclusions

The qualitative and quantitative analysis of ketamine plays an important role in clinical and forensic investigations. The clinical testing of ketamine involves a myriad of analytical techniques that are capable of analyzing different types of biological matrices. The optimal methodology and sample type to be used for a particular application will depend upon various factors, including (but not limited to) the urgency of the test request, the availability of technological and manpower resources, the level of sensitivity or accuracy required, and the desired detection window. When choosing an appropriate matrix type, it is also important to consider other issues such as subject privacy (e.g., collection under direct observation), sample substitution/adulteration, and analyte stability. At times, analysis using a combination of different methods and sample types may be required.

A growing concern upon the analysis of drugs of abuse is the appearance of novel psychoactive substances (e.g., methoxetamine). These compounds are constantly changing in chemical structure to bypass current legislation and are often difficult to identify by conventional drug testing methods. Failure to identify these novel drugs at an early stage will allow them to take root and become tremendous economic, social, and medical burdens to the society. Proactive development of analytical methods and continual surveillance of these novel drugs in the society are imperative in the fight against this rapidly expanding group of dangerous chemicals.

References

1. Domino, E.F., Taming the ketamine tiger. 1965, *Anesthesiology*, 113, 678, 2010.
2. Corazza, O., Schifano, F., Simonato, P., Fergus, S., Assi, S., Stair, J., Corkery, J., Trincas, G., Deluca, P., Davey, Z., Blaszko, U., Demetrovics, Z., Moskalewicz, J., Enea, A., di Melchiorre, G., Mervo, B., di Furia, L., Farre, M., Flesland, L., Pasinetti, M., Pezzolesi, C., Pisarska, A., Shapiro, H., Siemann, H., Skutle, A., Enea, A., di Melchiorre, G., Sferrazza, E., Torrens, M., van der Kreeft, P., Zummo, D. and Scherbaum, N., Phenomenon of new drugs on the Internet: The case of ketamine derivative methoxetamine, *Hum. Psychopharmacol.*, 27, 145, 2012.
3. Wu, A.H., Point-of-care drug testing, in *Principles and Practice of Point-of-Care Testing*, Kost, G., Ed., Lippincott Williams & Wilkins, Philadelphia, PA, 2002.
4. Tsui, T.K., Chan, A.S., Lo, C.W., Wong, A., Wong, R.C. and Ho, C.S., Performance of a point-of-care device for oral fluid ketamine evaluated by a liquid chromatography-tandem mass spectrometry method, *J. Anal. Toxicol.*, 36, 210, 2012.

5. Burtis, C.A., Ashwood, E.R. and Bruns, D.E., *Tietz Textbook of Clinical Chemistry and Molecular Diagnostics*, 5th ed., Elsevier Health Sciences, St. Louis, MO, 2012.

6. Cortez Diagnostics Inc., Kit insert for One Step Ketamine RapiDip InstaTest, available at http://www.rapidtest.com/pdf/Ketamine%20120301-1%20_09 -27-2010_.pdf (accessed December 16, 2013).

7. DIPRO Med, Product information and instructions: One step ketamine test device (urine), available at http://www.dipro.co.at/pdf/dipro_pdf _45ddb0b8c3d4c.pdf (accessed December 15, 2013).

8. Huang, M.H., Wu, M.Y., Wu, C.H., Tsai, J.L., Lee, H.H. and Liu, R.H., Performance characteristics of ELISAs for monitoring ketamine exposure, *Clin. Chim. Acta*, 379, 59, 2007.

9. Melanson, S.E., The utility of immunoassays for urine drug testing, *Clin. Lab. Med.*, 32, 429, 2012.

10. Tsai, J.S. and Lin, G.L., Drug-testing technologies and applications, in *Drug of Abuse Body Fluid Testing*, Wong, R.C., Tse, H.Y., Eds., Humana Press, New Jersey, 2005, Totowa, NJ, pp. 29–69.

11. Cheng, P.S., Fu, C.Y., Lee, C.H., Liu, C. and Chien, C.S., GC-MS quantifica-tion of ketamine, norketamine, and dehydronorketamine in urine specimens and comparative study using ELISA as the preliminary test methodology, *J. Chromatogr. B Analyt. Technol. Biomed. Life Sci.*, 852, 443, 2007.

12. Parkin, M.C., Turfus, S.C., Smith, N.W., Halket, J.M., Braithwaite, R.A., Elliott, S.P., Osselton, M.D., Cowan, D.A. and Kicman, A.T., Detection of ketamine and its metabolites in urine by ultra high pressure liquid chromatography-tandem mass spectrometry, *J. Chromatogr. B Analyt. Technol. Biomed. Life Sci.*, 876, 137, 2008.

13. Choi, H., Baeck, S., Jang, M., Lee, S., Choi, H. and Chung, H., Simultaneous analysis of psychotropic phenylalkylamines in oral fluid by GC-MS with automated SPE and its application to legal cases, *Forensic Sci. Int.*, 215, 81, 2012.

14. Harun, N., Anderson, R.A. and Miller, E.I., Validation of an enzyme-linked immunosorbent assay screening method and a liquid chromatography-tandem mass spectrometry confirmation method for the identification and quantification of ketamine and norketamine in urine samples from Malaysia, *J. Anal. Toxicol.*, 33, 310, 2009.

15. International Union of Pure and Applied Chemistry, available at http://www .iupac.org/publications/analytical_compendium/ (accessed December 16, 2013).

16. Adams, H., Weber, B., Bachmann-M, B., Guerin, M. and Hempelmann, G., The simultaneous determination of ketamine and midazolam using high pressure liquid chromatography and UV detection (HPLC/UV), *Anaesthesist*, 41, 619, 1992.

17. Niedorf, F., Bohr, H.H. and Kietzmann, M., Simultaneous determination of ketamine and xylazine in canine plasma by liquid chromatography with ultraviolet absorbance detection, *J. Chromatogr. B Analyt. Technol. Biomed. Life Sci.*, 791, 421, 2003.

18. Rofael, H.Z. and Abdel-Rahman, M.S., Development and validation of a high-performance liquid chromatography method for the determination of cocaine, its metabolites and ketamine, *J. Appl. Toxicol.*, 22, 123, 2002.

19. Raikos, N., Theodoridis, G., Alexiadou, E., Gika, H., Argiriadou, H., Parlapani, H. and Tsoukali, H., Analysis of anaesthetics and analgesics in human urine by headspace SPME and GC, *J. Sep. Sci.*, 32, 1018, 2009.
20. Xiong, J., Chen, J., He, M. and Hu, B., Simultaneous quantification of amphetamines, caffeine and ketamine in urine by hollow fiber liquid phase microextraction combined with gas chromatography-flame ionization detector, *Talanta*, 82, 969, 2010.
21. Chang, T. and Glazko, A.J., A gas chromatographic assay for ketamine in human plasma, *Anesthesiology*, 36, 401, 1972.
22. Davisson, J.N., Rapid gas chromatographic analysis of plasma levels of ketamine and major metabolites employing either nitrogen selective or mass spectroscopic detection, *J. Chromatogr.*, 146, 344, 1978.
23. Lo, J.N. and Cumming, J.F., Interaction between sedative premedicants and ketamine in man in isolated perfused rat livers, *Anesthesiology*, 43, 307, 1975.
24. Brown, S.D., Rhodes, D.J. and Pritchard, B.J., A validated SPME-GC-MS method for simultaneous quantification of club drugs in human urine, *Forensic Sci. Int.*, 171, 142, 2007.
25. Castro, A.L., Tarelho, S., Silvestre, A. and Teixeira, H.M., Simultaneous analysis of some club drugs in whole blood using solid phase extraction and gas chromatography-mass spectrometry, *J. Forensic Leg. Med.*, 19, 77, 2012.
26. Chen, C.Y., Lee, M.R., Cheng, F.C. and Wu, G.J., Determination of ketamine and metabolites in urine by liquid chromatography-mass spectrometry, *Talanta*, 72, 1217, 2007.
27. Chen, L.C., Hashimoto, Y., Furuya, H., Takekawa, K., Kubota, T. and Hiraoka, K., Rapid detection of drugs in biofluids using atmospheric pressure chemi/chemical ionization mass spectrometry, *Rapid Commun. Mass Spectrom.*, 23, 333, 2009.
28. Cheng, J.Y. and Mok, V.K., Rapid determination of ketamine in urine by liquid chromatography-tandem mass spectrometry for a high throughput laboratory, *Forensic Sci. Int.*, 142, 9, 2004.
29. Cheng, P.S., Lee, C.H., Liu, C. and Chien, C.S., Simultaneous determination of ketamine, tramadol, methadone, and their metabolites in urine by gas chromatography-mass spectrometry, *J. Anal. Toxicol.*, 32, 253, 2008.
30. Cheng, W.C., Yau, T.S., Wong, M.K., Chan, L.P. and Mok, V.K., A high-throughput urinalysis of abused drugs based on a SPE-LC-MS/MS method coupled with an in-house developed post-analysis data treatment system, *Forensic Sci. Int.*, 162, 95, 2006.
31. Chou, S.L., Yang, M.H., Ling, Y.C. and Giang, Y.S., Gas chromatography-isotope dilution mass spectrometry preceded by liquid–liquid extraction and chemical derivatization for the determination of ketamine and norketamine in urine, *J. Chromatogr. B Analyt. Technol. Biomed. Life Sci.*, 799, 37, 2004.
32. del Mar Ramirez Fernandez, M., Laloup, M., Wood, M., De Boeck, G., Lopez-Rivadulla, M., Wallemacq, P. and Samyn, N., Liquid chromatography-tandem mass spectrometry method for the simultaneous analysis of multiple hallucinogens, chlorpheniramine, ketamine, ritalinic acid, and metabolites, in urine, *J. Anal. Toxicol.*, 31, 497, 2007.
33. Dowling, G. and Regan, L., A new mixed mode solid phase extraction strategy for opioids, cocaines, amphetamines, and adulterants in human blood with hybrid liquid chromatography tandem mass spectrometry detection, *J. Pharm. Biomed. Anal.*, 54, 1136, 2011.

34. Dowling, G., Regan, L., Tierney, J. and Nangle, M., A hybrid liquid chromatography-mass spectrometry strategy in a forensic laboratory for opioid, cocaine and amphetamine classes in human urine using a hybrid linear ion trap-triple quadrupole mass spectrometer, *J. Chromatogr. A*, 1217, 6857, 2010.
35. Favretto, D., Vogliardi, S., Stocchero, G., Nalesso, A., Tucci, M., Terranova, C. and Ferrara, S.D., Determination of ketamine and norketamine in hair by micropulverized extraction and liquid chromatography-high resolution mass spectrometry, *Forensic Sci. Int.*, 226, 88, 2013.
36. Gentili, S., Cornetta, M. and Macchia, T., Rapid screening procedure based on headspace solid-phase microextraction and gas chromatography-mass spectrometry for the detection of many recreational drugs in hair, *J. Chromatogr. B Analyt. Technol. Biomed. Life Sci.*, 801, 289, 2004.
37. Harun, N., Anderson, R.A. and Cormack, P.A., Analysis of ketamine and norketamine in hair samples using molecularly imprinted solid-phase extraction (MISPE) and liquid chromatography-tandem mass spectrometry (LC-MS/MS), *Anal. Bioanal. Chem.*, 396, 2449, 2010.
38. Huang, M.K., Liu, C., Li, J.H. and Huang, S.D., Quantitative detection of ketamine, norketamine, and dehydronorketamine in urine using chemical derivatization followed by gas chromatography-mass spectrometry, *J. Chromatogr. B Analyt. Technol. Biomed. Life Sci.*, 820, 165, 2005.
39. Inagaki, S., Makino, H., Fukushima, T., Min, J.Z. and Toyo'oka, T., Rapid detection of ketamine and norketamine in rat hair using micropulverized extraction and ultra-performance liquid chromatography-electrospray ionization mass spectrometry, *Biomed. Chromatogr.*, 23, 1245, 2009.
40. Jen, H.P., Tsai, Y.C., Su, H.L. and Hsieh, Y.Z., On-line preconcentration and determination of ketamine and norketamine by micellar electrokinetic chromatography. Complementary method to gas chromatography/mass spectrometry, *J. Chromatogr. A.*, 1111, 159, 2006.
41. Kim, E.M., Lee, J.S., Choi, S.K., Lim, M.A. and Chung, H.S., Analysis of ketamine and norketamine in urine by automatic solid-phase extraction (SPE) and positive ion chemical ionization-gas chromatography-mass spectrometry (PCI-GC-MS), *Forensic Sci. Int.*, 174, 197, 2008.
42. Kim, J.Y., Cheong, J.C., Lee, J.I., Son, J.H. and In, M.K., Rapid and simple GC-MS method for determination of psychotropic phenylalkylamine derivatives in nails using micro-pulverized extraction, *J. Forensic Sci.*, 57, 228, 2012.
43. Kim, J.Y., In, M.K. and Kim, J.H., Determination of ketamine and norketamine in hair by gas chromatography/mass spectrometry using two-step derivatization, *Rapid Commun. Mass Spectrom.*, 20, 3159, 2006.
44. Kim, J.Y., Shin, S.H. and In, M.K., Determination of amphetamine-type stimulants, ketamine and metabolites in fingernails by gas chromatography-mass spectrometry, *Forensic Sci. Int.*, 194, 108, 2010.
45. Kim, J.Y., Shin, S.H., Lee, J.I. and In, M.K., Rapid and simple determination of psychotropic phenylalkylamine derivatives in human hair by gas chromatography mass spectrometry using micro-pulverized extraction, *Forensic Sci. Int.*, 196, 43, 2010.
46. Kuwayama, K., Inoue, H., Kanamori, T., Tsujikawa, K., Miyaguchi, H., Iwata, Y.T., Miyauchi, S. and Kamo, N., Analysis of amphetamine-type stimulants and their metabolites in plasma, urine and bile by liquid chromatography with a strong cation-exchange column-tandem mass spectrometry, *J. Chromatogr. B Analyt. Technol. Biomed. Life Sci.*, 867, 78, 2008.

47. Kwon, W., Kim, J.Y., Suh, S. and In, M.K., Rapid and simple determination of psychotropic phenylalkylamine derivatives in urine by direct injection liquid chromatography-electrospray ionization-tandem mass spectrometry, *Biomed. Chromatogr.*, 27, 88, 2013.

48. Lee, H.H., Lee, J.F., Lin, S.Y., Chen, P.H. and Chen, B.H., Simultaneous determination of HFBA-derivatized amphetamines and ketamines in urine by gas chromatography-mass spectrometry, *J. Anal. Toxicol.*, 35, 162, 2011.

49. Legrand, T., Roy, S., Monchaud, C., Grondin, C., Duval, M. and Jacqz-Aigrain, E., Determination of ketamine and norketamine in plasma by micro-liquid chromatography-mass spectrometry, *J. Pharm. Biomed. Anal.*, 48, 171, 2008.

50. Lendoiro, E., Quintela, O., de Castro, A., Cruz, A., Lopez-Rivadulla, M. and Concheiro, M., Target screening and confirmation of 35 licit and illicit drugs and metabolites in hair by LC-MSMS, *Forensic Sci. Int.*, 217, 207, 2012.

51. Lin, H.R., Lin, H.L., Lee, S.F., Liu, C. and Lua, A.C., A fast screening procedure for ketamine and metabolites in urine samples with tandem mass spectrometry, *J. Anal. Toxicol.*, 34, 149, 2010.

52. Lin, H.R. and Lua, A.C., Simultaneous determination of amphetamines and ketamines in urine by gas chromatography/mass spectrometry, *Rapid Commun. Mass Spectrom.*, 20, 1724, 2006.

53. Lua, A.C. and Lin, H.R., A rapid and sensitive ESI-MS screening procedure for ketamine and norketamine in urine samples, *J. Anal. Toxicol.*, 28, 680, 2004.

54. Melent'ev, A., Determination of promedol (trimeperidine) and ketamine in blood using gas chromatography-mass spectrometry, *J. Anal. Chem.*, 59, 972, 2004.

55. Merola, G., Gentili, S., Tagliaro, F. and Macchia, T., Determination of different recreational drugs in hair by HS-SPME and GC/MS, *Anal. Bioanal. Chem.*, 397, 2987, 2010.

56. Miyaguchi, H. and Inoue, H., Determination of amphetamine-type stimulants, cocaine and ketamine in human hair by liquid chromatography/linear ion trap-Orbitrap hybrid mass spectrometry, *Analyst*, 136, 3503, 2011.

57. Moore, K.A., Sklerov, J., Levine, B. and Jacobs, A.J., Urine concentrations of ketamine and norketamine following illegal consumption, *J. Anal. Toxicol.*, 25, 583, 2001.

58. Olmos-Carmona, M.L. and Hernandez-Carrasquilla, M., Gas chromatographic-mass spectrometric analysis of veterinary tranquillizers in urine: Evaluation of method performance, *J. Chromatogr. B Biomed. Sci. Appl.*, 734, 113, 1999.

59. Rodriguez Rosas, M.E., Patel, S. and Wainer, I.W., Determination of the enantiomers of ketamine and norketamine in human plasma by enantioselective liquid chromatography-mass spectrometry, *J. Chromatogr. B Analyt. Technol. Biomed. Life Sci.*, 794, 99, 2003.

60. Sergi, M., Bafile, E., Compagnone, D., Curini, R., D'Ascenzo, G. and Romolo, F.S., Multiclass analysis of illicit drugs in plasma and oral fluids by LC-MS/MS, *Anal. Bioanal. Chem.*, 393, 709, 2009.

61. Sergi, M., Compagnone, D., Curini, R., D'Ascenzo, G., Del Carlo, M., Napoletano, S. and Risoluti, R., Micro-solid phase extraction coupled with high-performance liquid chromatography-tandem mass spectrometry for the determination of stimulants, hallucinogens, ketamine and phencyclidine in oral fluids, *Anal. Chim. Acta*, 675, 132, 2010.

62. Strano-Rossi, S., Anzillotti, L., Castrignano, E., Felli, M., Serpelloni, G., Mollica, R. and Chiarotti, M., UHPLC-ESI-MS/MS method for direct analysis of drugs of abuse in oral fluid for DUID assessment, *Anal. Bioanal. Chem.*, 401, 609, 2011.

63. Tabernero, M.J., Felli, M.L., Bermejo, A.M. and Chiarotti, M., Determination of ketamine and amphetamines in hair by LC/MS/MS, *Anal. Bioanal. Chem.*, 395, 2547, 2009.

64. Wang, K.C., Shih, T.S. and Cheng, S.G., Use of SPE and LC/TIS/MS/MS for rapid detection and quantitation of ketamine and its metabolite, norketamine, in urine, *Forensic Sci. Int.*, 147, 81, 2005.

65. Wohlfarth, A., Weinmann, W. and Dresen, S., LC-MS/MS screening method for designer amphetamines, tryptamines, and piperazines in serum, *Anal. Bioanal. Chem.*, 396, 2403, 2010.

66. Wu, Y.H., Lin, K.L., Chen, S.C. and Chang, Y.Z., Simultaneous quantitative determination of amphetamines, ketamine, opiates and metabolites in human hair by gas chromatography/mass spectrometry, *Rapid Commun. Mass Spectrom.*, 22, 887, 2008.

67. Xiang, P., Shen, M. and Zhuo, X., Hair analysis for ketamine and its metabolites, *Forensic Sci. Int.*, 162, 131, 2006.

68. Xiang, P., Sun, Q., Shen, B. and Shen, M., Disposition of ketamine and norketamine in hair after a single dose, *Int. J. Legal Med.*, 125, 831, 2011.

69. Zancanaro, I., Limberger, R.P., Bohel, P.O., dos Santos, M.K., De Boni, R.B., Pechansky, F. and Caldas, E.D., Prescription and illicit psychoactive drugs in oral fluid—LC-MS/MS method development and analysis of samples from Brazilian drivers, *Forensic Sci. Int.*, 223, 208, 2012.

70. Zhu, K.Y., Leung, K.W., Ting, A.K., Wong, Z.C., Fu, Q., Ng, W.Y., Choi, R.C., Dong, T.T., Wang, T., Lau, D.T. and Tsim, K.W., The establishment of a highly sensitive method in detecting ketamine and norketamine simultaneously in human hairs by HPLC-Chip-MS/MS, *Forensic Sci. Int.*, 208, 53, 2011.

71. Chang, T. and Glazko, A.J., Biotransformation and disposition of ketamine, *Int. Anesthesiol. Clin.*, 12, 157, 1974.

72. Pedersen, A.J., Dalsgaard, P.W., Rode, A.J., Rasmussen, B.S., Muller, I.B., Johansen, S.S. and Linnet, K., Screening for illicit and medicinal drugs in whole blood using fully automated SPE and ultra-high-performance liquid chromatography with TOFMS with data-independent acquisition, *J. Sep. Sci.*, 36, 2081, 2013.

73. Watson, J.T. and Sparkman, O.D., *Introduction to Mass Spectrometry: Instrumentation, Applications and Strategies for Data Interpretation*, 4th ed., John Wiley & Sons Ltd, West Sussex, England, 2007.

74. CLSI, *Mass Spectrometry in the Clinical Laboratory: General Principles and Guidance; Approved Guideline*, CLSI document C50-A. Clinical and Laboratory Standards Institute, Wayne, PA, 2007.

75. Ardrey, R.E., *Liquid Chromatography-Mass Spectrometry: An Introduction*, John Wiley & Sons Ltd, West Sussex, England, 2003.

76. Snyder, L.R., Kirkland, J.J. and Dolan, J.W., *Introduction to Modern Liquid Chromatography*, 3rd ed., John Wiley & Sons Inc., Hoboken, NJ, 2010.

77. Niessen, W.M., *Liquid Chromatography-Mass Spectrometry*, 3rd ed., CRC Press, Boca Raton, FL, 2006.

78. Willoughby, R., Sheehan, E. and Mitrovich, S., *A Global View of LC/MS: How to Solve Your Most Challenging Analytical Problems*, 2nd ed., Global View Publishing, Pittsburgh, PA, 2002.
79. Dass, C., *Principles and Practice of Biological Mass Spectrometry*, Wiley-Interscience, New York, 2001.
80. Commission decision of 12 August 2002 implementing Council Directive 96/23/EC concerning the performance of analytical methods and the interpretation of results (2002/657/EC), *Official Journal of the European Communities*, August 17, 2002.
81. Society of Forensic Toxicologists Inc. and American Academy of Forensic Sciences, SOFT/AAFS Forensic Laboratory Guidelines, 2006.
82. Ghassabian, S., Moosavi, S.M., Valero, Y.G., Shekar, K., Fraser, J.F. and Smith, M.T., High-throughput assay for simultaneous quantification of the plasma concentrations of morphine, fentanyl, midazolam and their major metabolites using automated SPE coupled to LC-MS/MS, *J. Chromatogr. B Analyt. Technol. Biomed. Life Sci.*, 903, 126, 2012.
83. Idder, S., Ley, L., Mazellier, P. and Budzinski, H., Quantitative on-line preconcentration-liquid chromatography coupled with tandem mass spectrometry method for the determination of pharmaceutical compounds in water, *Anal. Chim. Acta*, 805, 107, 2013.
84. Marinova, M., Artusi, C., Brugnolo, L., Antonelli, G., Zaninotto, M. and Plebani, M., Immunosuppressant therapeutic drug monitoring by LC-MS/MS: Workflow optimization through automated processing of whole blood samples, *Clin. Biochem.*, 46, 1723, 2013.
85. Hopley, C., Bristow, T., Lubben, A., Simpson, A., Bull, E., Klagkou, K., Herniman, J. and Langley, J., Towards a universal product ion mass spectral library reproducibility of product ion spectra across eleven different mass spectrometers, *Rapid Commun. Mass Spectrom.*, 22, 1779, 2008.
86. Wissenbach, D.K., Meyer, M.R., Weber, A.A., Remane, D., Ewald, A.H., Peters, F.T. and Maurer, H.H., Towards a universal LC-MS screening procedure—Can an LIT LCMS(n) screening approach and reference library be used on a quadrupole-LIT hybrid instrument?, *J. Mass Spectrom.*, 47, 66, 2012.
87. McMaster, M., *GC/MS: A Practical User's Guide*, 2nd ed., John Wiley & Sons Inc., Hoboken, NJ, 2008.
88. Apollonio, L.G., Pianca, D.J., Whittall, I.R., Maher, W.A. and Kyd, J.M., A demonstration of the use of ultra-performance liquid chromatography-mass spectrometry (UPLC/MS) in the determination of amphetamine-type substances and ketamine for forensic and toxicological analysis, *J. Chromatogr. B*, 836, 111, 2006.
89. Cheng, W.C., Ng, K.M., Chan, K.K., Mok, V.K.K. and Cheung, B.K.L., Roadside detection of impairment under the influence of ketamine—Evaluation of ketamine impairment symptoms with reference to its concentration in oral fluid and urine, *Forensic Sci. Int.*, 170, 51, 2007.
90. Fitzgibbon, E.J., Hall, P., Schroder, C., Seely, J. and Viola, R., Low dose ketamine as an analgesic adjuvant in difficult pain syndromes: A strategy for conversion from parenteral to oral ketamine, *J. Pain Symptom Manage.*, 23, 165, 2002.

91. Moll, V., Clavijo, C., Cohen, M., Christians, U. and Galinkin, J., eds., A analytical method to determine ketamine and norketamine levels in dried blood spots, *Proceedings of the Annual Meeting of the American Society of Anesthesiologists (ASA)*, New Orleans, LA, 2009.

92. Adamowicz, P. and Kala, M., Urinary excretion rates of ketamine and norketamine following therapeutic ketamine administration: Method and detection window considerations, *J. Anal. Toxicol.*, 29, 376, 2005.

93. Yanagihara, Y., Kariya, S., Ohtani, M., Uchino, K., Aoyama, T., Yamamura, Y. and Iga, T., Involvement of CYP2B6 in N-demethylation of ketamine in human liver microsomes, *Drug Metab. Dispos.*, 29, 887, 2001.

94. Hijazi, Y. and Boulieu, R., Contribution of CYP3A4, CYP2B6, and CYP2C9 isoforms to N-demethylation of ketamine in human liver microsomes, *Drug Metab. Dispos.*, 30, 853, 2002.

95. Clements, J. and Nimmo, W., Pharmacokinetics and analgesic effect of ketamine in man, *Br. J. Anaesth.*, 53, 27, 1981.

96. Idvall, J., Ahlgren, I., Aronsen, K. and Stenberg, P., Ketamine infusions: Pharmacokinetics and clinical effects, *Br. J. Anaesth.*, 51, 1167, 1979.

97. Malinovsky, J., Servin, F., Cozian, A., Lepage, J. and Pinaud, M., Ketamine and norketamine plasma concentrations after iv, nasal and rectal administration in children, *Br. J. Anaesth.*, 77, 203, 1996.

98. Hijazi, Y., Bodonian, C., Bolon, M., Salord, F. and Boulieu, R., Pharmacokinetics and haemodynamics of ketamine in intensive care patients with brain or spinal cord injury, *Br. J. Anaesth.*, 90, 155, 2003.

99. Grant, I., Nimmo, W. and Clements, J., Pharmacokinetics and analgesic effects of im and oral ketamine, *Br. J. Anaesth.*, 53, 805, 1981.

100. Moffat, A.C., Osselton, D.M. and Widdop, B., *Clarke's Analysis of Drugs and Poisons: In Pharmaceuticals, Body Fluids, and Postmortem Material*, Pharmaceutical Press, London, 2004.

101. Lum, G. and Mushlin, B., Urine drug testing: Approaches to screening and confirmation testing, *Lab. Med.*, 35, 368, 2004.

102. Feng, N., Vollenweider, F., Minder, E., Rentsch, K., Grampp, T. and Vonderschmitt, D., Development of a gas chromatography-mass spectrometry method for determination of ketamine in plasma and its application to human samples, *Ther. Drug Monit.*, 17, 95, 1995.

103. Cruces-Blanco, C. and Garcia-Campana, A.M., Capillary electrophoresis for the analysis of drugs of abuse in biological specimens of forensic interest, *Trends Analyt. Chem.*, 31, 85, 2012.

104. Maquille, A., Guillarme, D., Rudaz, S. and Veuthey, J.-L., High-throughput screening of drugs of abuse in urine by supported liquid–liquid extraction and UHPLC coupled to tandem MS, *Chromatographia*, 70, 1373, 2009.

105. Bush, D.M., The US mandatory guidelines for federal workplace drug testing programs: Current status and future considerations, *Forensic Sci. Int.*, 174, 111, 2008.

106. Wu, A.H., Bristol, B., Sexton, K., Cassella-McLane, G., Holtman, V. and Hill, D.W., Adulteration of urine by "Urine Luck," *Clin. Chem.*, 45, 1051, 1999.

107. Urry, F.M., Komaromy-Hiller, G., Staley, B., Crockett, D.K., Kushnir, M., Nelson, G. and Struempler, R.E., Nitrite adulteration of workplace urine drug-testing specimens I. Sources and associated concentrations of nitrite in urine and distinction between natural sources and adulteration, *J. Anal. Toxicol.*, 22, 89, 1998.

108. George, S. and Braithwaite, R., The effect of glutaraldehyde adulteration of urine specimens on Syva EMIT II drugs-of-abuse assays, *J. Anal. Toxicol.*, 20, 195, 1996.
109. Arndt, T., Urine-creatinine concentration as a marker of urine dilution: Reflections using a cohort of 45,000 samples, *Forensic Sci. Int.*, 186, 48, 2009.
110. Jaffee, W.B., Trucco, E., Levy, S. and Weiss, R.D., Is this urine really negative? A systematic review of tampering methods in urine drug screening and testing, *J. Subst. Abuse Treat.*, 33, 33, 2007.
111. Cody, J.T. and Valtier, S., Effects of stealth. adulterant on immunoassay testing for drugs of abuse, *J. Anal. Toxicol.*, 25, 466, 2001.
112. Warner, A., Interference of common household chemicals in immunoassay methods for drugs of abuse, *Clin. Chem.*, 35, 648, 1989.
113. Tsai, J.S., ElSohly, M.A., Tsai, S.-F., Murphy, T.P., Twarowska, B. and Salamone, S.J., Investigation of nitrite adulteration on the immunoassay and GC-MS analysis of cannabinoids in urine specimens, *J. Anal. Toxicol.*, 24, 708, 2000.
114. Substance Abuse and Mental Health Services Administration, Mandatory Guidelines for Federal Workplace Drug Testing Programs, 2008, pp. 1–50.
115. Judson, B.A., Himmelberger, D.U. and Goldstein, A., Measurement of urine temperature as an alternative to observed urination in a narcotic treatment program, *Am. J. Drug Alcohol Abuse*, 6, 197, 1979.
116. Schuttler, J., Stanski, D.R., White, P.F., Trevor, A.J., Horai, Y., Verotta, D. and Sheiner, L.B., Pharmacodynamic modeling of the EEG effects of ketamine and its enantiomers in man, *J. Pharmacokinet. Biopharm.*, 15, 241, 1987.
117. Krystal, J.H., Petrakis, I.L., Webb, E., Cooney, N.L., Karper, L.P., Namanworth, S., Stetson, P., Trevisan, L.A. and Charney, D.S., Dose-related ethanol-like effects of the NMDA antagonist, ketamine, in recently detoxified alcoholics, *Arch. Gen. Psychiatry*, 55, 354, 1998.
118. Burch, H.J., Clarke, E.J., Hubbard, A.M. and Scott-Ham, M., Concentrations of drugs determined in blood samples collected from suspected drugged drivers in England and Wales, *J. Forensic Leg. Med.*, 20, 278, 2013.
119. Immunalysis, Product catalogue, September 2012, available at http://www.immunalysis.com/images/productcatalogsept2012.pdf (accessed December 15, 2013).
120. Randox clinical diagnostic solutions, Ketamine ELISA and Complementary Products, 2009.
121. Svensson, J.-O. and Gustafsson, L.L., Determination of ketamine and norketamine enantiomers in plasma by solid-phase extraction and high-performance liquid chromatography, *J. Chromatogr. B Biomed. Sci. Appl.*, 678, 373, 1996.
122. Mueller, C., Weinmann, W., Dresen, S., Schreiber, A. and Gergov, M., Development of a multi-target screening analysis for 301 drugs using a QTrap liquid chromatography/tandem mass spectrometry system and automated library searching, *Rapid Commun. Mass Spectrom.*, 19, 1332, 2005.
123. Hijazi, Y., Bolon, M. and Boulieu, R., Stability of ketamine and its metabolites norketamine and dehydronorketamine in human biological samples, *Clin. Chem.*, 47, 1713, 2001.
124. Langman, L.J., The use of oral fluid for therapeutic drug management, *Ann. N.Y. Acad. Sci.*, 1098, 145, 2007.
125. Aps, J.K. and Martens, L.C., Review: The physiology of saliva and transfer of drugs into saliva, *Forensic Sci. Int.*, 150, 119, 2005.

126. Dams, R., Choo, R.E., Lambert, W.E., Jones, H. and Huestis, M.A., Oral fluid as an alternative matrix to monitor opiate and cocaine use in substance-abuse treatment patients, *Drug Alcohol Depend.*, 87, 258, 2007.
127. Drummer, O.H., Drug testing in oral fluid, *Clin. Biochem. Rev.*, 27, 147, 2006.
128. Pai, A. and Heining, M., Ketamine, *Contin. Educ. Anaesth. Crit. Care Pain*, 7, 59, 2007.
129. Marleen, L., Nele, S. and Gert, D.B., Oral fluid toxicology, in *Wiley Encyclopedia of Forensic Science*, Jamieson, A., Moenssens, A., Eds., John Wiley & Sons Ltd, West Sussex, UK, 2009.
130. Mucklow, J.C., Bending, M.R., Kahn, G.C. and Dollery, C.T., Drug concentration in saliva, *Clin. Pharmacol. Ther.*, 24, 563, 1978.
131. Crouch, D.J., Oral fluid collection: The neglected variable in oral fluid testing, *Forensic Sci. Int.*, 150, 165, 2005.
132. Gallardo, E. and Queiroz, J., The role of alternative specimens in toxicological analysis, *Biomed. Chromatogr.*, 22, 795, 2008.
133. Wylie, F., Torrance, H., Anderson, R. and Oliver, J., Drugs in oral fluid: Part I. Validation of an analytical procedure for licit and illicit drugs in oral fluid, *Forensic Sci. Int.*, 150, 191, 2005.
134. Venture Labs Inc., Ketamine ELISA, available at http://biosellsolutions.com /userfiles/Kits/ELISA/Bio%20Ketamine.pdf (accessed December 16, 2013).
135. Wong, R.C., Tran, M. and Tung, J.K., Oral fluid drug tests: Effects of adulterants and foodstuffs, *Forensic Sci. Int.*, 150, 175, 2005.
136. Jenkins, A.J., *Drug Testing in Alternate Biological Specimens*, Humana Press, Totowa, NJ, 2008.
137. Curtis, J. and Greenberg, M., Screening for drugs of abuse: Hair as an alternative matrix: A review for the medical toxicologist, *Clin. Toxicol. (Phila.)*, 46, 22, 2008.
138. Kintz, P., Cirimele, V. and Ludes, B., Pharmacological criteria that can affect the detection of doping agents in hair, *Forensic Sci. Int.*, 107, 325, 2000.
139. Kidwell, D.A., Lee, E.H. and DeLauder, S.F., Evidence for bias in hair testing and procedures to correct bias, *Forensic Sci. Int.*, 107, 39, 2000.
140. Kronstrand, R., Nystrom, I., Strandberg, J. and Druid, H., Screening for drugs of abuse in hair with ion spray LC-MS-MS, *Forensic Sci. Int.*, 145, 183, 2004.
141. Musshoff, F. and Madea, B., New trends in hair analysis and scientific demands on validation and technical notes, *Forensic Sci. Int.*, 165, 204, 2007.
142. Baumgartner, W.A. and Hill, V.A., Sample preparation techniques, *Forensic Sci. Int.*, 63, 121, 1993.
143. Negrusz, A., *Detection of "Date-Rape" Drugs in Hair and Urine*, Final Report, National Institute of Justice, 2001, available at https://www.ncjrs.gov/pdffiles1 /nij/grants/201894.pdf (accessed December 6, 2013).
144. Miki, A., Katagi, M., Kamata, T., Zaitsu, K., Tatsuno, M., Nakanishi, T., Tsuchihashi, H., Takubo, T. and Suzuki, K., MALDITOF and MALDIFTICR imaging mass spectrometry of methamphetamine incorporated into hair, *J. Mass Spectrom.*, 46, 411, 2011.
145. Yaemsiri, S., Hou, N., Slining, M.M. and He, K., Growth rate of human fingernails and toenails in healthy American young adults, *J. Eur. Acad. Dermatol. Venereol.*, 24, 420, 2010.
146. McCarthy, D.J., Anatomic considerations of the human nail, *Clin. Podiatr. Med. Surg.*, 21, t477, 2004.

147. Madry, M.M., Steuer, A.E., Binz, T.M., Baumgartner, M.R. and Kraemer, T., Systematic investigation of the incorporation mechanisms of zolpidem in fingernails, *Drug Test. Anal.*, 6, 533, 2014.
148. Kobayashi, Y., Komatsu, T., Sumi, M., Numajiri, S., Miyamoto, M., Kobayashi, D., Sugibayashi, K. and Morimoto, Y., In vitro permeation of several drugs through the human nail plate: Relationship between physicochemical properties and nail permeability of drugs, *Eur. J. Pharm. Sci.*, 21, 471, 2004.
149. Palmeri, A., Pichini, S., Pacifici, R., Zuccaro, P. and Lopez, A., Drugs in nails: Physiology, pharmacokinetics and forensic toxicology, *Clin. Pharmacokinet.*, 38, 95, 2000.
150. Liu, H.C., Liu, R.H., Ho, H.O. and Lin, D.L., Development of an information-rich LC.MS/MS database for the analysis of drugs in postmortem specimens, *Anal. Chem.*, 81, 9002, 2009.
151. Wills, S., *Drugs of Abuse*, 2nd ed., Pharmaceutical Press, Cornwall, UK, 2005.
152. Badawi, N., Simonsen, K.W., Steentoft, A., Bernhoft, I.M. and Linnet, K., Simultaneous screening and quantification of 29 drugs of abuse in oral fluid by solid-phase extraction and ultraperformance LC-MS/MS, *Clin. Chem.*, 55, 2004, 2009.
153. Hegstad, S., Hermansson, S., Betner, I., Spigset, O. and Falch, B.M., Screening and quantitative determination of drugs of abuse in diluted urine by UPLC-MS/MS, *J. Chromatogr. B Analyt. Technol. Biomed. Life Sci.*, 947–948, 83, 2014.
154. Paterson, S., Cordero, R., McCulloch, S. and Houldsworth, P., Analysis of urine for drugs of abuse using mixed-mode solid-phase extraction and gas chromatography-mass spectrometry, *Ann. Clin. Biochem.*, 37 (Pt 5), 690, 2000.
155. Tsai, I.L., Weng, T.I., Tseng, Y.J., Tan, H.K., Sun, H.J. and Kuo, C.H., Screening and confirmation of 62 drugs of abuse and metabolites in urine by ultra-high-performance liquid chromatography-quadrupole time-of-flight mass spectrometry, *J. Anal. Toxicol.*, 37, 642, 2013.
156. Davis, G.G., Drug abuse: Newly-emerging drugs and trends, *Clin. Lab. Med.*, 32, 407, 2012.
157. Gibbons, S., 'Legal highs'—Novel and emerging psychoactive drugs: A chemical overview for the toxicologist, *Clin. Toxicol. (Phila.)*, 50, 15, 2012.
158. Meyer, M.R. and Peters, F.T., Analytical toxicology of emerging drugs of abuse—An update, *Ther. Drug Monit.*, 34, 615, 2012.
159. Rosenbaum, C.D., Carreiro, S.P. and Babu, K.M., Here today, gone tomorrow...and back again? A review of herbal marijuana alternatives (K2, Spice), synthetic cathinones (bath salts), kratom, Salvia divinorum, methoxetamine, and piperazines, *J. Med. Toxicol.*, 8, 15, 2012.
160. European Monitoring Centre for Drugs and Drug Addiction (EMCDDA) and Europol, New drugs in Europe, 2012. EMCDDA-Europol 2012 Annual Report on the implementation of Council Decision 2005/387/JHA, 2012.
161. Hofer, K.E., Grager, B., Muller, D.M., Rauber-Luthy, C., Kupferschmidt, H., Rentsch, K.M. and Ceschi, A., Ketamine-like effects after recreational use of methoxetamine, *Ann. Emerg. Med.*, 60, 97, 2012.
162. Ward, J., Rhyee, S., Plansky, J. and Boyer, E., Methoxetamine: A novel ketamine analog and growing health-care concern, *Clin. Toxicol. (Phila.)*, 49, 874, 2011.
163. Wood, D.M., Davies, S., Puchnarewicz, M., Johnston, A. and Dargan, P.I., Acute toxicity associated with the recreational use of the ketamine derivative methoxetamine, *Eur. J. Clin. Pharmacol.*, 68, 853, 2012.

164. Archer, J.R., Dargan, P.I., Hudson, S. and Wood, D.M., Analysis of anony-mous pooled urine from portable urinals in central London confirms the sig-nificant use of novel psychoactive substances, *QJM*, 106, 147, 2013.

165. Peters, F.T. and Martinez-Ramirez, J.A., Analytical toxicology of emerging drugs of abuse, *Ther. Drug Monit.*, 32, 532, 2010.

166. Tang, M.H.Y., Ching, D.C.K., Lam, Y.H., Lee, C.Y.W., Chong, C.Y.K., Wong, W.W.S. and Mak, T.W.L., Surveillance of emerging drugs of abuse in sub-stance abusers—Final report to the Beat Drugs Fund Association (Narcotics Division, Security Bureau of the government of HKSAR). Project no. BDF101021, 2013.

167. Tang, M.H.Y., Ching, D.C.K., Lee, C.Y.W., Lam, Y.H. and Mak, T.W.L., Simultaneous detection of 93 conventional and emerging drugs of abuse and their metabolites in urine by UHPLC-MS/MS, *J. Chromatogr.* B, 969, 272, 2014.

168. Tang, M.H.Y., Ching, C.K., Tse, M.L., Ng, C., Lee, C., Chong, C., Wong, W. and Mak, T.W.L., Surveillance of emerging drugs of abuse in Hong Kong: Validation of an analytical tool, *Hong Kong Med. J.*, Submitted manuscript, 2014.

Index

Page numbers followed by f and t denotes figures and tables, respectively.